FRONTIERS IN NEUROSCIENCE

Series Editors
Sidney A. Simon, Ph.D.
Miguel A.L. Nicolelis, M.D., Ph.D.

Published Titles

Apoptosis in Neurobiology
Yusuf A. Hannun, M.D., Professor of Biomedical Research and Chairman/Department of Biochemistry and Molecular Biology, Medical University of South Carolina
Rose-Mary Boustany, M.D., tenured Associate Professor of Pediatrics and Neurobiology, Duke University Medical Center

Methods for Neural Ensemble Recordings
Miguel A.L. Nicolelis, M.D., Ph.D., Professor of Neurobiology and Biomedical Engineering, Duke University Medical Center

Methods of Behavioral Analysis in Neuroscience
Jerry J. Buccafusco, Ph.D., Alzheimer's Research Center, Professor of Pharmacology and Toxicology, Professor of Psychiatry and Health Behavior, Medical College of Georgia

Neural Prostheses for Restoration of Sensory and Motor Function
John K. Chapin, Ph.D., Professor of Physiology and Pharmacology, State University of New York Health Science Center
Karen A. Moxon, Ph.D., Assistant Professor/School of Biomedical Engineering, Science, and Health Systems, Drexel University

Computational Neuroscience: Realistic Modeling for Experimentalists
Eric DeSchutter, M.D., Ph.D., Professor/Department of Medicine, University of Antwerp

Methods in Pain Research
Lawrence Kruger, Ph.D., Professor of Neurobiology (Emeritus), UCLA School of Medicine and Brain Research Institute

Motor Neurobiology of the Spinal Cord
Timothy C. Cope, Ph.D., Professor of Physiology, Emory University School of Medicine

Nicotinic Receptors in the Nervous System
Edward D. Levin, Ph.D., Associate Professor/Department of Psychiatry and Pharmacology and Molecular Cancer Biology and Department of Psychiatry and Behavioral Sciences, Duke University School of Medicine

Methods in Genomic Neuroscience
Helmin R. Chin, Ph.D., Genetics Research Branch, NIMH, NIH
Steven O. Moldin, Ph.D, Genetics Research Branch, NIMH, NIH

Methods in Chemosensory Research
Sidney A. Simon, Ph.D., Professor of Neurobiology, Biomedical Engineering, and Anesthesiology, Duke University
Miguel A.L. Nicolelis, M.D., Ph.D., Professor of Neurobiology and Biomedical Engineering, Duke University

The Somatosensory System: Deciphering the Brain's Own Body Image
Randall J. Nelson, Ph.D., Professor of Anatomy and Neurobiology, University of Tennessee Health Sciences Center

The Superior Colliculus: New Approaches for Studying Sensorimotor Integration
William C. Hall, Ph.D., Department of Neuroscience, Duke University
Adonis Moschovakis, Ph.D., Institute of Applied and Computational Mathematics, Crete

New Concepts in Cerebral Ischemia
Rick C. S. Lin, Ph.D., Professor of Anatomy, University of Mississippi Medical Center

DNA Arrays: Technologies and Experimental Strategies
Elena Grigorenko, Ph.D., Technology Development Group, Millennium Pharmaceuticals

Methods for Alcohol-Related Neuroscience Research
Yuan Liu, Ph.D., National Institute of Neurological Disorders and Stroke, National Institutes of Health
David M. Lovinger, Ph.D., Laboratory of Integrative Neuroscience, NIAAA

***In Vivo* Optical Imaging of Brain Function**
Ron Frostig, Ph.D., Associate Professor/Department of Psychobiology, University of California, Irvine

Primate Audition: Behavior and Neurobiology
Asif A. Ghazanfar, Ph.D., Primate Cognitive Neuroscience Lab, Harvard University

Methods in Drug Abuse Research: Cellular and Circuit Level Analyses
Dr. Barry D. Waterhouse, Ph.D., MCP-Hahnemann University

Functional and Neural Mechanisms of Interval Timing
Warren H. Meck, Ph.D., Professor of Psychology, Duke University

Biomedical Imaging in Experimental Neuroscience
Nick Van Bruggen, Ph.D., Department of Neuroscience Genentech, Inc., South San Francisco
Timothy P.L. Roberts, Ph.D., Associate Professor, University of Toronto

The Primate Visual System
John H. Kaas, Department of Psychology, Vanderbilt University
Christine Collins, Department of Psychology, Vanderbilt University

Neurosteroid Effects in the Central Nervous System
Sheryl S. Smith, Ph.D., Department of Physiology, SUNY Health Science Center

Modern Neurosurgery: Clinical Translation of Neuroscience Advances
Dennis A. Turner, Department of Surgery, Division of Neurosurgery, Duke University Medical Center

Sleep: Circuits and Functions
Pierre-Hervé Luoou, Université Claude Bernard Lyon I, Lyon, France

Methods in Insect Sensory Neuroscience
Thomas A. Christensen, Arizona Research Laboratories, Division of Neurobiology, University of Arizona, Tucson, AZ

Motor Cortex in Voluntary Movements
Alexa Riehle, INCM-CNRS, Marseille, France
Eilon Vaadia, The Hebrew University, Jeruselum, Israel

NEURAL PLASTICITY IN ADULT SOMATIC SENSORY-MOTOR SYSTEMS

Edited by

Ford F. Ebner

Vanderbilt University
Department of Psychology
Nashville, TN

Taylor & Francis
Taylor & Francis Group

Boca Raton London New York Singapore

A CRC title, part of the Taylor & Francis imprint, a member of the Taylor & Francis Group, the academic division of T&F Informa plc.

Published in 2005 by
CRC Press
Taylor & Francis Group
6000 Broken Sound Parkway NW, Suite 300
Boca Raton, FL 33487-2742

Printed in the United States of America on acid-free paper
10 9 8 7 6 5 4 3 2 1

International Standard Book Number-10: 0-8493-1521-2 (Hardcover)
International Standard Book Number-13: 978-0-8493-1521-3 (Hardcover)
Library of Congress Card Number 2004058571

Library of Congress Cataloging-in-Publication Data

Neural plasticity in adult somatic sensory-motor system / edited by Ford F. Ebner.
p. cm. -- (Frontiers in neuroscience)
ISBN 0-8493-1521-2 (alk. paper)
1. Sensorimotor cortex. 2. Neuroplasticity. I. Ebner, Ford F. II. Frontiers in neuroscience (Boca Raton, Fla.)

QP383.15.N475 2005
612.8'252--dc22 2004058571

T&F informa
Taylor & Francis Group
is the Academic Division of T&F Informa plc.

Visit the Taylor & Francis Web site at
http://www.taylorandfrancis.com

and the CRC Press Web site at
http://www.crcpress.com

Preface

Neural plasticity is now well accepted as a universal property of multi-cellular nervous systems. Plasticity has been studied in particular detail in the mammalian cerebral cortex. The word "plasticity" has been applied to a wide variety of cortical changes, so an initial question is always: what metric has been used to conclude that a plastic event has occurred? The chapters in this book illustrate important examples in which the metric for plasticity is physiological alterations in neuronal response properties or changes in behavioral skills. The locus of these changes is in the somatic sensory pathways to and within sensory cortex or motor cortex in response to a variety of challenges. The initial chapters discuss issues relevant to modifications in sensory processing.

Although controversial and easy to ignore, an increasing number of investigators are convinced that silent neurons need further study. In somatic sensory cortex the silent neuron idea is linked to a 1988 paper by Robert Dykes and Yves Lamour in which they showed that a large fraction of cortical cells did not fire action potentials in response to tactile stimuli, even though the cells seemed healthy and responded vigorously to locally applied glutamate. Their hypothesis that the silent neurons become wired into cortical circuits during learning was too novel, and arrived too early, to be embraced by other workers in the field without additional lines of evidence. Strong evidence for the existence of silent neurons has since appeared, and the chapter by Michael Brecht and his colleagues in this book poses important questions about the silent neurons' role in cortical function. The specific contribution of these neurons to cortical plasticity is a particularly important ongoing idea that remains to be clarified.

Another fascinating dimension of sensory transduction is that rats may use the whiskers on their face to listen to vibrations in the world. Rats and mice are known to use their whiskers as a main source of sensory information. Christopher Moore and Mark Andermann describe how the resonance properties of the whiskers, like in the cochlea of the human ear, may allow rodents to amplify signals and help rats detect small vibrations present in the sensory world. These vibrations could be crucial to a rodent's ability to perceive the subtle texture properties of a solid surface, which generate these small vibrations when a whisker is swept across. They further provide evidence that rodent whiskers could even be used to "hear" sounds. Beyond just being an amplifier, the whiskers are organized in an orderly way, such that the shorter whiskers near the snout amplify higher frequency inputs than the longer whiskers further back. This arrangement of the whiskers, like the strings of a harp, creates a systematic map of tuning across the rat's face. This orderly map in the periphery creates an orderly neural representation in the primary somatic sensory cortex, a map of frequency embedded within the well-described body map representation. These authors also provide evidence for further subdivisions of this representation into isofrequency columns, modular groups of cells that all respond best to the same amplified frequency. These novel findings are considered with regard to classical theories of how resonance facilitates perception in other sensory systems, ranging from the cockroach to the human ear, and also consider how these principles of the biomechanical transduction of information may provide lessons for understanding the optimal use of tools by humans.

Continuing the coding theme more centrally, Mathew Diamond then discusses the role of modular, maplike cortical organization in the processing of sensory information, including the functional significance of cortical maps, as well as the individual modules that create the topographic framework for spatial coding in primary sensory cortex. These spatial rules for barrel cortex plasticity co-exist with temporal fluctuations in excitability (temporal coding), characterized in anesthetized rats by bursts of spikes that are synchronized across the entire barrel cortex. The bursts appear to briefly open a plasticity gate allowing incoming sensory inputs to modify the efficacy of the activated intracortical circuits. During the time between bursts the plasticity gate is closed and incoming inputs have no long-term effect on intracortical circuits. These modifications by sensory input patterns during discrete intervals provide a theoretical basis for understanding barrel cortex changes in awake, exploring rats because rhythmic oscillations occur in awake rat cortex as well.

The isolation of neural codes related to perception and learning is another important issue discussed in this series by Ranulfo Romo and his colleagues. The underlying premise is that unraveling the sensory code from the periphery to cortical processing is key to understanding initial perceptual processes. They use the ideas of Vernon Mountcastle and colleagues who quantified the relationship between action potentials in cutaneous, primary afferents and mechanical (especially flutter) stimuli applied to the skin. By combining human psychophysics with single unit analysis in monkeys, they looked for the psychophysical link between stimulus and sensation. Using this approach, it should be possible to identify neural codes for simple stimuli in early stages of cortical processing that can be compared with the psychophysical responses. However, even the simplest cognitive task may engage many cortical areas, and each one might represent sensory information using a different code, or combine new inputs with stored signals representing past experience. Romo and and his colleagues explore these ideas in primary somatic sensory (SI) cortex of primates. Starting with optimal conditions for flutter discrimination, they studied the neuronal responses in SI cortex, and correlated them with psychophysical performance. The evoked neuronal responses in SI could be shown to correlate well with correct or incorrect responses, even when they bypassed the usual sensory pathway by electrical activation of neuronal clusters in SI to produce an artificial perceptual input to SI cortex that could be used by the animals to guide their behavior.

In Krish Sathian's studies on human perception, he and his colleagues used a variety of stimuli and tasks to study the transfer of perceptual learning between fingers and hands. They employed periodic gratings actively stroked by the subjects where the task was to discriminate between gratings that varied either in their groove width or in their ridge width. Initial training was carried out with one index finger, and progressed to the index or middle finger of the other hand. Learning was reflected in improved performance, and transfer of learning occurred between fingers, and was substantial between the two hands, presumably based on interhemispheric connections. In subsequent studies, these findings were extended to a variety of tactile stimuli and tasks leading to the conclusion that transfer of tactile learning appears to be a general rule. It is interesting to speculate that interhemispheric transfer of tactile learning may relate to intermanual referral of tactile sensations following amputation or stroke. The mechanisms of perceptual learning are relevant to the perceptual improvements that are observed in spared modalities following sensory deprivation in a particular modality, such as improved tactile skills in people with very low vision.

Examples of somatic sensory processing after early postnatal sensory deprivation has identified a number of ways in which activity is needed to develop normal sensory processing in cortex. Ford Ebner and Michael Armstrong-James describe the nature of cortical impairments induced by low activity during the early postnatal period in the somatic sensory system in rats and mice after they mature to normal-looking adults. The literature shows that both excitatory and inhibitory processes are affected by sensory deprivation, with the severity of effects depending upon the time of onset, the duration of the deprivation, and the length of the recovery period after deprivation ends. Intracortical circuit dynamics are most severely affected. Neural transmission from cortical layer IV to more superficial layers II/III is a major site of synaptic dysfunction. Trimming all whiskers produces a more uniform down-regulation of sensory transmission than trimming a subset of whiskers presumably because restricted deprivation creates competition between active and inactive inter-connected cell groups. Activity-based changes in function can be induced by altered tactile experience throughout life, but early postnatal deprivation degrades neuronal plasticity, and interferes with the animal's ability to learn subtle tactile discriminations throughout life.

The remaining chapters deal with the motor side of sensory-motor transformations.

John Chapin and his colleagues discuss the mechanisms by which the brain transforms sensory inputs into motor outputs. The rules for such sensory-motor conversions have proven elusive, and the authors suggest that this is due to the multiplicity of "bridges" between these systems in the CNS. Moreover, while the development and maintenance of the sensorimotor transformation machinery must involve some sort of plasticity, it is not yet clear how or where this plasticity occurs. They then offer specific recommendations for studying these issues in awake animals performing behaviors that involve sensory-motor transformations, an area in which they have made significant contributions.

The plastic responses of neurons in motor cortex after stroke-like lesions have clinical as well as basic science relevance. Randy Nudo and his colleagues have been studying the mutability of sensory, motor and premotor maps of the mature cerebral cortex following experimental lesions of cortex to document the mechanisms of neuroplasticity in the adult brain. They use direct brain stimulation (ICMS) in layer V of motor cortex to elicit muscle or joint movement before and after motor skill training. The maps are composed of various digit and arm movements. An initial result was that monkeys trained to use their digits to retrieve food pellets from a food board showed an increase in the size of representations of the digits used in the task. Further, multijoint responses to ICMS were infrequent before training, but were found in abundance after digit training. The implication is that simultaneous movements may become associated in the cortex through Hebbian synaptic mechanisms in which horizontal fibers connecting two areas become strengthened through associated repetitive activation. When spontaneous recovery was studied at 3 to 5 months after a hand area motor cortex lesion, skilled use of the hand returned, but roughly half of the digit movement representation was still replaced by shoulder and elbow. However, if squirrel monkeys were trained to retrieve food pellets from food wells, and then re-trained after a motor cortex lesion using the less affected hand (ipsilateral to a small infarct), the monkeys returned to baseline levels on the most difficult food-well task. In this case, motor skill training saved the remaining pre-infarct distal hand representation from the expected takeover by surrounding inputs. The implication of these results is that physical rehabilitation after stroke can drive physiological changes in the cortex associated with recovering skilled hand use, if the conditions are optimized.

Jon Kaas then discusses how motor experience rebalances dynamic systems to reveal latent neural circuit properties. Short term changes emerge over a time period ranging from seconds to hours due to a range of activity-dependent cellular mechanisms that affect synaptic strengths. Over somewhat longer periods of days to weeks, anatomical circuits may be lost or gained as local circuits grow and rearrange. Over a time period of weeks to months, considerable new growth of axons and synapses can occur that considerably alter the functional organization of sensory and motor systems, sometimes in ways that promote behavioral recovery, and sometimes in ways that do not promote such recovery.. One goal of research on sensory-motor plasticity is to understand the mechanisms of change and how to manipulate them in order to maximize recovery after sensory and motor loss. This chapter focuses on changes in the motor system that are the result of a particularly severe type of motor system damage— the loss of an entire forelimb or hindlimb. In humans, badly damaged limbs might require amputation, and it is important to determine what happens to the somatosensory and motor systems as a result of the loss of both the sensory afferents from the limb and the motor neuron outflow to the muscles of that limb.

Leonardo Cohen and colleagues focus on central nervous system adaptations to environmental challenges or lesions. Understanding the mechanisms underlying cortical plasticity can provide clues to enhance neurorehabilitative efforts. Upper limb amputation (e.g., at the elbow level) results in an increase in the excitability of body part representations in the motor cortex near the deafferented zone in the form of decreased motor thresholds, larger motor maps and a lateral shift of the center of gravity with transcranial magnetic stimulation. This increased excitability appears to be predominantly cortical in origin. The mechanisms underlying these reorganizational changes are incompletely understood, however, intracortical inhibition in the motor cortex contralateral to an amputated limb is decreased relative to healthy subjects suggesting that GABAergic inhibition may be reduced. Another issue is phantom limb pain, a condition characterized by the presence of painful perceptions referred to the missing limb. Phantom limb pain is associated with profound changes in cortical and subcortical organization. Reorganization in the primary somatosensory cortex has been demonstrated to be strongly correlated with the magnitude of phantom limb pain. Interestingly, phantom pain was more prominent in patientsin whom the motor representations of face muscles were displaced medially, possibly reflecting an invasion of the face motor representation in motor cortex.

In the last chapter the behavioral basis of focal hand dystonia is discussed by Nancy Byl as a form of aberrant learning in the somatic sensory cortex. The cause of this disabling movement disorder has remained elusive. It is common in productive, motivated individuals, such as musicians, who perform highly repetitive, intensive hand tasks., Their studies document degradation of the cortical somatosensory representation of the hand characterized by large receptive fields overlapped across adjacent digits, overlap of glabrous-hairy surfaces, persistence of digital receptive fields across broad cortical distances, high ratio of amplitude to latency in somatic sensory evoked field responses, and abnormal digit representation. Challenging, rewarded, repetitive behavioral tasks that require high speed, high force, precision and intense work cycles with minimal breaks accelerate the onset and severity of dystonia. The development of dystonia may be minimized if individuals use the hands in a functional, mid-range position, take frequent breaks, work at variable speeds for short durations, attend to sensory-motor feedback, and initiate digital movements with the intrinsic muscles. The central theme is that attended, progressive, rewarded, learning-based sensory-motor training consistent with the principles of neuroplasticity, can facilitate recovery of task-specific motor control.

All of the examples in this book suggest that our understanding of neural plasticity and its mechanisms is increasing at a rapid rate, and that the knowledge will modify many of the procedures now in place to improve perceptual and motor skills after brain damage.

Ford Ebner

Editor

Ford F. Ebner, Ph.D., was raised in the American Pacific Northwest where he attended Washington State University (WSU). After receiving a B.S. in biology and a D.V.M. degree at WSU, he spent 2 years as a veterinary officer in the US Army at the Walter Reed Army Medical Center, and the Armed Forces Institute of Pathology in Washington, D.C. He worked with Dr. Ronald Myers at the Walter Reed Army Institute of Research and continued to study the transfer of learned information through the corpus callosum under the sponsorship of Dr. Vernon Mountcastle at the Johns Hopkins School of Medicine Department of Physiology. Dr. Ebner returned to graduate school to earn his Ph.D. with Dr. Walle Nauta, a neuroanatomist at Walter Reed, who was affiliated with the University of Maryland School of Medicine in Baltimore. After 2 years on the University of Maryland faculty he moved to Brown University in Providence, Rhode Island, where he remained for two decades teaching medical neuroscience and continuing research on cortical function. During this period Dr. Ebner received a Javits Neuroscience Investigator Award from the NIH to support his research. In 1991 he moved to Vanderbilt University in Nashville, Tennessee as director of the Institute for Developmental Neuroscience in the John F. Kennedy Center at Vanderbilt University. He is currently Professor of Psychology and Cell Biology at Vanderbilt where he continues cutting-edge research on cortical plasticity. His experience and expertise were instrumental in drawing together the very talented group of investigators who contributed to this book.

Contributors

Mark L. Andermann
Massachusetts Institute of Technoolgy, McGovern Insitutte of Brain Research and Department of Brain and Cognitive Sciences
Harvard University Program in Biophysics
Massachusetts General Hospital
Boston, Massachusetts

Michael Armstrong-James
Department of Physiology
Vanderbilt University
Nashville, Tennessee

Scott Barbay
University of Kansas Medical Center
Department of Molecular and Integrative Physiology
Kansas City, Kansas

Nancy N. Byl
Department of Physical Therapy and Rehabilitation Science
University of California, San Francisco
School of Medicine
San Francisco, California

Michael Brecht
Max Planck Institute for Medical Research
Department of Cell Physiology
Heidelberg Germany

John K. Chapin
Department of Physiology and Pharmacology
SUNY Downstate Medical Center
Brooklyn, New York

Leonardo G. Cohen
Human Cortical Physiology Section
National Institute of Neurological Disorders and Stroke
National Institutes of Health
Bethesda, Maryland

Mathew E. Diamond
Cognitive Neuroscience Sector
International School for Advanced Studies
Trieste, Italy

Ford F. Ebner
Department of Psychology
Vanderbilt University
Nashville, Tennessee

Christian Gerloff
Cortical Physiology Research Group, Department of Neurology and Hertie Institute for Clinical Brain Research
Eberhard Karls University
Tuebingen, Germany

Linda Hermer-Vazquez
Department of Psychology and McKnight Brain Institute
University of Florida
Gainesville, Florida

Raymond Hermer-Vazquez
Department of Psychology and McKnight Brain Institute
University of Florida
Gainesville, Florida

Adrián Hernandez
Instituto de Fisiología Celular
Universidad Nacional Autonoma de México
México City, Mexico

Friedhelm Hummel
Human Cortical Physiology Section
National Institute of Neurological Disorders and Stroke
National Institutes of Health
Bethesda, Maryland
and
Cortical Physiology Research Group, Department of Neurology and Hertie Institute for Clinical Brain Research
Eberhard Karls University
Tuebingen, Germany

Jon H. Kaas
Department of Physiology
Vanderbilt University
Nashville, Tennessee

Victor de Lafuente
Instituto de Fisiología Celular
Universidad Nacional Autonoma de México
México City, Mexico

Luis Lemus
Instituto de Fisiología Celular
Universidad Nacional Autonoma de México
México City, Mexico

Rogelio Luna
Instituto de Fisiología Celular
Universidad Nacional Autonoma de México
México City, Mexico

Ian D. Manns
Max Planck Institute for Medical Research
Department of Cell Physiology
Heidelberg Germany

Christoper I. Moore
Massachusetts Institute of Technology, McGovern Insitute of Brain Research and Department of Brain and Cognitive Sciences
Harvard University Program in Biophysics
Massachusetts General Hospital
Boston, Massachusetts

Randolph J. Nudo
University of Kansas Medical Center
Department of Molecular and Integrative Physiology
Kansas City, Kansas

Ranulfo Romo
Instituto de Fisiología Celular
Universidad Nacional Autonoma de México
México City, Mexico

K. Sathian
Department of Neurology
Emory University School of Medicine
Atlanta, Georgia

Miriam Schnieider
Max Planck Institute for Medical Research
Department of Cell Physiology
Heidelberg Germany

Antonio Zainos
Instituto de Fisiología Celular
Universidad Nacional Autonoma de México
México City, Mexico

Elena Zoubina
University of Kansas Medical Center
Department of Molecular and Integrative Physiology
Kansas City, Kansas

Contents

Chapter 1 Silent Neurons in Sensorimotor Cortices: Implications for Cortical Plasticity 1

Michael Brecht, Miriam Schneider, and Ian D. Manns

Chapter 2 The Vibrissa Resonance Hypothesis 21

Christopher Moore and Mark L. Andermann

Chapter 3 Spatial and Temporal Rules Underlying Rat Barrel Cortex Plasticity 61

Mathew E. Diamond

Chapter 4 Probing the Cortical Evidence of Somatosensory Discrimination 79

Ranulfo Romo, Adrián Hernández, Antonio Zainos, Luis Lemus, Victor de Lafuente, and Rogelio Luna

Chapter 5 Perceptual Learning and Referral in the Tactile System 97

K. Sathian

Chapter 6 The Effects of Sensory Deprivation on Sensory Function of SI Barrel Cortex 109

Ford F. Ebner and Michael Armstrong-James

Chapter 7 Role of Plasticity in Sensorimotor Transformations 139

Linda Hermer-Vazquez, Raymond Hermer-Vazquez, and John K. Chapin

Chapter 8 Neural Plasticity in Adult Motor Cortex 155

Scott Barbay, Elena Zoubina and Randolph J. Nudo

Chapter 9 Reorganization of Motor Cortex after Damage to the Motor System 189

Jon H. Kaas

Chapter 10 Modulation of Cortical Function and Plasticity in the Human Brain 207

Friedhelm Hummel, Christian Gerloff, and Leonardo G. Cohen

Chapter 11 Behavioral Basis of Focal Hand Dystonia: Aberrant Learning in the Somatosensory Cortex 227

Nancy N. Byl

Index 263

1 Silent Neurons in Sensorimotor Cortices: Implications for Cortical Plasticity

Michael Brecht, Miriam Schneider, and Ian D. Manns

CONTENTS

I. Evidence for a Predominance of Silent Cortical Neurons in Sensorimotor Cortices 2
   A. Evaluation of Presynaptic Activity in the Somatosensory Cortex 3
      1. Results from Extracellular Unit Recording 3
      2. Results from Sharp Microelectrode Recordings 3
      3. Results from Whole-Cell Recordings in the Vibrissae Barrel Cortices of Anesthetized Animals 4
      4. Results from Whole-Cell Recordings in the Vibrissae Barrel Cortices of Awake Animals 4
      5. Inconsistencies and Caveats from Whole-Cell Studies 6
      6. Results from Derivatives of the Whole-Cell Recordings Technique 6
      7. All Techniques Agree that Firing Rates of Cortical Neurons are Very Heterogeneous 7
   B. Evaluation of Postsynaptic Activty in the Somatosensory Cortex 7
      1. Experimental Agreement on the Amplitude of Subthreshold Signals 7
      2. The Synaptic Composition of Postsynaptic Responses Is Controversial 8
      3. An Attempt to Quantitatively Determine the Synaptic Composition of a Cortical Sensory Response Suggests Very Low Presynaptic Activity 8

C. Evaluation of the Motor Effects of APs in the Primary Motor Cortex 10
1. Low Firing Rates Imply a High Efficacy of Cortical APs 10
2. Small Numbers of APs in Single Cells of the Primary Motor Cortex Can Evoke Movements 10
3. Effective Cells Greatly Outnumber Active Cells in M1 12
II. Silent Neurons and Cortical Plasticity 12
A. Cortical Plasticity and Alternative Hypotheses for Silent Neurons 12
1. Silent Cortical Cells as a Corollary of Metabolic Demands 12
2. Contribution of Silent Cells to Learning and Cortical Plasticity 12
3. Silent Cells May Signal by Non AP-Dependent Signaling 13
4. Silent Cells May Function by Sparse AP Activity 13
B. Data on an Involvement of Silent Cells in Plasticity are Largely Absent 13
C. Silent Neurons and Synaptic Learning Rules 13
III. Conclusion 15
References 16

Silence is music, too.

Miles Davis

Evidence indicates that cortical neurons are mostly silent. Early in the 1970s, Barlow hypothesized that few neurons may be sufficient for a sensory representation.[1] This was clarified 15 yr ago by extracellular recordings combined with glutamate application that indicated a large fraction of cortical cells do not fire spikes in response to tactile stimuli.[2] An important implication is that if silent cells, are "desilenced," they could profoundly contribute to cortical plasticity.

This chapter will review the following: 1. Evidence from new experimental approaches indicates that only a very small fraction of cortical cells do fire APs. 2. While a potential role for these enigmatically silent cortical neurons in cortical plasticity is an attractive hypothesis, very little evidence for this is provided. 3. Examining the contribution of silent cortical neurons to cortical plasticity poses conceptual and experimental challenges.

I. EVIDENCE FOR A PREDOMINANCE OF SILENT CORTICAL NEURONS IN SENSORIMOTOR CORTICES

The observation that most cortical neurons do not discharge APs was made using a wide range of experimental paradigms. It is beyond the scope of this chapter to review all evidence. Instead, we will mention some landmark studies and point out that very diverse techniques lead to that conclusion. Attention will be paid to our own studies on the quantitative description of neural activity in a rat's barrel cortex. We will refer to cells as silent neurons, if they discharge less then 0.1 APs for an optimized stimulus (in our case a strong backward deflection of the best whisker).

A. Evaluation of Presynaptic Activity in the Somatosensory Cortex

1. Results from Extracellular Unit Recording

Extracellular unit recording was the first and is still the most common technique to quantify cellular activity in the somatosensory cortex. The general impression of most of these studies is that neurons in the somatosensory cortex discharge APs when the appropriate tactile stimulus is applied.[3] Results have been obtained not only in the cat[3] and monkey[4] primary somatosensory cortex (S1), but also in the vibrissae barrel cortices of rats that were both anesthetized[5] and awake.[6] In studies like this, when responses to controlled deflections of the (best) principal whisker (PW) are quantified, values around 1 AP per PW stimulus were reported.[6-9] These studies also report a considerable level of spontaneous AP activity of around 1Hz.[10,11]

However, not all studies that analyzed S1 activity by unit recordings came to these conclusions. In a series of influential papers, Dykes and colleagues argued that most neurons in the somatosensory cortex could not be driven by conventional stimuli.[2,12] What made these studies so compelling, was the deliberate effort to analyze every AP discharge in order to minimize sampling biases, and even more so, the use of iontophoretic injections of glutamate and other neurotransmitters, which uncovered the existence of previously unresponsive neurons around the recording electrode. Few researchers fully agree to the idea of a majority of unresponsive cells in the S1 cortex. Nonetheless the technical elegance of the work of Dykes and colleagues has made it clear that unit recordings result in enormous sampling biases against neurons with low levels of AP activity. Swadlow supported this idea in a series of studies on various cortical areas. In these studies, sampling biases were minimized by antidromic identification of recorded units.[13-16] Apart from exceptions like corticofugally projecting layer V neurons, most identified neurons in these studies were found to have very low spontaneous and evoked AP activity. Similarly, some recent unit recording studies on the vibrissae barrel cortex report rather low rates of AP activity.[17]

2. Results from Sharp Microelectrode Recordings

Sharp microelectrode recordings have been applied in a wide variety of preparations (and it is beyond the scope of this chapter to review all this evidence). In the barrel cortex, it has been observed that sharp microelectrode recordings report slightly higher AP rates than extracellular unit recordings[18]. In particular, spontaneous AP activity can be very high in these recordings and may even exceed 10 Hz.[19] This observation suggests that neurons under these recording conditions are close to the AP initiation threshold. Control experiments in frog spinal neurons appear that leaks introduced by the impalement of the cell are responsible for more depolarized membrane potentials and high firing rates observed in sharp microelectrode recordings.[20]

3. Results from Whole-Cell Recordings in the Vibrissae Barrel Cortices of Anesthetized Animals

The whole-cell recording technique has been used for about ten years for *in vivo* recordings.[21,22] A substantial number of studies were conducted in the vibrissae barrel cortex of anesthetized rats and most of them came to similar conclusions with respect to AP activity. As first reported, for urethane-anesthetized rats by Moore and Nelson 1998[23] and confirmed by Zhu and Connors 1999,[24] most neurons in the barrel cortex of animals anesthetized with barbiturates do not show evoked AP responses. In our laboratory, we conducted a series of recording studies under urethane anesthesia on identified neurons in the vibrissae region of the ventral posterior medial (VPM) thalamus and the barrel cortex. In the VPM we observed a mean of 0.5 APs per (6°) PW deflection,[25] a result that is only two-fold lower than the findings of unit recordings in the VPM, which report about ~1 AP per PW stimulus.[26,27] In contrast, for layer IV barrel cortex neurons we observed a mean of 0.14 APs per PW deflection,[28] a response that is five- to ten-fold smaller than what has been reported for layer IV neurons by unit recordings under the same anesthesia.[9,29] In layers II/III, we observed evoked AP rates of only 0.031 APs per PW stimulus,[30] which is about 40-fold less than what has been reported by unit recordings.[29] Thus, the low AP rate estimate of whole-cell recordings versus that of extracellular unit recordings is in line with the idea that unit recordings bias against cells with low firing rates. Indeed, if the firing rate estimates of whole-cell recordings for layer II/III are correct, most of these cells could not possibly be detected by unit recordings because they do not fire APs.

4. Results from Whole-Cell Recordings in the Vibrissae Barrel Cortices of Awake Animals

A major unknown in the studies discussed above is the effect of anesthesia on AP activity. To address this issue we performed whole-cell recordings in layer II/III and layer IV of the barrel cortex of awake head-fixed animals.[22] Results from this recording are shown in Figure 1.1. This layer IV cell (Figure 1.1A) was recorded for 20 min and discharged APs upon current injection (Figure 1.1B). However, besides the current evoked APs, only four further APs were observed during that time. No APs were observed when the animal was resting (Figure 1.1C), when it was alert and whisking (Figure 1.1C), and upon tactile stimulation of the appropriate whiskers (data not shown). Thus, it is hard to believe that the absence of APs in this cell was artifactual. Even in awake animals, AP rates are very low. For healthy cells with stable resting membrane potentials, spontaneous AP was around 0.1 Hz, which is higher than what one observes with whole-cell recordings in urethane anesthesia, but much lower than what has been reported with unit recordings.[31]

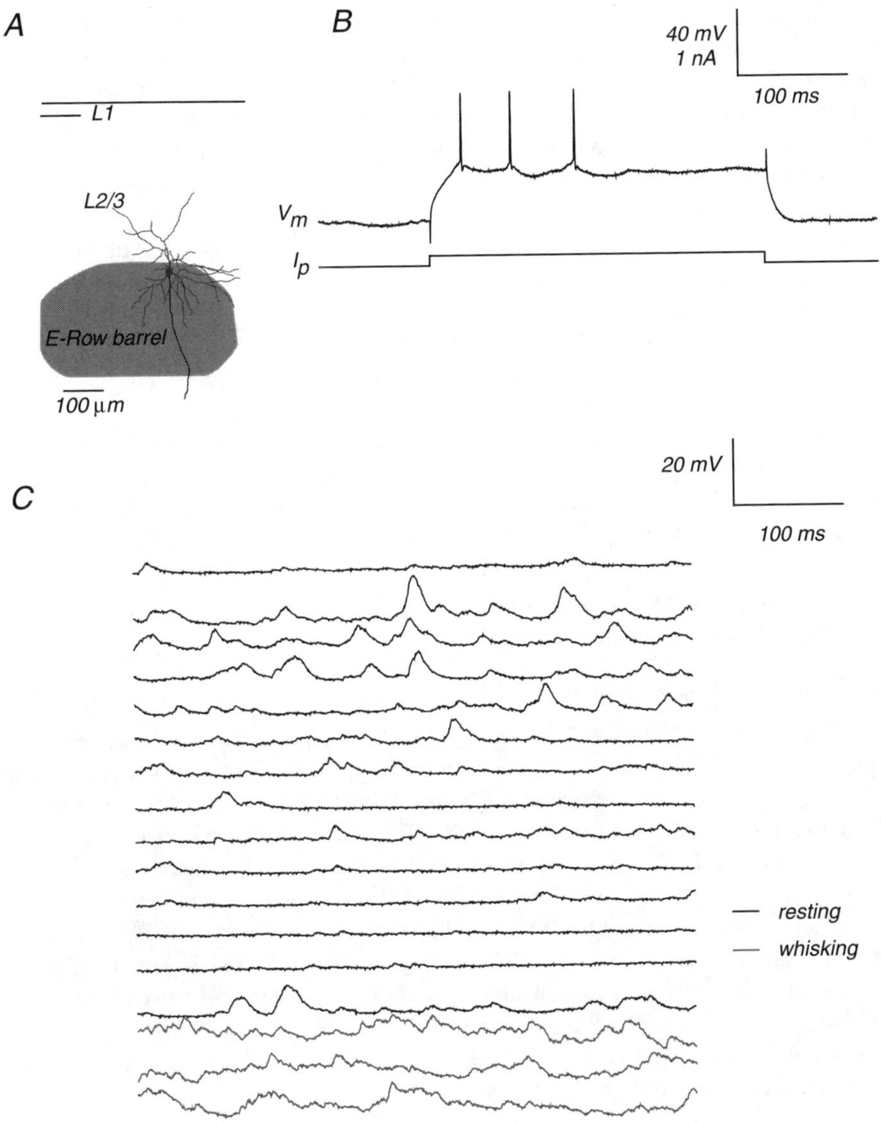

FIGURE 1.1 Activity of a star-shaped pyramidal layer IV neuron in an awake rat A, coronal section through the barrel cortex with topographic position, and the dendritic, axonal arbor of the stimulated pyramid in L4. B, Current injection reveals a regular spiking pattern. C, Ongoing activity of the neuron, while the animal is resting and while it is whisking. Note the absence of APs.

5. Inconsistencies and Caveats from Whole-Cell Studies

For obtaining whole-cell recordings, pressure applied to the pipette interior prevents pipette occlusion while cells are approached. As a consequence, intracellular high-potassium solution is pushed into the tissue, and this depolarizes neurons and leads to a transient depression of neuron firing. We therefore compared the results of whole-cell recordings obtained from recordings where we minimized the spill of internal solution (by patching cells with minimal pressure applied to the pipette interior and the first pipette in the experiment) to recordings with massive spillover (patching cells with high pressures after numerous electrode penetrations). With high spillover of potassium, postsynaptic activity can be suppressed during the first 1 to 2 minutes of the recording. This was not the case in recordings without large potassium spillover. However, after less than 5 minutes, recordings under the two conditions were indistinguishable. It is unlikely that spillover of internal solution is a major contributor to the low firing rates observed with whole-cell recordings. Another potentially confounding factor for whole-cell recordings is dialysis of cells by the recording pipette as described below.

Observations in the barrel cortex tend to indicate very low firing rates with whole-cell recordings.[23,24,28,30] One study however, reports substantially higher rates of AP activity (spontaneous AP rates of about 1 AP/s,[32] The origin of this difference unfortunately is unknown.

6. Results from Derivatives of the Whole-Cell Recordings Technique

a. *Cell-Attached Recordings*

As already mentioned,, dialysis of recorded neurons with intracellular solution may distort the results of whole-cell recordings. To address this issue we performed sequential cell-attached and whole-cell recordings of AP activity from neurons in the barrel cortices of anesthetized animals.[22] To exclude any unintended dialysis, we tested before and after the cell-attached recordings to see whether a giga-seal between pipette and recorded neuron was established. We observed low firing rates (spontaneous AP rates less than 0.1Hz, and less than 0.1 evoked AP per PW stimulus) in both cell-attached and whole-cell recordings and found that AP activity slightly increased in neurons after establishing the whole-cell configuration. Cell-attached recordings are single-cell extracellular recordings, selected for seal formation and not for AP activity, suggesting high firing rates observed with extracellular unit recordings are a result of sampling biases.

b. *Targeted Whole-Cell Recordings*

Since basically all techniques for recording cellular cortical activity *in vivo* rely on blind sampling, the question arises to what extent are the recorded neurons representative of the neuronal group or population. We recently developed a targeted recording technique based on two-photon-microscopy (two-photon targeted patching, TPTP) and applied it to fluorescently labeled layer II/III interneurons *in vivo*.[33] In line with other results from whole-cell recordings, we found that such interneurons display low levels of evoked AP activity (0.3 APs per PW stimulus). This estimate of AP activity is lower than what was reported from most unit studies of putative interneurons.[34-36] As neurons are optically selected by this technique, the possibility of non-representative sampling from a blind approach biased by firing rate is ruled out.

c. *Transneuronal Recordings of Spikelet Activity*

It has become clear that cortical interneurons are mutually coupled by electric synapses. In the case of strong coupling, presynaptic APs result in an AP-like waveform in postsynaptic interneurons called a spikelet.[37,38] As predicted, such spikelets are observed *in vivo* in recordings from interneurons.[33] These recordings reveal that large spikelets (>2mV) occur at low rates between 0.2 to 2.7 Hz (mean ca. 0.5 Hz). Since spikelet-events are likely to reflect APs in one or more electrically coupled presynaptic cells, this infers that most interneurons discharge at low rates <1Hz. It is important to note that AP rates inferred from transneuronal recordings reflect the activity of cells that are not directly recorded and thought to be unaffected by factors such as dialysis of intracellular solution. Thus, a large number of factors that could distort AP counts in conventional recordings can be excluded here. Once again, the AP rates are much lower than those reported from extracellular studies.[34-36]

7. All Techniques Agree that Firing Rates of Cortical Neurons are Very Heterogeneous

As discussed above, different recording techniques lead to different quantitative assessments of cortical activity. Still most researchers agree that spontaneous and evoked AP rates can be very diverse and may vary by ×10 to ×100 between cells.

B. Evaluation of Postsynaptic Activty in the Somatosensory Cortex

1. Experimental Agreement on the Amplitude of Subthreshold Signals

In sensory cortices, almost all cells show postsynaptic responses upon sensory stimulation. Signals from simultaneous intracellular and local field potential recordings are well correlated (Brecht, unpublished observations), and simultaneous whole-cell recordings and voltage-sensitive dye recordings also show an excellent correspondence.[39] Sharp microelectrode recordings and whole-cell recordings in a variety of cortical areas report similar amplitudes of evoked responses of 5 to 25 mV. In barrel cortex, postsynaptic potential (PSP) amplitudes of sensory responses recorded with sharp microelectrodes[18,40] and whole-cell recordings[23,24,28,30] usually differ by a factor of less than two when the average of the population of recorded cells is considered. Such consistency is very remarkable if one considers the fact that estimates of AP activity may differ by more than two orders of magnitude, i.e., they range from 0.1 Hz recorded by whole-cell recording in the barrel cortices of awake animals to >10 Hz recorded by sharp microelectrodes in various cortical areas of cats.[19]

2. The Synaptic Composition of Postsynaptic Responses Is Controversial

Given general agreement on the amplitude of PSP responses, opinions diverge on the composition of synaptic events that underlie such responses. Some authors favor the idea of a large number of synaptic excitatory and inhibitory inputs that balance each other out to create much smaller net PSPs. Such hypotheses are referred to as high input regimes or synaptic bombardment scenarios.[41,42] We have suggested that this may not be the case and that cortical responses are generated by a few carefully selected inputs designated the selective input regime.[28]

Cortical neurons make and receive a large number of synaptic contacts. For example a layer II/III neuron may receive and allocate about 10,000 (thousand) terminals.[43] Since neuron connections between neurons often consist of around five terminals,[44,45] each neuron will form connections with a few thousand pre- and postsynaptic cells. In the barrel cortex, it is clear that the average amplitude of single unitary connections is often in the size range of 10% of a sensory evoked PSP. Thus, layer IV PSPs for PW stimuli are on average around 14 mV in size, whereas layer IV to layer IV unitary connections are around 1.6 mV in amplitude.[44] In layer II/III neurons, sensory evoked PSPs are around 9 mV in amplitude and layer II/III to layer II/III unitary connections are around 0.8 mV[46] in amplitude. In layer V neurons sensory evoked PSPs are on average 5 mV,[47] while unitary connections are found to be 0.3 mV.[48]

From these numbers it is clear that few out of the several thousand presynaptic inputs could underlie sensory responses in the barrel cortex, but a balanced excitation-to-inhibition scenario is also possible. This balanced scenario suggests that huge inhibitory and excitatory inputs are hidden in net response. Informal testing of such scenarios by current injection experiments fail to uncover such hidden inputs (Brecht, unpublished data). For slightly suboptimal stimuli, one often observes unitary-response-like synaptic events and total response failures (Brecht and Sakmann, unpublished observations). Such observations are difficult to reconcile with the idea that responses are generated by hundreds or thousands of balanced excitatory and inhibitory synaptic inputs.

3. An Attempt to Quantitatively Determine the Synaptic Composition of a Cortical Sensory Response Suggests Very Low Presynaptic Activity

We tried to quantitatively determine the synaptic composition of a sensory response of a layer II/III pyramidal cell by combining voltage clamp experiments and compartmental modeling. In the ideal voltage clamp experiment, the contribution of excitation and inhibition could be directly estimated from the current responses at different clamping potentials. However, the real experiment is subject to voltage clamp and space-clamp errors. Thus, to understand the currents actually active during a sensory response we tried to correct for these errors using compartmental modeling of the recorded neuron. We simulated a sensory response recorded from a morphologically reconstructed layer II/III neuron (Figure 1.2A) using a multicompartmental neuron model[49] of a morphologically reconstructed layer II/III neuron. Passive properties of the model, such as membrane resistance, membrane capacitance, and axial resistance were adjusted to appropriate values by comparing the voltage responses of the simulated neuron to those of the recorded neuron.

To simulate the real synapses from layer II/III and layer IV, synaptic conductance changes were placed onto the model neuron. Geometric synaptic distribution and unitary synaptic strength for these synapse-models were taken from *in vitro* measurements.[46,50,51] To mimic realistic temporary synaptic distributions, presynaptic spike times measured *in vivo* in layer II/III and layer IV were convolved with typical PSP response time course measured *in vitro*.

The response analysed here (Figure 1.2B) was small (approximately 100 pA amplitude), but in the range of other PSC amplitudes observed for PW stimulation in these experiments (30pA – 180pA, n = 7). Analysis of this realistic model and comparison of its sensory responses to the recorded sensory response revealed that 10 to 30 active excitatory unitary connections and 1 to 10 inhibitory unitary connections best reproduce the clamping behaviour of the recorded sensory responses (Figure 1.2C). More precisely in this scenario, which reproduced the real measurement best, 50 excitatory and 7 inhibitory active synaptic terminals corresponding to ~11 excitatory and ~2 inhibitory presynaptic neurons were active (taking 4.5 as the average number of synaptic terminals per unitary connection[46]). From our simulations one can reject a high input scenario (Figure 1.2D), because the voltage clamp

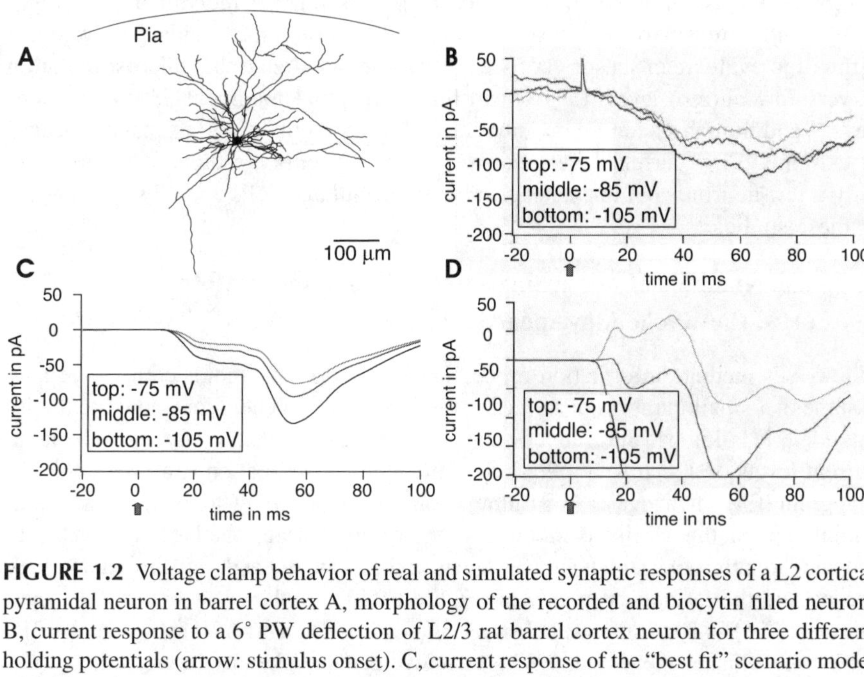

FIGURE 1.2 Voltage clamp behavior of real and simulated synaptic responses of a L2 cortical pyramidal neuron in barrel cortex A, morphology of the recorded and biocytin filled neuron. B, current response to a 6° PW deflection of L2/3 rat barrel cortex neuron for three different holding potentials (arrow: stimulus onset). C, current response of the "best fit" scenario model with 50 excitatory and 7 inhibitory synaptic terminals, which is reproducing the real experiment most closely. D, current response of the "synaptic bombardment" scenario model with ~ 500 excitatory and ~ 350 inhibitory synaptic contacts.

behavior of such responses is entirely different from the measured ones (Figure 1.2B). Our analysis was designed to take into account space clamp problems, high access resistance, etc. Nonetheless, it is likely that multiple errors such as mistakes in estimates of the chloride reversal potential, access resistance or geometric distribution of the synapses, will distort our estimate of synaptic inputs. Still, our data seem to rule out the possibility that massive inhibitory inputs mask excitation in layer II/III cells, because such inputs target proximal regions of the neuron and would have undoubtedly been detected in our somatic clamp experiments. In summary, these data suggest that only <100 out of about 10,000 terminals or only <50 unitary connections out of several 1,000 inputs are active during a sensory response to an optimized stimulus (a large PW deflection). Given the experimental uncertainties, one should treat these data as an order of magnitude estimate rather than an ultimate count of synaptic inputs. Nevertheless, this evaluation is a further indication that most cortical synapses are silent during whisker stimulation.

C. Evaluation of the Motor Effects of APs in the Primary Motor Cortex

1. Low Firing Rates Imply a High Efficacy of Cortical APs

If the aforementioned assessments of cortical AP activity by whole-cell recordings were correct, one would conclude that few APs carry out cortical processing. Thus, contrary to mass action views of cortical processing[41,42,] individual APs might significantly impact on the result of cortical computations. This possibility is difficult to evaluate in sensory cortices. It was demonstrated that microstimulation at very low current levels (5 μA) can bias perceptual judgments.[52] Rats,[53] monkeys,[54] and humans[55] can report intracortical microstimulation in sensory cortices at extremely low current levels (<2 μA). Such currents are thought to stimulate only a few neurons, but the exact number of stimulated cells and APs is unknown in these studies.[56]

2. Small Numbers of APs in Single Cells of the Primary Motor Cortex Can Evoke Movements

If few APs mediate information processing in the primary motor cortex, one may assume that small numbers of APs can affect movement generation. Consistent with this idea, it is known that intracortical microstimulation can evoke movements at current levels of 1–2 μA.[57,58] Similarly, intracellular stimulation can evoke electromyogram (EMG) changes.[59] We followed up this evidence by applying intracellular stimulation in the vibrissae motor cortices of lightly anesthetized rats. Whisker movements offer special advantages because they can be easily quantified and they escape anesthesia-induced paralysis (Figure 1.3A). As shown in Figure 1.3, small numbers of APs (10 in this case, Figure 1.3C) initiated in a layer VI cell (Figure 1.3B) can evoke whisker movements. Whisker movements evoked by intracellular stimulation were usually of small amplitudes (on the average about 0.5°). Movements evoked by intracellular stimulation were complex because they consisted of long sequences of movements (a duration of many seconds, Figures 1.3D and E) and

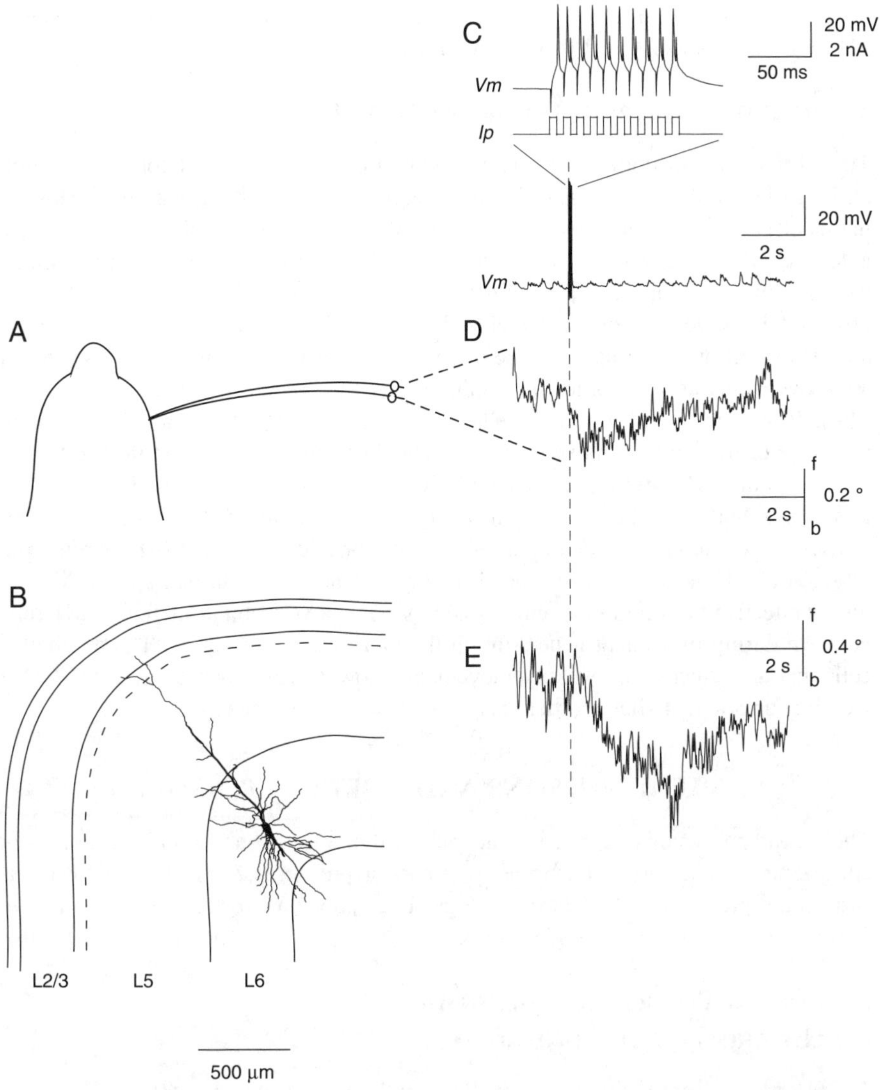

FIGURE 1.3 Whisker movements evoked by intracellular stimulation of an L6 pyramidal neuron. A, dorsal view of a rat's snout. top: two whisker positions observed in the intracellular stimulation experiment are illustrated. Only the C2 whisker, which carries a reflex foil label, is drawn. B, coronal section through M1 with topographic position, and the morphology of the stimulated pyramid in L6. The cell was recorded close to a site where extracellular stimulation evoked backward movement of C and B whisker rows. C, membrane potential recordings and injection pattern of depolarizing current steps during intracellular stimulation (10 action potentials (APs) at 100 Hz).D, position of whisker C2 during the intracellular stimulation trial. E, movement average of 15 single-cell stimulations. The dashed lines in C–E indicate the onset of AP initation.

always involved multiple whiskers. The direction of evoked movements depended on the frequency of initiated APs. Thus, it seems that small numbers of APs in single M1 cells could specify motor programs for whisker movements.[60]

3. Effective Cells Greatly Outnumber Active Cells in M1

The vibrissae motor cortex takes up a large part of the rat's frontal cortex and based on its surface area, we estimate that it contains 1 to 1.5 million neurons.[61] Most of these cells are layer V and layer VI neurons, and in about 20% of these cells we were able to evoke movements by intracellular stimulation.[60] A 0.5° movement amplitude would seem small in an awake animal, where whisker movements of up 100° are observed. However, it is of considerable size if one takes into account that movements are very small under anesthesia. The ongoing whisker movements in the case shown above were only about 1° or less in amplitude (see the single trial data in Figure 1.3D). Thus, 10 APs in one of a million cells can evoke movements with amplitude within the range of the ongoing movements. Intracellular stimulation is even more effective in awake animals.[60] Given that intracellularly evoked movement amplitudes were often around 0.5° and up to 2 to 3°, it seems likely if there is a linear relationship, that 100° movements could be mediated by a few hundred or a few thousand M1 neurons, i.e., by <1% of M1 neurons. Taken together, the evidence suggests that APs in M1 are highly effective in evoking movements and that only a very small fraction of M1 cells is active during movement generation in the vibrissae motor cortex. The number of cells that are effective in evoking movements, however, is much larger (20%). One would conclude that silent cortical neurons are not simply ineffective.

II. SILENT NEURONS AND CORTICAL PLASTICITY

The second part of this chapter is concerned with the functional significance of silent cortical neurons. If >90% of neurons are silent, the presence of so many silent neurons is a central problem of cortical physiology. The major goal of this section will be to frame questions that could guide research on the significance of silent cortical cells.

A. Cortical Plasticity and Alternative Hypotheses for Silent Neurons

1. Silent Cortical Cells as a Corollary of Metabolic Demands

Theory holds that it is implausible that many cortical cells can be active at any given time due to energetic costs of firing APs.[63] This theory concludes that less than 5% of cortical neurons could be active even within specifically activated cortical regions (e.g. visual cortex during visual stimulus presentation).

2. Contribution of Silent Cells to Learning and Cortical Plasticity

What if the animal was posed with learning a novel situation where it could adapt, learn, and perform? An attractive hypothesis for silent cortical cells places them in mechanisms of learning and plasticity. They are conspicuously quiet targets for transformation during plasticity into substrates for nascent neural activity. A conversion of a silent to a spiking neuron is thought to occur as a consequence of a plastic change/learning process and may enhance the ability of the cortical network to adapt to changing demands. A hypothesis like this was expressed (among others) by Dykes and colleagues[2] and by Moore and Nelson 1998[23] which suggested that the large subthreshold RFs of cortical cells serve this purpose. Tenuous support for an idea like this is that silent cortical cells seem to be anatomically integrated into the normal network of cortical neurons and evidence implicates networks in learning and memory. Studies on expression of immediate early genes show that large numbers of cortical neurons change gene expression during cortical plasticity; these data thus point to activity in (putatively the erstwhile) silent cells.[64]

3. Silent Cells May Signal by Non AP-Dependent Signaling

Not all neural communication among cortical neurons depends on AP generation. Thus, pre- and postsynaptic neurons exchange complex molecular signals and communicate electrically via miniature PSPs in the absence of APs. These forms of neural communication deserve special attention and may be important for maintaining synaptic strengths in cortical and particularly silent cortical cells.

4. Silent Cells May Function by Sparse AP Activity

Although most cortical cells in sensorimotor cortices fire APs only rarely, even such sparse AP activity, if it is close temporal relation with other sparsely firing neurons (ie. AP rates <0.1 AP per optimal stimulus), could have functional significance.

B. Data on an Involvement of Silent Cells in Plasticity are Largely Absent

Critical experiments on the significance of silent cortical neurons have not been done yet. Evidently, recording 0.5 to 2 h from a neuron without AP activity does not reveal the cell's function. An involvement of silent cells in cortical plasticity would best be revealed by monitoring sub- and suprathreshold activity of identified cells over extended time periods. Chronic recordings of unit activity are possible, but even with tetrode techniques it will be difficult to verify that previously non-spiking cells join the population of active cells. The most promising techniques to confront the problem of silent cells might lie in two photon microscopy-based optical measurements of neural activity. Visualization of neural activity based on dye injection[65] or genetic encoded indicators can be done with cellular resolution and individual cells can be identified over extended time periods.[66]

C. Silent Neurons and Synaptic Learning Rules

Here, we consider synaptic mechanisms that could lead to a recruitment of previously silent cells into the population of spiking neurons. Potential mechanisms of this change will be constrained by the cortical activity's quantitative aspects which are

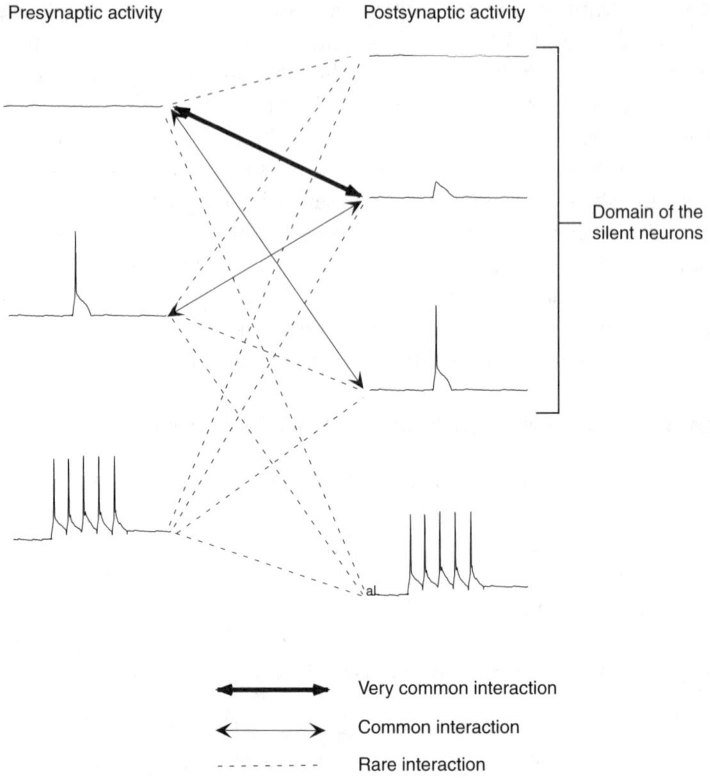

FIGURE 1.4 Pre- and postsynaptic activity patterns during sensory stimulation. Left side, presynaptic activity patterns: 0 APs, Top; Sparse APs, middle; Bursts of APs bottom. Effects of subthreshold presynaptic activity on axon terminals are neglected. Right side, postsynaptic activity patterns: no subthreshold activity, Top; subthreshold activity, second from top; subthreshold activity and sparse APs, third from top; subthreshold activity and bursts of APs, bottom.

worth recalling.. For optimized stimuli (a strong PW deflection), a large fraction (>90%) of neurons in a cortical column and most neurons in neighboring columns show only subthreshold responses. Thus, in layer II/III of the barrel cortex of an anesthetized rat, we estimate that about 40,000 neurons show subthreshold responses whereas only 200 to 300 APs are evoked in layer II/III.[30] So, in a scenario like this what type of information is available to silent cells and their synapses? Figure 1.4 illustrates the most common pre- and postsynaptic activity patterns that are likely to occur under strong sensory stimulation: The most common pre- and postsynaptic activity pattern will be between a silent presynaptic site and a silent postsynaptic site, which is exposed to substantial subthreshold postsynaptic activity. Less frequently pre- and postsynaptic activity pattern should consist of a silent presynaptic sites and postsynaptic sites with suprathreshold activity and backpropagating APs. A further pattern of pre- and postsynaptic activity consists of sparse presynaptic activity and postsynaptic sites with subthreshold activity. Other patterns of pre- and postsynaptic activity would seem to be rare.

From these considerations it becomes clear that many of the classic plasticity protocols that rely on strong pre- and postsynaptic activity cannot predict the types of changes that one may expect in silent cells. Not only in LTP type experiments, but also most paradigms involving spike–timing-dependent plasticity bursts of pre- and postsynaptic activity are applied.[67,68] It must be emphasized that most *in vitro* plasticity experiments are also done in the absence of neuromodulators, which are known to facilitate cortical plasticity *in vivo* and may be critical in silent cells. Learning rules which would seem to be important for understanding synaptic change in silent neurons are those uncovered by experiments on homeostatic mechanisms.[69] In these experiments, adjustments were observed in not only the AP thresholds but in the synaptic strength of cells in the absence of AP activity.[70,71] These synaptic changes may involve heterosynaptic learning rules and could be the mechanism by which silent cells use subthreshold activity to adjust synaptic strength at their largely inactive synapses. This is not to say that conventional forms of synaptic plasticity which involve strong pre- and postsynaptic are irrelevant. Most likely such types of plasticity will occur between the few highly active cortical neurons and will fine tune the properties of these most important cells.

We speculate that spike-based and nonspike-based learning rules set up a continuous competition: a large number of cortical neurons compete to generate a small number of APs that best represent sensory inputs and motor outputs.

III. CONCLUSION

While measurements of cortical subthreshold activity by sharp microelectrodes, whole-cell recordings, local field potential recordings, and voltage-sensitive dye imaging agree, there is no consensus about AP activity in sensorimotor cortices. Thus, we do not know yet whether >90% or >99% of neurons in sensorimotor cortices are silent. Evidence from whole-cell recordings, cell-attached recordings, postsynaptic estimates of incoming inputs, and intracellular stimulation experiments suggests a predominance of silent cortical neurons. Most but not all of the evidence suggesting that cortical neurons are generally active comes from extracellular unit recordings; this technique cannot detect nonspiking cells and therefore has an inherently poor quantitative performance in networks with silent cells. Resolving the discrepancies in our estimates of cortical AP activity will be fundamental for understanding cortical activity. Until then it might be wise to not treat the result of any one recording technique as representative. The most attractive hypothesis for the significance of silent cortical neurons is they can become converted to spiking neurons by forms of learning and they can thereby enhance the cortical ability to adapt to changing demands. Evidence for this hypothesis is lacking, however, and synaptic learning rules that adjust synaptic strength under conditions of low or no AP activity are not well characterized. We speculate that spike-based and non-spike-based learning rules constrain cortical learning where a large number of cortical neurons compete to generate a small number of APs that best represent the inner and outer world.

REFERENCES

1. Barlow, H.B., Single units and sensation: a neuron doctrine for perceptual psychology? *Perception*, 1, 371, 1972.
2. Dykes, R. and Lamour, Y., Neurons without demonstrable receptive fields outnumber neurons having receptive fields in samples from the somatosensory cortex of anesthetized or paralyzed cats and rats. *Brain Res.*, 440,133, 1988.
3. Mountcastle, V.B., Modality and topographic properties of single neurons of cat's somatosensory cortex. *J. Neurophysiol.*, 20, 403, 1957.
4. Mountcastle, V.B., Talbot, W.H., Sakata, H. and Hyvarinen J., Cortical neuronal mechanisms in flutter-vibration studied in unanesthetized monkeys. Neuronal periodicity and frequency discrimination. *J. Neurophysiol.*, 32, 452-84, 1969.
5. Welker, C. Receptive fields of barrels in the somatosensory neocortex of the rat. *J. Comp. Neurol.*, 166, 173, 1976.
6. Simons, D.J., Response properties of vibrissa units in rat SI somatosensory neocortex. *J. Neurophysiol.*, 41, 798, 1978.
7. Armstrong-James, M. and Fox, K. Spatio-temporal divergence and convergence in the rat "barrel" cortex. *J. Comp. Neurol.*, 263, 265, 1987.
8. Armstrong-James, M., Fox, K. and Das-Gupta, A. Flow of excitation within rat barrel cortex on striking a single vibrissa. *J. Neurophysiol.*, 68, 1345, 1992.
9. Diamond, M.E, Armstrong-James, M., and Ebner, F.F., Experience-dependent plasticity in the barrel cortex of adult rats. *Proc. Natl. Acad. Sci. USA* 90, 2602, 1993.
10. Huang, W., Armstrong-James, M., Reama, V., Diamond, M.E. and Ebner, F.F., Contribution of supragranular layers to sensory processing and plasticity in adult rat barrel cortex. *J. Neurophysiol.*, 80, 3261, 1998.
11. Armstrong-James, M., Diamond, M. E. and Ebner, F.F., An innocuous bias in whisker use in adult rats modifies receptive fields of barrel cortex neurons. *J. Neurosci.* 14, 6978-6991, 1994.
12. Warren R.A. and Dykes R.W., Population analysis of single neurons in cat somatosensory cortex. *Somatosens. Mot. Res.*, 4, 297, 1992.
13. Swadlow, H.A., Efferent neurons and suspected interneurons in motor cortex of the awake rabbit: axonal properties, sensory receptive fields, and subthreshold synaptic inputs. *J. Neurophysiol.*, 71, 437, 1994.
14. Swadlow, H.A., Efferent neurons and suspected interneurons in second somatosensory cortex of the awake rabbit: receptive fields and axonal properties. *J. Neurophysiol.*, 66, 1392, 1991.
15. Swadlow, H.A., Efferent neurons and suspected interneurons in S-1 forelimb representation of the awake rabbit: receptive fields and axonal properties. *J. Neurophysiol.*, 63, 1477, 1990.
16. Swadlow, H.A., Efferent neurons and suspected interneurons in S-1 vibrissa cortex of the awake rabbit: receptive fields and axonal properties. *J. Neurophysiol.*, 62, 288, 1989.
17. Shimegi S., Ichikawa, T., Akasaki, T., and Sato, H., Temporal characteristics of response integration evoked by multiple whisker stimulations in the barrel cortex of rats. *J. Neurosci.*, 19, 10164, 1999.
18. Carvell, G.E. and Simons, D.J. Membrane potential changes in rat SmI cortical neurons evoked by controlled stimulation of mystacial vibrissae. *Brain Res.*, 448, 186, 1988.

19. Timofeev, I., Grenier, F. and Steriade, M., Disfacilitation and active inhibition in the neocortex during the natural sleep-wake cycle: an intracellular study. *Proc. Natl. Acad. Sci. USA.*, 98, 1924, 2001.
20. Li, W.C., Soffe, S.R. and Roberts A. Direct comparison of whole cell patch and sharp electrodes by simultaneous recording from single spinal neurons in frog tadpoles. *J. Neurophysiol.*, 92 380, 2004.
21. Pei, X., Volgushev, M., Vidyasagar, T.R. and Creutzfeldt, O.D., Whole cell recording and conductance measurements in cat visual cortex *in vivo, Neuroreport*, 2, 485, 1991.
22. Margrie, T. W., Brecht, M. and Sakmann, B., *In vivo* low resistance whole-cell recordings from neurons in the anaesthetized and awake mammalian brain. *Pflügers Archiv. Eur. J. Physiol.* 444,491, 2002.
23. Moore, C.I. andNelson, S.B., Spatio-temporal subthreshold receptive fields in the vibrissa representation of rat primary somatosensory cortex. *J. Neurophysiol.*, 80, 2882, 1998.
24. Zhu, J.J. and Connors, B.W. Intrinsic firing patterns and whisker-evoked synaptic responses of neurons in the rat barrel cortex. *J. Neurophysiol.*, 81, 1171, 1999.
25. Brecht, M. and Sakmann, B., Whisker maps of neuronal subclasses in the rat ventral posterior medial (VPM) thalamus identified by whole-cell voltage recording and morphological reconstruction. *J. Physiol.*, 538, 495, 2002.
26. Armstrong, M. and Callahan, C., Thalamocortical mechanisms in the formation of receptive fieldsof rat barrel cortex neurones. II. spatiotemporal convergence in the thalamic ventroposterior medial nucleus (VPm) and its relevance to generation of receptive fields of S1 cortical "barrel" neurons. *J. Comp. Neurol.*, 303, 211, 1991.
27. Diamond, M.E., Armstrong-James, M., Budway,M.J. and Ebner, F.F., Somatic sensory responses in the rostral sector of the posterior group (POm) and in the ventral posterior medial nucleus (VPM) of the rat thalamus: dependence on the barrel field cortex. *J. Comp. Neurol.*, 319, 66, 1992.
28. Brecht, M. and Sakmann, B., Dynamic representation of whisker deflection by postsynaptic potentials in morphologically reconstructed spiny stellate and pyramidal cells in the barrels and septa of layer IV in rat somatosensory cortex. *J. Physiol.*, 543, 49, 2002.
29. Armstrong-James, M., The nature and plasticity of sensory processing within adult rat barrel cortex. In The Barrel Cortex of Rodents, Jones, E. G. and Diamond, I. T., Eds., Plenum Press, New York, 1995, pp. 333-374.
30. Brecht, M., Sakmann B., Dynamic receptive fields of reconstructed pyramidal cells in layers I/III of rat somatosensory barrel cortex, *J. Physiol.*, 553:243, 2003.
31. Fanselow, E.E. and Nicolelis, M.A., Behavioral modulation of tactile responses in the rat somatosensory system. *J. Neurosci.*, 19, 7603, 1999.
32. Chung, S., Li, X. and Nelson, S.B., Short-term depression at thalamocortical synapses contributes to rapid adaptation of cortical sensory responses *in vivo. Neuron*, 34, 437, 2002.
33. Margrie, T.W., Meyer, A.H., Caputi, A., Monyer, H., Hasan M.T., Schaefer, A.T., Denk, W., and Brecht, M., Targeted whole-cell recordings in the mammalian brain *in vivo. Neuron*, 39, 911, 2003.
34. Simons, D.J., Temporal and spatial integration in the rat SI vibrissa cortex. *J. Neurophysiol.* 54, 615, 1985.
35. Simons, D.J. and Carvell, G.E., Thalamocortical response transformations in the rat vibrissa / barrel system. *J. Neurophysiol.*, 61, 311, 1989.

36. Simons, D.J., Neuronal integration in the somatosensory whisker/barrel cortex. In The Barrel Cortex of Rodents, Jones, E. G. and Diamond, I. T., Eds., Plenum Press, New York, 1995, pp. 262-298.
37. Galarreta, M. and Hestrin, S., Electrical synapses between GABA-releasing interneurons. *Nat. Rev. Neurosci.*, 2 425, 2001.
38. Gibson, J.R., Beierlein, M. and Connors, B.W., Two networks of electrically coupled inhibitory neurons in neocortex. *Nature*, 402, 75, 1999.
39. Petersen, C.C., Grinvald, A. and Sakmann B., Dynamics of sensory responses in layer II/III of rat barrel cortex measured *in vivo* by voltage-sensitive dye imaging combined with whole-cell voltage recordings and neuron reconstructions. *J. Neurosci.*, 23, 1298, 2003.
40. Stern, E.A., Maravall, M., Svoboda, K. Rapid development and plasticity of layer II/III maps in rat barrel cortex *in vivo. Neuron.* 31, 305, 2001.
41. Shadlen, M.N. and Newsome, W.T. The variable discharge of cortical neurons: implications for connectivity, computation, and information coding. *J. Neurosci.* 18, 3870, 1998.
42. Destexhe, A. and Pare, D., Impact of network activity on the integrative properties of neocortical pyramidal neurons *in vivo. J. Neurophysiol.* 81, 1531, 1999.
43. DeFelipe, J. and Farinas, I., The pyramidal neuron of the cerebral cortex: morphological and chemical characteristics of the synaptic inputs. *Neurobiology*, 39, 563, 1992.
44. Feldmeyer, D., Egger, V., Lübke, J. and Sakmann, B., Reliable synaptic connections between pairs of excitatory layer IV neurones within a single 'barrel' of developing rat somatosensory cortex. *J. Physiol.*, 521, 169, 1999.
45. Markram, H., Lübke J., Frotscher, M., Roth, A. and Sakmann, B., Physiology and anatomy of synaptic connections between thick tufted pyramidal neurones in the developing rat neocortex. *J. Physiol.*, 500, 409, 1997.
46. Thomson, A.M., Activity-dependent properties of synaptic transmission at two classes of connections made by rat neocortical pyramidal axons *in vitro*, *J. Physiol.*, 502/1, 131, 1997.
47. Manns, I.D., Sakmann B. and Brecht M., 2004 Sub- and suprathreshold receptive field properties of pyramidal neurons in layers VA and VB of rat somatosensory barrel cortex. *J. Physiol.*, 556, 601, 2004.
48. Reyes, A. and Sakmann, B. Developmental switch in the short-term modification of unitary EPSPs evoked in layer II/III and layer V pyramidal neurons of rat neocortex. *J. Neurosci.*, 19, 3827, 1999.
49. Hines, M., NEURON — A Program for Simulation of Nerve Equations, aus \newline Eeckman F., Neural Systems: Analysis and Modeling, Kluwer Academic Publishers, 127, 1993.
50. Feldmeyer, D., Lübke, J., Silver, R.A. and Sakmann, B., Synaptic connections between Layer IV spiny neurons — layer II/III pyramidal cell pairs in juvenile rat barrel cortex: physiology and anatomy of inter-laminar signaling within a cortical column, *J. Physiol.*, 538/3, 803, 2002.
51. Gupta, A., Wang, Y. and Markram, H., Organizing principles for a diversity of GABAergic nterneurons and synapses in the neocortex, *Science*, 287, 273, 2000.
52. Murasugi, C.M., Salzman, C.D., Newsome, W.T., Microstimulation in visual area MT: effects of varying pulse amplitude and frequency. *J. Neurosci.*, 13, 1719, 1993.
53. Krauss, A., Manns, I., and Brecht, M., Laminar distribution of detection thresholds for electric stimulation in rat barrel cortex. SFN *Abstr.*, 2003.

54. Tehovnik, E.J., Slocum, W.M. and Schiller P.H., Saccadic eye movements evoked by microstimulation of striate cortex. *Eur. J. Neurosci.*, 17, 870, 2003.
55. Schmidt, E.M., Bak, M.J., Hambrecht, F.T., Kufta, C.V., O'Rourke, D.K and Vallabhanath P. Feasibility of a visual prosthesis for the blind based on intracortical microstimulation of the visual cortex. *Brain*, 119, 507, 1996.
56. Tehovnik, E.J., Electrical stimulation of neural tissue to evoke behavioral responses. *J. Neurosci. Methods*, 65, 1, 1996.
57. Asanuma, H., Stoney, S.D. Jr and Abzug, C., Relationship between afferent input and motor outflow in cat motorsensory cortex. *J. Neurophysiol.*, 31, 670, 1968.
58. Stoney, S.D. Jr, Thompson, W.D. and Asanuma, H., Excitation of pyramidal tract cells by intracortical microstimulation: effective extent of stimulating current. *J. Neurophysiol.*, 31, 659, 1968.
59. Woody, C.D. and Black-Cleworth, P., Differences in excitability of cortical neurons as a function of motor projection in conditioned cats. *J. Neurophysiol.*, 36, 1104, 1973.
61. Brecht, M., Schneider, M., Sakmann, B. and Margrie, T.W., Whisker movements evoked by stimulation of single pyramidal cells in rat motor cortex. *Nature* 427, 704, 2004.
62. Rockel, A.J., Hiorns, R.W. and Powell, T.P., The basic uniformity in structure of the neocortex. *Brain*, 103, 221, 1980.
63. Lennie, P., The cost of cortical computation. *Curr. Biol.*, 13,493, 2003.
64. Bisler, S., Schleicher, A., Gass, P., Stehle, J.H., Zilles, K., and Staiger, J.F., Expression of c-Fos, ICER, Krox-24 and JunB in the whisker-to-barrel pathway of rats: time course of induction upon whisker stimulation by tactile exploration of an enriched environment. *J. Chem. Neuroanat.*, 23,187, 2002.
65. Stosiek, C., Garaschuk, O., Holthoff, K. and Konnerth A., *In vivo* two-photon calcium imaging of neuronal networks. *Proc Natl Acad Sci USA*, 100, 7319, 2003.
66. Trachtenberg, J.T., Chen, B.E., Knott, G.W., Feng, G., Sanes, J.R., Welker, E. and Svoboda, K. Long-term *in vivo* imaging of experience-dependent synaptic plasticity in adult cortex. *Nature*, 420, 788, 2002.
67. Markram H, Lübke J, Frotscher M, Sakmann B., Regulation of synaptic efficacy by coincidence of postsynaptic APs and EPSPs. *Science*, 275, 213, 1997.
68. Sjostrom, P.J., Nelson, S.B., Spike timing, calcium signals and synaptic plasticity. *Curr. Opin. Neurobiol.*, 12, 305, 2002.
69. Desai, N.S., Rutherford, L.C. and Turrigiano, G.G., Plasticity in the intrinsic excitability of cortical pyramidal neurons. *Nat. Neurosci.*, 2, 515, 1999.
70. Leslie, K.R., Nelson, S.B., Turrigiano, G.G., Postsynaptic depolarization scales quantal amplitude in cortical pyramidal neurons. *J. Neurosci.*, 21, RC170, 2001.
71. Turrigiano, G.G., Nelson, S.B., Hebb and homeostasis in neuronal plasticity. *Curr. Opin. Neurobiol.* 10, 358, 2000.

2 The Vibrissa Resonance Hypothesis

Christopher Moore and Mark L. Andermann

CONTENTS

I. Overview 22
II. Introduction 23
A. The Behavioral Relevance of High Frequency Somatosensory Perception 23
B. Sensory Capabilities of the Vibrissa Sensory System 24
C. The Vibrissa Resonance Hypothesis 25
1. Resonance and Frequency Encoding in Other Sensory Systems 25
2. The Vibrissa Resonance Hypothesis 27
III. Vibrissa Resonance: Rate Coding and Place Coding of Frequency-Specific Tactile Information 28
A. Vibrissa Resonance 28
B. Neural Correlates of Vibrissa Resonance 30
1. Sinusoidal Stimuli 30
2. Complex Stimuli 33
C. Somatotopic Frequency Mapping and Isofrequency Columns 33
D. Velocity Sensitivity and Vibrissa Resonance 38
1. Frequency Tuning versus Velocity Tuning: Excitatory versus Inhibitory Neural Tuning in Rat SI 38
2. Velocity Sensitivity: Impact on the Representation of Specific Frequency Bands 38
E. Higher Harmonics: Implications for the Vibrissa Resonance Hypothesis 39
IV. Active Sensing and Vibrissa Resonance 41
A. Modulation of Whisking Velocity 41
B. Modulation of Vibrissa Damping 42
V. Temporal Coding and Vibrissa Resonance 42
A. Dynamic Evolution of Vibrissa Resonance Tuning: 'Contact' versus Frequency Coding during Different Epochs of the Response 42
B. Vibrissa Resonance and Fine Timing of Action Potential Activity 43

1. Vibrissa Resonance and Trigeminal Temporal Coding 43
2. Vibrissa Resonance and the Volley Principle 47
3. Vibrissa Resonance and SI Temporal Coding 50
C. Vibrissa Resonance and Intrinsic Neural Frequency Tuning 51
VI. Summary and Human Implications 54
Acknowledgments 55
References 56

I. OVERVIEW

The vibrissa sensory system has recently emerged as a predominant model for investigation of mammalian sensory processing. The anatomical markers of representation throughout the system, most notably the barrels in primary somatosensory cortex (SI),[1] make this system attractive for studies of neural structure and function. Further, rats can employ their vibrissae for a wide variety of perceptual tasks, ranging from shape perception to the fine discrimination of textured surfaces. Despite the growing interest in the vibrissa sensory system, until recently the transduction of information into neural activity by the vibrissae themselves has received relatively little attention.

In this chapter, we propose the vibrissa resonance hypothesis. We suggest that one of the intrinsic biomechanical properties of the vibrissae — their propensity to resonate at specific frequencies — plays an important role in information processing. Specifically, we propose that the amplification of high frequency inputs by vibrissa resonance should enhance the detection of small amplitude high frequency stimuli (e.g., of sound or of surface roughness) and that the frequency tuning within single vibrissae and the orderly organization of these properties across multiple vibrissae facilitates the discrimination of frequency specific stimuli. Vibrissa resonance may also be crucial to the temporal coding of sensory input on broad and fine time scales.

In the second section of this chapter, we provide an overview of the importance of frequency encoding in somatosensory processing. We then provide a review of other sensory systems in which resonance is proposed to play a role in signal transduction, most notably the auditory system. We conclude with a description of the vibrissa resonance hypothesis.

In the third section of this chapter, we review recent findings demonstrating that vibrissae resonate in response to sinusoidal and naturalistic stimuli, including sound pressure waves. We then describe the neural correlates of vibrissa resonance. Vibrissa resonance is translated into a frequency-specific increase in the rate of action potential activity demonstrated by peripheral and SI neurons. Because vibrissa resonance properties vary systematically across the face of a rat, with lower frequencies represented more posteriorly and higher frequencies represented more anteriorly, a map of frequencies is observed across SI with a system of isofrequency columns extending along arcs of vibrissae. These findings support the hypothesis that a place code exists for the representation of frequency information in this system, by which an increase in mean firing rate in a specific position within a map indicates the frequency of stimulation.

In the fourth section of this chapter, we discuss how active sensory processes may impact perception employed by vibrissa resonance. Specifically, we discuss how modification of the velocity of whisking of the vibrissae against and over objects may be crucial to engagement of vibrissa resonances, and we discuss the possibility that damping in the vibrissa follicle may play a role in modulating the expression of vibrissa resonance.

In the fifth section of this chapter, we provide evidence that vibrissa resonance not only contributes to a mean firing rate code and place code, but may also contribute to temporal coding of frequency information, considered over broad and fine temporal scales. Following the onset of a stimulus, vibrissa resonance has a relatively slow rise time, requiring tens of milliseconds to reach the full amplification of motion. Correspondingly, more optimal neural frequency tuning is observed at longer latencies after stimulus initiation (e.g., >25 msec post-onset). This finding suggests that initial neural activity may encode object contact and the somatotopic position of vibrissa deflection, while longer-latency activity may convey frequency specific information. We provide evidence from peripheral and SI recordings that the high frequencies, in the hundreds of Hertz (Hz), transmitted by vibrissae at their fundamental resonance frequency can drive high fidelity neural activity. These findings suggest that a volley principle may exist in the transmission of high resolution temporal signals from the periphery. Specifically, the sensitivity of this system to high velocity vibrissa motion, and the high degree of direction tuning observed in many trigeminal neurons, suggest that an inherent frequency doubling may occur through the convergence of signals from direction-tuned peripheral neurons.

In the final section of this chapter, we summarize the main findings and give a brief discussion of how the principles described here may relate to human tactile perception.

II. INTRODUCTION

A. The Behavioral Relevance of High Frequency Somatosensory Perception

Frequency information approximately tens of Hz to over a kHz is essential to accurate sensory processing in the auditory, somatosensory, and visual domains. The somatosensory system employs frequency-specific information in a variety of perceptual contexts. Vibratory stimuli transmitted through solid media provide an important sensory input to a variety of mammalian species. For example, blind moles are known to seek out conspecifics and termite colonies using primarily vibratory cues transmitted through the ground.[2] Elephants similarly are believed to detect seismic waves generated up to tens of kilometers away and may produce these signals as a means of communication.[3] Considering the natural arboreal environments of many primates, monkeys may detect vibrations resulting from motion occurring in the tree they are occupying, a potentially important alerting cue.

High frequency, temporally varying information also directly benefits human tactile perception. Surface perception in humans may employ temporally varying signals related to vibration perception. While spatially intensive cues dominate the judgment of roughness for surfaces with a periodicity in the range of ~1–3 mm,[4,5] temporal frequency information likely contributes to the judgment of spatial frequency patterns when microgeometric surfaces, with periodicity on the order of <1 mm, are perceived.[6,8] Vibratory stimuli are also an essential feature of the percept of spatio-temporally varying information. For example, robust tactile apparent motion illusions can be evoked when specific frequencies of stimulation, typically ≥50 Hz, are applied either sequentially across a series of three or more contact points or in bursts applied to ≥2 contact points.[9-11]

Although less commonly discussed, airborne media may also transmit high frequency information to the somatosensory system. Cockroaches can discriminate between laminar airflow of the type generated by wind from turbulent airflow of the type generated by an overhead predator using their cercal sensillae,[12,13] and specialized mechanical organs in the leg may be used to respond to sound stimuli.[14] Deaf individuals may also 'hear' sound through somatic receptors in what is likely a combination of vibration of the body cavity and perception of vibrations maintained in surrounding structures (e.g., the floor).[15] Airborne signals may be particularly important for animals that live in enclosed spaces such as tunnels, where sudden changes in air pressure or the interruption of air flow may indicate the entry of another animal into the system of tunnels or other important environmental changes.

B. Sensory Capabilities of the Vibrissa Sensory System

Different kinds of mammals, ranging from seals to chinchillae, employ long facial hairs (vibrissae) to obtain sensory information. In rats and mice, the lateral posterior surface of the face is covered with an orderly array of rows and arcs of vibrissae that are typically identified by letters (rows) and numbers (arcs). The longest vibrissae are located most posterior and are referred to as the macrovibrissae, and smaller microvibrissae are grouped more anteriorly in a dense patch. Although microvibrissae are likely important for perception,[16,17] especially of objects a rat is about to attempt to eat, they have received markedly less study. Because of this, references to vibrissae throughout the chapter will indicate the macrovibrissae only, although many of the observations described below should also apply to microvibrissae.

As suggested by the name *vibrissa*, mammals are capable of performing high-resolution frequency-related tasks with these sensors. With regards to airborne stimuli, rats can discriminate between different frequencies presented with an oscillating air stream.[18] In aqueous environments, seals can use their vibrissae to track moving objects through the water, a task likely accomplished by the detection of high frequency standing vortices generated by target movement.[19] Species of rats that gather food in aqueous environments have specifications of the vibrissa follicle and of the pattern of vibrissa innervation that suggest that they employ their vibrissae for navigation and foraging in opaque underwater environments.[19] Close parallels to vibrissa perception also exist in the lateral line system in fish.[20]

Rats can use their vibrissae for the detection and discrimination of solid textured surfaces,[21-25] tasks that likely requires the encoding of high frequency vibrissa vibrations generated by the interaction of the vibrissae with surface features.[21,26-28] * Using

* Several recent robotics projects employing vibrissa-like sensors have also chosen texture identification as an initial target problem.[29-31]

only their macrovibrissae, rats can detect the presence of a grating, as compared to a smooth surface, when the grooves in the texture are as small as 90 μm. This task requires only the presence of a single posterior vibrissa. Rats can also use their vibrissae to discriminate between periodic gratings with a spacing of 1.00 vs. 1.06 mm. For this kind of discrimination, two vibrissae, isolated to the same row, were minimally required.[21,22] Vibrissa-based texture discrimination capabilities also extend to more complex textures, like sandpaper[24,25] which contain a variety of complex frequency components[27-29] (see Figures 2.2, 2.5 and 2.6).

C. The Vibrissa Resonance Hypothesis

Sensory stimuli transmitted through airborne, liquid and solid media interacting with the vibrissae likely generate high frequencies of vibrissa vibration. In the case of the textures employed in behavioral tasks, estimates ranging from 200 Hz to over 1 kHz have been suggested for the rate of vibration generated when a rat sweeps its vibrissae over these surfaces.[21,27,28] An open question remains how the vibrissae can encode these high frequency signals. As described in this chapter, we hypothesize that vibrissa resonance may play a crucial role in frequency-specific transduction of tactile sensory information. In the following section, we provide background on other systems that are believed to employ resonance in the transduction of sensory input, followed by a more explicit statement of the vibrissa resonance hypothesis.

1. Resonance and Frequency Encoding in Other Sensory Systems

For any given object, there exists a specific set of frequencies that, when applied, will cause the object to demonstrate a larger relative amplitude of vibration. This property is known as resonance and can be observed in several common contexts. For example, when pushing a pendulum such as a child's swing, one can get the largest motion with the smallest amount of applied force if one pushes repetitively at the natural frequency of the swing (i.e., pushing the swing once for each cycle of motion). The largest resonance amplification is observed when stimuli are applied at the fundamental resonance frequency, but other higher harmonics are also observed and also amplify object motion. Larger amplitudes of motion applied at a given frequency also create larger velocities an important consideration for processing in the vibrissa sensory system. See Figure 2.1 for an example of a vibrissa's propensity to resonate.

The vibrissa resonance hypothesis follows a long line of similar proposals made for other sensory transduction systems. According to Wever, the first suggestion that the resonance properties of a transduction organ facilitate frequency-specific representation was made by Bauhin in 1605, with the proposal that the cavities of the inner ear resonate to enable sound representation.[30] Several other systems are thought to employ resonance in frequency-specific perception. Cockroaches possess cercal hairs that are capable of encoding different frequency components of airborne stimuli. This frequency-specific encoding is thought to result in part from the resonance tuning of the sensillae located on the cerci, whose frequency tuning varies inversely as a function of their length.[31,32] Variation in human sensory biomechanics

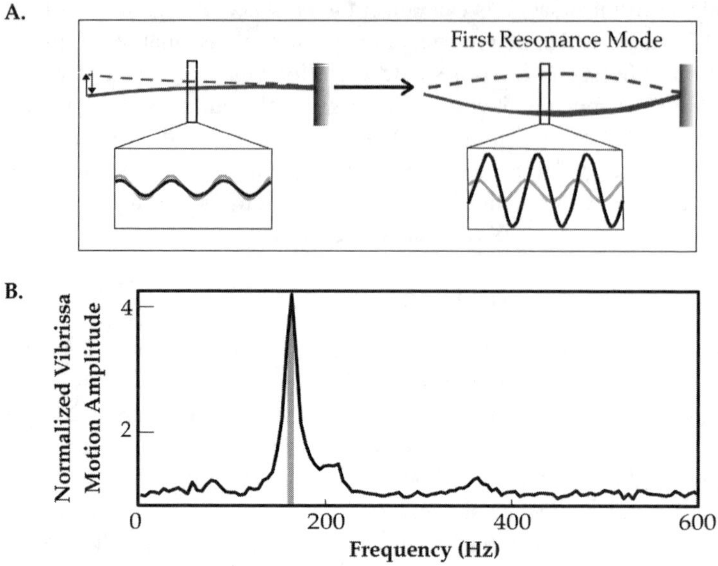

FIGURE 2.1 An example of vibrissa resonance tuning. A. When equal amplitude stimuli are applied to the tip (or base) of a thin elastic beam, a significantly greater amplitude of motion is observed at the fundamental resonance frequency of the vibrissa and at higher harmonics. In this diagram, the amplitude, frequency and phase of the stimulus input are shown in light gray, while the amplitude of the beam motion (at the midpoint along the beam) is shown in black. The image on the left depicts motion at a non-resonant frequency below the fundamental (first) resonance mode, and the image on the right depicts motion at the fundamental resonance frequency. B. An example of a vibrissa resonance frequency tuning curve. A constant amplitude stimulus was applied to the vibrissa tip at frequencies from 0–600 Hz, and the relative change in vibrissa motion amplitude recorded using an optical sensor placed at the mid-point of the vibrissa length. The fundamental resonance frequency in this example occurred at ~160 Hz (grey vertical bar). (Adapted from Neimark et al., *J. Neurosci.* 23, 2003. With permission.)

across the body surface may also play a role in the perception of vibratory stimuli.[33,34] Different regions of the hand have different mechanical impedances as a function of the constitution of substrates beneath the skin, for example, the relative concentration of bone, muscle, or fat[33,35].

The most famous proposal that resonance enhances frequency specific processing was made by Helmholtz, who suggested that the resonance properties of the cochlea generated a spatial segregation of frequency bands similar to the separation of frequencies on the strings of a piano, with low frequencies localized at one end of the cochlea and higher frequencies at the other.[36] While the precise mechanisms by which cochlear biomechanics transduce sound information have been substantially modified and updated for mammalian hearing since his initial prediction,* the basic principle proposed by Helmholtz has proven correct: The cochlea oscillates in specific regions of its extent as a function of the frequency of the stimulus presented. As such, the biomechanics of the cochlea form a spatial tonotopic map in which vibration, in response to a given frequency, is translated into increased firing rate in a specific subset of neurons in the auditory nerve that innervate around the point of largest cochlear motion. The spatial organization of these auditory fibers is preserved in an ascending system of neural tonotopic maps. This position-specific frequency amplification is believed to provide a place code in which the increased mean firing rate at a given position within a central auditory map is employed by an animal to make sound frequency judgments. While cochlear place coding is an important component of stimulus representation, several findings suggest that the fine timing of neural activity evoked by auditory stimuli may also play an essential role in stimulus representation and perception.[30,39,40] In support of this view, the place code and frequency tuning of the mean firing rate derived from cochlear amplification are thought to lack the specificity required to explain perceptual skill in lower frequency pitch judgments <500 Hz.[40]

2. The Vibrissa Resonance Hypothesis

We and others have recently discovered that the biomechanical properties of vibrissae may play a central role in frequency transduction.[26,28,37,38] Specifically, vibrissae resonate in response to stimulation at a relatively narrow range of frequencies generating frequency tuning in tactile sensory transduction that is translated into neural tuning in the periphery and the SI. Further, because the optimal frequencies for inducing resonance vary inversely with the length of the vibrissae, a map of frequency tuning is present across the rat's face. This map is also translated into a central frequency map in SI neural activity.[26]

These findings have led us to propose the vibrissa resonance hypothesis: that the amplification of frequency information by vibrissa resonance plays a meaningful role in the *detection* and discrimination of high frequency stimuli. There are three central components to this hypothesis. The first prediction is that vibrissa resonance enhances the detection of small amplitude high frequency stimuli because mechanical amplification of these signals generates a detectable increase in the neural mean firing rate. The second prediction is that vibrissa resonance facilitates the *discrimination* or identification of high frequency information in part through the tuning of individual vibrissae and their corresponding neural representation and in part through a place code provided by the systematic mapping of frequency preference across the face and in central somatic representations. The third prediction is that vibrissa resonance contribute to temporal coding of high frequency information by enabling the recruitment of precise temporal neural activity, most importantly in response to otherwise subthreshold small amplitude stimuli. Within this framework we predict that vibrissa resonance enhances the detection and discrimination of high frequency information generated in a variety of contexts, including airborne stimuli and textured surfaces.

---

* Alligator lizards have a population of free-standing inner hair cells whose lengths vary as a function of position along the basilar membrane and whose mechanical resonances largely determine their frequency tuning.[37,38] This kind of resonance mechanism for auditory tuning and mapping is in closer agreement with the initial Helmholtz proposal, and with the model of vibrissa sensory encoding proposed below.

III. VIBRISSA RESONANCE: RATE CODING AND PLACE CODING OF FREQUENCY-SPECIFIC TACTILE INFORMATION

A. VIBRISSA RESONANCE

Vibrissae resonance demonstrates oscillations of greater amplitude when stimulated at a narrow range of frequencies (Figure 2.1). As their shape suggests vibrissae can be modeled as a thin elastic conical beam.[28,37,38] As predicted by this model, the resonance tuning of a vibrissa is inversely dependent on its length, a factor that changes systematically from the back of the face (longer vibrissae, lower frequency tuning) to the front (shorter vibrissae, higher frequency tuning). As such, an anterior-posterior map of vibrissa frequency tuning is present across the face.[28]

Vibrissae demonstrate resonance tuning *ex vivo* (plucked from the face) or *in vivo* (attached to an anesthetized or behaving rat). and under a variety of boundary conditions at the vibrissa tip — when the tip is free and a sharp deflection is applied, when the tip or shaft of the vibrissa is pressed and moved over a surface (Figures 2.2 and 2.5), or when the tip is firmly attached to a stimulator (Figures 2.1 and 2.4).[28,38] Active sensing mechanisms, such as the control of vibrissa motion velocity and/or damping in the vibrissa follicle, may dictate the expression of vibrissa resonance in the behaving animal. In support of this suggestion there is initial supporting evidence that resonance may not be expressed in behaving animals when discrete contact of a vibrissa is made onto a vertical bar.[38]

Although the spatial frequencies present in the surface textures that make up a natural scene for vibrissa-based perception are not well documented and likely vary as a function of the many different environments that rodents inhabit. However the existing data suggest that the range of fundamental resonance frequencies present in a set of adult rat vibrissae may be well positioned to encode this aspect of their surrounding environment. Costa (2000) used a laser displacement sensor to show that natural surfaces, specifically 20 mm and 10 mm samples of concrete blocks and sandpaper, respectively, follow a power law distribution of spatial frequencies, $1/f^b$ (b = 2.68 for concrete and 100 grit sandpaper, b = 2.22 for 60 grit sandpaper).[41] These characteristics hold for spatial frequencies from 0.5 to 50 mm^{-1}, which span a range that rats can detect and discriminate with their posterior vibrissae, as assessed in studies using gratings.[21.22]

To study how vibrissae are employed to identify textures, it is important to understand the nature of the interaction between vibrissae and surfaces. As suggested,[27,28] depending on surface friction, the stiffness of the vibrissa, and the force of contact, vibrissa-texture interactions will fall on a continuum between a slip-stick behavior and a more reliable gliding movement (like that of a record player needle). In all cases, natural surfaces are expected to generate high frequency (>100 Hz) mechanical vibrations: Slip-stick behavior should lead to high frequency vibrations due to the impulsive, spectrally dispersed forces on the vibrissa, and the amplitude of these vibrations may be texture-specific. Whereas record player behavior should transduce the features of the surface more faithfully. We have observed a one-to-one translation of spatial frequencies to temporal frequencies

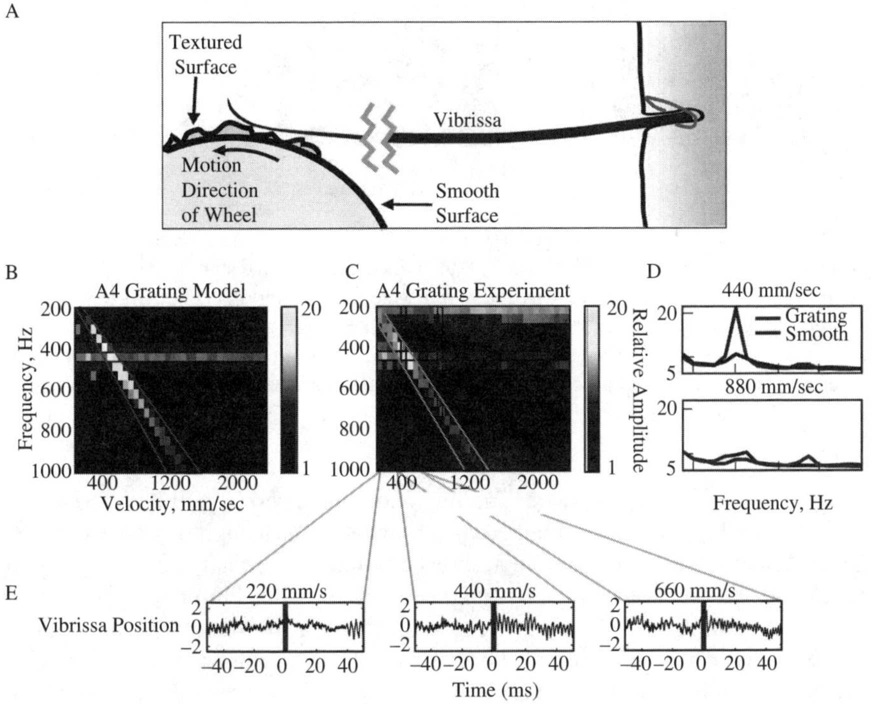

FIGURE 2.2 (See color figure following page 78) Vibrissae resonate when driven by complex natural stimuli. A. Half of a smooth wheel was covered with a textured surface, either a grating or sandpaper, and was moved at different velocities while contacting a vibrissa. This motion in turn drove vibrations in the vibrissa that were monitored with an optical sensor. B. and C. Plots comparing the power spectra of vibrissa oscillations driven by a grating at different speeds of wheel motion. Color scale indicates the relative amplitude of vibrissa motion. In C, increasing wheel speed caused an increased rate of vibrissa vibration as predicted by a one-to-one translation of the predominant frequencies of the grating as a function of wheel velocity (increased diagonal band of activation bounded by green lines). This signal was amplified when the grating drove the vibrissa at its fundamental resonance frequency (~400 Hz, horizontal band of activation bounded by black lines). A model of the vibrissa as a thin elastic beam[28] predicted this pattern of resonance amplification (B). D. The relative amplification of the vibrissa was shown for the grating surface and for a smooth surface for two speeds of wheel motion. The grating surface evoked peak amplification of vibrissa motion, although a small increase was observed at the vibrissa fundamental resonance frequency at ~400 Hz in both instances. E. Traces of vibrissa oscillation are shown for three distinct wheel speeds, with the red bar indicating the point at which the textured region of the wheel surface came into contact with the vibrissa. These traces show the amplification of vibrissa motion observed when the grating was moved at 440 mm/sec. (Adapted from Neimark et al., *J. Neurosci.* 23, 2003. With permission.) (continued)

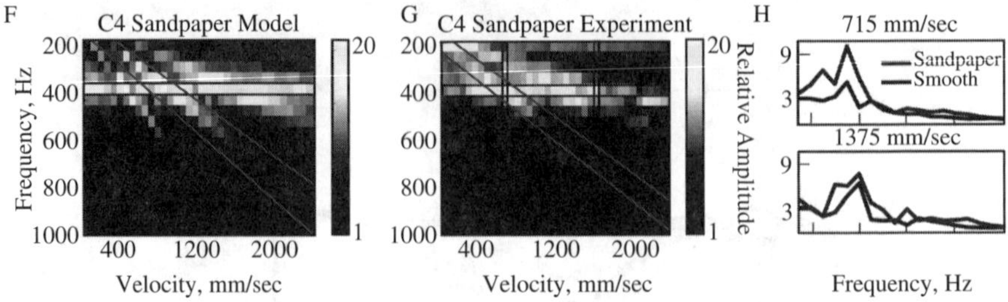

FIGURE 2.2 (CONTINUED) (See color figure following page 78) Vibrissae resonate when driven by complex natural stimuli. F. and G. Data plotted as in B. and C. for the vibrissa response. When 80-grit sandpaper was used as the textured stimulus. Note that both the smooth and textured surfaces drove increased vibrissa resonance at ~375 Hz, and that the textured stimulus was again more effective at driving motion at the fundamental resonance frequency when the wheel speed generated an optimal driving frequency. See Reference number 28 for further details and Figure 2.5 for an example of neural frequency tuning evoked under parallel stimulus conditions. (Adapted from Neimark et al., *J. Neurosci.* 23, 2003. With permission).

during surface interactions with gratings and sandpaper (e.g., Figures 2.2 and 2.5).[26,28,42] Preliminary data obtained using high-speed videography also suggests that the friction between vibrissae and surfaces like glass, sandpaper and gratings is high and that the drag on a vibrissa against a surface, in addition to its frequency response, could prove an essential sensory cue;[42] (J. Ritt, M. Andermann, and C. Moore, unpublished observation). Further recordings of vibrissa interactions with surfaces during behavior will be required to determine which interactions predominate during different perceptual tasks.

The range of fundamental resonance frequencies covered by the vibrissae also provides a complement to the audible range of perception in rodents. Cochlear transduction in rats initiates at a relatively high frequency, beginning between 250 Hz and 1 kHz, and hearing is typically poor below 2 kHz.[44,45] The vibrissa fundamental resonance frequencies observed in adult rats for the posterior macrovibrissae span from ~30 Hz to 1 kHz[28,38] and are likely higher for the microvibrissae. As such, vibrissae may extend the range of oscillatory encoding by amplifying airborne oscillations that would not otherwise be perceived or may be poorly represented by hearing in the rodent. In support of this prediction, vibrissa resonance amplification can be driven by sound stimuli and can demonstrate precise frequency tuning (Figure 2.3).

B. Neural Correlates of Vibrissa Resonance

1. Sinusoidal Stimuli

Vibrissa resonance amplification is translated into increased mean firing rate in neurons at multiple levels of the vibrissa sensory system. In our initial studies of the neural correlates of vibrissa resonance, we employed sinusoidal stimuli used to facilitate the precise characterization of frequency responses, and small amplitude tip deflections to reflect subtle surface features or airborne displacements. These

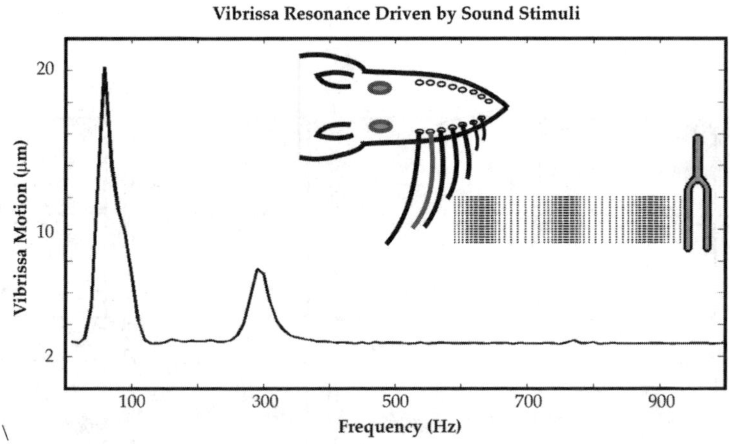

FIGURE 2.3 Vibrissae show precise frequency tuning in response to sound stimuli. A speaker was placed ~5 mm from an *ex vivo* C1 vibrissa fixed at the base. Pure sinusoidal tones were presented from 0–600 Hz (–1 to 1 volt input) while vibrissa motion was recorded using a photodiode sensor placed at the mid-point of the vibrissa length.[26,28] Under these conditions, vibrissa resonance at the fundamental resonance frequency (~55 Hz) led to frequency-specific amplification of vibrissa motion by approximately an order of magnitude, from 2 μm to 20 μm oscillations at the fundamental resonance frequency.

small amplitude stimuli also place the system in a range where amplification has an opportunity to show a differential impact on neural activity. In this context, the frequency-specific amplification of vibrissa motion generated by vibrissa resonance is translated into increased neural firing rate in peripheral neurons.[26] Among trigeminal neurons, 88% demonstrated frequency tuning associated with vibrissa resonance (Figures 2.4 and 2.11). Given the small (80 μm) vibrations that were applied at the vibrissa tip for these trigeminal recordings (a motion that subtends typically less than ~0.5 degrees of arc), the class of trigeminal neurons recorded in these experiments were most likely the slowly adapting (SA) neurons[46,47] that were previously reported to demonstrate the most sensitive receptive fields and the most robust sustained responses.

Cortical neurons also demonstrated resonance-related frequency tuning. Regular-spiking units (RSUs: putative excitatory neurons) and fast-spiking units (FSUs: putative inhibitory neurons) in SI barrels showed a relatively high probability of resonance-related frequency tuning (55% and 58%, respectively[26]; Figures 2.4 and 2.11). Beyond the demonstration of tuning, vibrissa resonance was often necessary to elicit any significant evoked activity. In these SI experiments we employed 160 μm deflections at the vibrissa tip, a stimulus that subtended less than 1 degree of arc in all cases. As can be seen in Figure 2.4, frequencies above or below the range of vibrissa resonance did not evoke any increase in evoked neural activity. This finding highlights the potential importance of vibrissa resonance for the detection of small amplitude sensory stimuli, as inputs of this amplitude would not have evoked neural activity without this amplification.

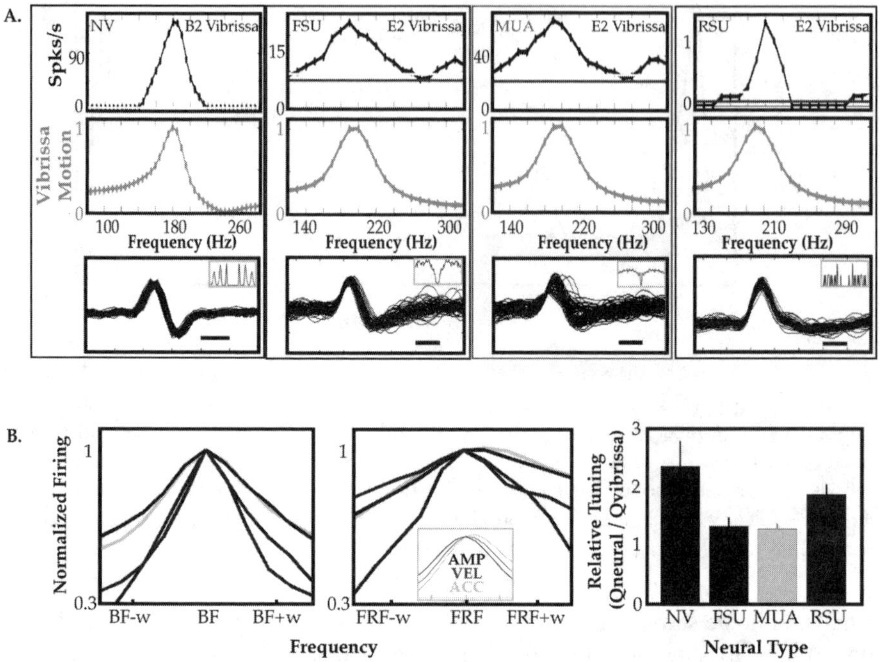

FIGURE 2.4 (See color figure following page 78) Vibrissa resonance tuning is translated into neural frequency tuning in somatosensory peripheral and cortical neurons. A. Vibrissa resonance tuning curves (gray lines, middle row of boxes) and corresponding neural frequency tuning curves (black lines, top row) are shown for peripheral and cortical recordings. The bottom row shows the corresponding spike traces for trigeminal (NV), fast spiking unit (FSU), regular spiking unit (RSU) and multi-unit activity (MUA). Green horizontal lines indicate the spontaneous firing rate; yellow lines indicate the threshold for significant evoked activity. Note that off-resonance stimuli were unable to evoke a significant increase in neural activity, demonstrating the potential importance of resonance for the amplification of sensory information. B. *Left and Center Boxes* Average neural tuning curves are shown for all four types of neural recording. In the graph on the left, average neural activity was normalized to peak firing rate and centered on the best frequency (BF), the frequency that drove the greatest increase in mean firing rate. On the right, average neural activity was normalized to peak firing rate and centered on the fundamental resonance frequency (FRF), the frequency that drove the greatest increase in the amplitude of vibrissa motion. All four classes of neural recording showed frequency tuning. *Right* The quality of neural frequency tuning (Qneural) normalized by the quality of vibrissa frequency tuning (Qvibrissa) for all four neural recording types. As seen in the BF- and FRF-centered average tuning curves, RSUs (red curve) and trigeminal neurons (green) demonstrated more refined tuning than FSUs (blue) and MUA (purple) for both averaging approaches.[26] See Figure 2.11. (Adapted from Andermann et al., *Neuron* 42, 2004. With permission.)

A consistent difference was observed in the specificity of tuning between recorded cell classes. The excitatory neuron classes in the periphery and SI, trigeminal neurons and RSUs, showed significantly more precise tuning than the FSUs (Figures 2.4 and 2.11). The greater precision observed in the frequency tuning of RSU responses parallels the tuning precision observed for other features in SI, as RSUs also tend to have more precise spatial tuning and more precise direction-of-deflection representations.[48,49] Several factors likely contribute to the sharper tuning of RSUs including inhibition by FSUs and more robust synaptic depression of thalamocortical inputs.[49-52] These neural mechanisms are probably not operating through a center-surround organization—inhibitory subthreshold potentials are for example, co-extensive with excitatory potentials throughout a canonical SI receptive field.[53-55] These mechanisms likely act to lower the total firing rate of RSUs in response to a wide range of stimuli, differentially suppressing weaker inputs and creating sharper tuning.

2. Complex Stimuli

We also recorded the neural response in a subset of neurons evoked by natural and complex stimuli. Figure 2.5 shows an example of a trigeminal recording maintained while a sandpaper stimulus was rolled over the primary vibrissa. As the speed of sandpaper motion was increased, the relative frequency of vibration of the vibrissa tip was correspondingly increased. Larger vibrissa oscillations were observed when the speed of the wheel combined with the prominent spatial frequencies of the sandpaper to drive the vibrissa at its resonance. In accordance, a greater mean firing rate was also observed in this recording when the vibrissa was driven at its resonance. These stimulus-generated velocities of vibrissa surface interaction are within the physiological range[21,56] suggesting that increases in mean firing rate should be observed in the awake behaving animal as they explore textured surfaces. In further experiments in the periphery and SI, we created white noise stimuli that were the sum of all sinusoids from 0–600 Hz, where each sinusoid was calibrated to 80 μm amplitude and randomly phase shifted. These stimuli were then notch filtered by selectively removing frequencies that would have driven vibrissa resonance while compensating for the total power in the stimulus across frequencies. When complex stimuli were applied without notching, a robust vibrissa resonance was observed and correspondingly, a significant increase in the mean firing rate (Figure 2.6). When the majority of frequencies amplified by vibrissa resonance were removed, resonance was not observed and the mean firing rate was also significantly lower. These examples demonstrate that the neural correlates of vibrissa resonance are observed even when more natural and/or complex stimuli are presented to the vibrissa.

C. Somatotopic Frequency Mapping and Isofrequency Columns

A potentially important consequence of the discovery of vibrissa resonance was the simultaneous discovery of a novel frequency map laid across the somatotopic map of the face.[28] As described above, the longer vibrissae located more posterior have

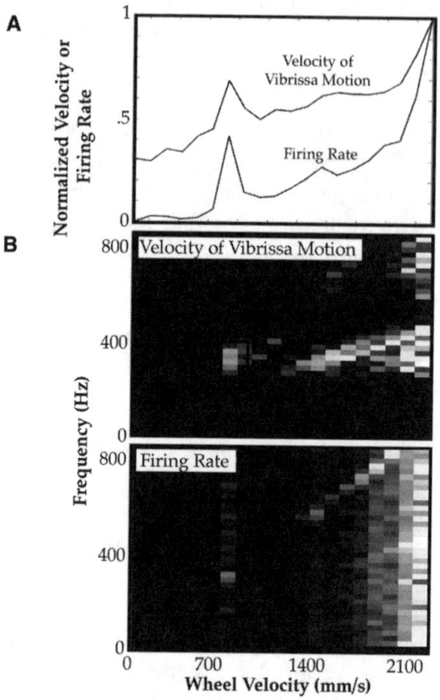

FIGURE 2.5 (See color figure following page 78) Vibrissa resonance evokes increased neural activity when natural complex (sandpaper) stimuli are applied. A. Multi-unit activity was recorded in the trigeminal ganglion while a stimulus wheel covered in 80-grit sandpaper was rolled against the primary vibrissa (see Figure 2.2 for a parallel example). As the wheel velocity increased, so did the vibrissa oscillation velocity (blue line). Vibrissa resonance amplification can be observed in the spike in vibrissa velocity (P(f)*f, *top*) at a wheel speed of 800 mm/s. Neural activity also showed a spike in mean firing rate at this velocity (green line). Neural activity also demonstrated a thresholded sensitivity to the increasing velocity of vibrissa oscillation at higher frequencies (≥ a wheel speed of 2000 mm/sec; see also Figure 2.8). B. Power spectra showing increased velocity of vibrissa motion or increased amplitude of neural activity (*bottom*) as a function of oscillation frequency and wheel speed. In the top panel, the peak in velocity signal at ~350 Hz (global increase in power) reflects the increased velocity of vibrissa motion generated when the wheel speed drove the predominant spatial frequency present in the texture (shown in the diagonal bands) at the vibrissa resonance (blue box). The increased mean firing rate in the associated neural response is indicated by the vertical band of increased activity observed at a wheel speed of 800 mm/s in the bottom panel. Note also that a peak is present in MUA power spectrum at the vibrissa resonance (~350 Hz), indicating fine temporal fidelity of spiking activity in response to a complex stimulus presentation (see also Figures 2.12–2.14).

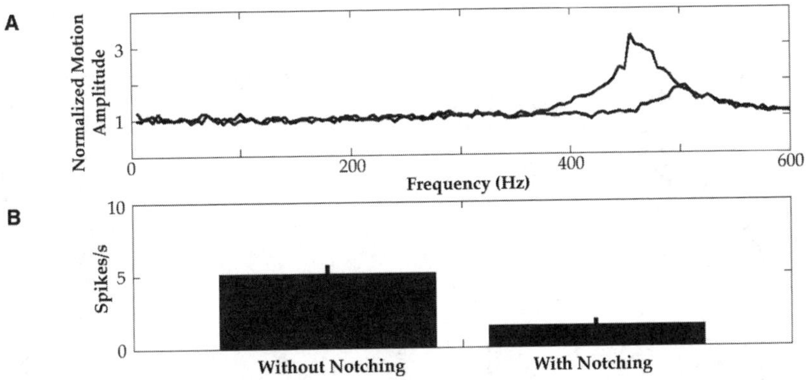

FIGURE 2.6 (See color figure following page 78) Vibrissa resonance evokes increased neural activity when synthesized complex stimuli are applied. A. White noise stimuli constructed as the sum of phase-shifted sinusoids from 0–600 Hz were presented through a piezoelectric stimulator to the vibrissa. A notched stimulus was also created in which the fundamental resonance and surrounding frequencies (400–500 Hz) were removed from the stimulus and the power adjusted across remaining frequencies. Vibrissa oscillations showed a resonance amplification at ~450 Hz when white noise stimuli were applied (green line) that is not present when notched stimuli were applied (blue line). B. These complex stimuli were presented while recording from a trigeminal ganglion single unit. Average neural activity was summed over the stimulation period (500 msec). As predicted by the differential increase in vibrissa motion, greater mean firing rate was evoked by the non-notched (green bar) than the notched stimulus (blue bar) (N = 37 trials, mean ± SE).

lower fundamental resonance frequencies than the shorter vibrissae located more anterior. This systematic peripheral gradient is translated into a frequency map in SI. As predicted by the consistent agreement between neural tuning and vibrissa tuning (Figure 2.4), the preferred vibratory frequency is higher in the barrels representing more anterior vibrissae (Figure 2.7). Because vibrissae in the same arc possess similar lengths, they also possess similar resonance tuning, creating a band of vibrissae with similar frequency encoding properties and a systematic network of putative cortical isofrequency columns.[26] This redundancy of frequency information across multiple vibrissae in an isofrequency column may provide robustness to the frequency code, diminishing the impact of the loss of or damage to a single vibrissa (Figure 2.7). Vibrissae along an arc may, in several perceptual contexts, encode substantially different kinds of inputs. The E row, located most ventral, may play a specific role in judging properties of the ground surface and in maintaining balance. The A row, which will seldom come into contact with the ground, could play a more specific role in the encoding of airborne frequency-varying signals (e.g., sound or changes in wind currents). A somatosensory isofrequency column may be subtended by neural regions biased to perform different processing tasks.

The frequency map and isofrequency columns discovered in SI mirror the tonotopic map and isofrequency columns described in primary auditory cortex (AI).[57,58] In both cases, the biomechanical properties of frequency transduction generate a systematic spatial mapping of frequency that is translated into a spatial position within the

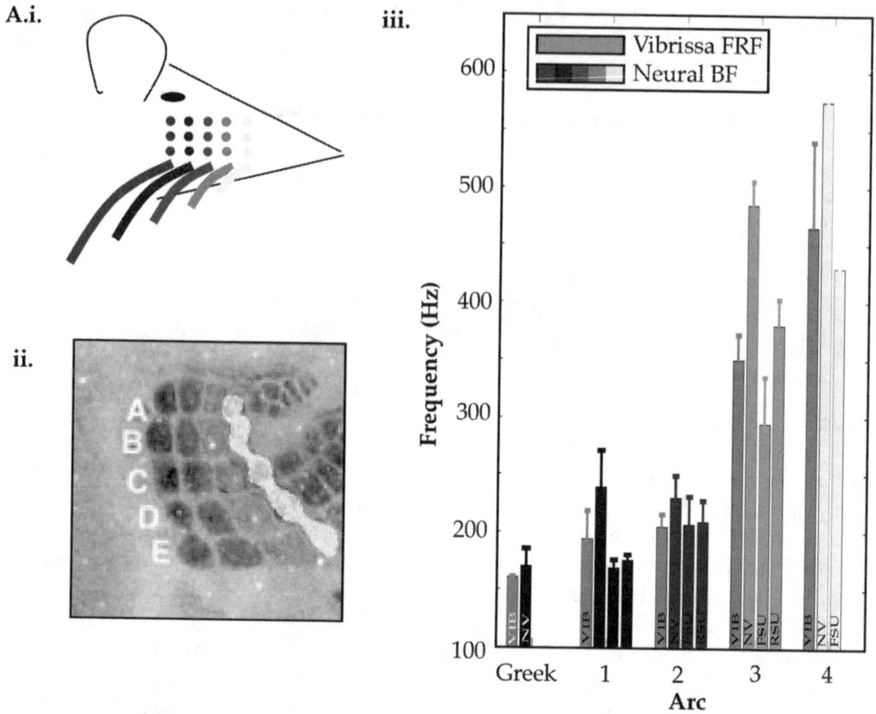

FIGURE 2.7 (See color figure following page 78) Vibrissa resonance creates a somatotopic frequency map and isofrequency columns in Si. A. i. A cartoon of the rat face, showing decreasing vibrissa length in more anterior vibrissae. ii. The similar lengths of vibrissae within an arc predict the existence of isofrequency columns, spanning multiple vibrissa representations. This prediction is shown on a cytochrome oxidase stain of the SI barrel map (anatomy from http://www.neurobio.pitt.edu/barrels). iii. Vibrissa fundamental resonance frequencies (FRF: gray bars) and the neural best frequency (BF: colored bars) increased as a function of arc position of the stimulated vibrissa.[26] (Adapted from Andermann et al., *Neuron* 42, 2004. With permission.) (continued)

cortical map. This similarity extends to the structure of the isofrequency columns in both systems as the elongated isofrequency lines observed in AI and SI show a similar aspect ratio (i.e., are asymmetrically extensive across the cortical sheet). An interesting nuance to the progression of isofrequency columns across the frequency map in SI is that the i and ii arcs tend to have vibrissae with similar lengths.[26,28] This characteristic of the system could provide a wider column of information in the approximately 100–300 Hz frequency band, potentially enhancing frequency judgments in this range.

The mapping of frequency across the somatotopic map and its parallels to the auditory map strongly suggest the existence of place coding in the processing of tactile information. Activation in a specific region of the SI map could be used to identify the frequency of stimulation applied across multiple vibrissae. This prediction does not necessarily imply that rats can use vibrissa resonance and a place code for perfect pitch identification of an isolated frequency presentation (a sensory capability that is also rare in the auditory system). The length of vibrissae varies over the life of the animal and changes on a daily basis making this kind of

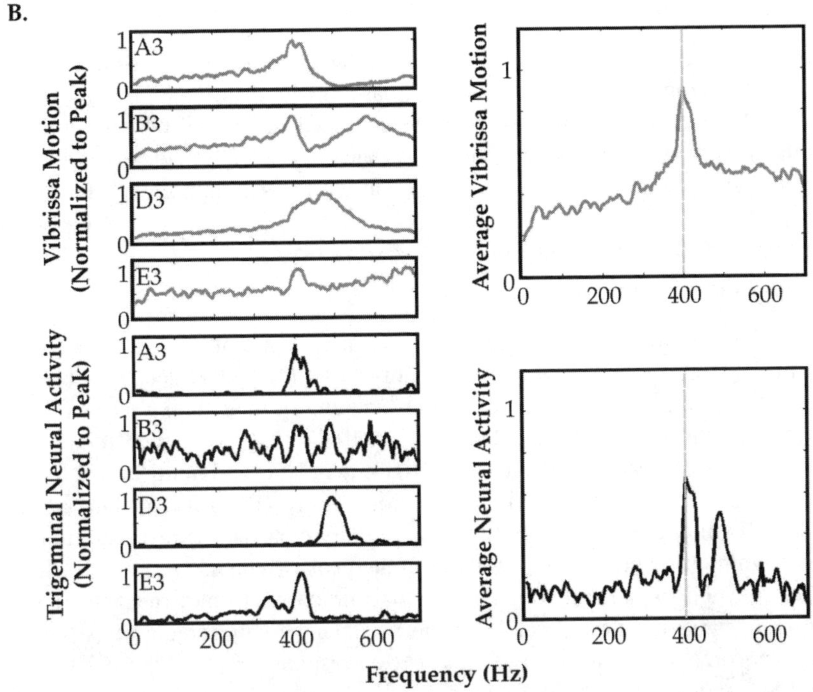

FIGURE 2.7 (CONTINUED) (See color figure following page 78) Vibrissa resonance creates a somatotopic frequency map and isofrequency columns in Si. B. *Left* Examples of four trigeminal single unit recordings obtained during primary vibrissa stimulation. Recordings were made from the same arc of vibrissae from one animal. Vibrissa frequency tuning curves are shown in the upper panels (gray), and neural frequency tuning curves in the lower panels (black). *Right* When responses were normalized and summed across the arc, a peak in vibrissa amplitude and neural activity was observed at ~400 Hz. This example highlights the coding of isofrequency information within an arc of vibrissae.

precision unlikely. Rather, the importance of the frequency map likely derives from the relative position of activation of columns of neurons in response to different frequency inputs — more posterior activation indicating relatively lower frequencies of presentation.

One classical argument for the utility of a place code, based on increased mean firing rate, is that it circumvents the need for fine temporal following of high frequency input. Mammalian auditory systems can routinely process signals of several kHz or higher. The refractory period of a typical neuron prevents the following of signals in this range. It remains an open question whether the vibrissae encounter and discriminate signals in the kHz frequency range, where place coding might be essential for adequate information transfer, and whether and how temporal encoding may contribute to high frequency perception. Even if the vibrissae only encode information accurately in the range of hundreds of Hz, a mean firing rate place code relieves the need to encode high frequency information in the precise timing of individual neurons.

D. Velocity Sensitivity and Vibrissa Resonance

Neurons throughout the vibrissa sensory system show robust velocity sensitivity.[59–64] This sensitivity strongly suggests that the increased mean firing rate observed as a product of vibrissa resonance results from the increased velocity of vibrissa motion and not the increased amplitude of motion.[26] As such, consideration of how velocity sensitivity interacts with resonance-related neural tuning is an important question.*

1. Frequency Tuning vs. Velocity Tuning: Excitatory vs. Inhibitory Neural Tuning in Rat SI

For sinusoidal stimuli, velocity increases as (frequency x amplitude). If neurons are sensitive to the increased velocity of vibrissa motion generated by vibrissa resonance, then they should demonstrate a broadening of their tuning curve on its high frequency side and/or a shift in the peak of neural tuning to a higher frequency. This behavior was observed in the average trigeminal, FSU and multi-unit activity (MUA) tuning curves recorded in our previous study (Figures 2.4 and 2.11). However, RSU tuning curves did not show a velocity bias, as their average tuning curve was symmetric with a peak centered on the peak of vibrissa resonance. This symmetry may be explained as the result of intracortical sharpening by FSU activity – FSUs respond strongly and at a shorter latency to high frequency stimuli, and are, therefore, well positioned to inhibit RSU responses[26] (see Figure 2.11 for relative FSU and RSU latencies). Whatever the mechanism, excitatory neuron activity in an SI barrel reflects the frequency tuning of vibrissa motion amplitude, and not velocity tuning, potentially providing relatively precise frequency coding that reflects vibrissa amplification.

2. Velocity Sensitivity: Impact on the Representation of Specific Frequency Bands

Diamond and colleagues recently presented a subset of high-frequency stimuli to multiple vibrissae simultaneously while recording MUA activity in SI.[63] These authors found that neurons were more sensitive to high frequency stimuli, and explained this trend as the product of velocity sensitivity in SI. We similarly observed that, for a given amplitude of vibrissa stimulation, many single neuron and MUA responses demonstrated high-pass sensitivity in addition to vibrissa resonance tuning (Figure 2.8)†. High-pass frequency/velocity sensitivity has several potential implications for the neural representation of vibrissa resonance. First, it suggests that those vibrissae that have a resonance tuning peak closer to the velocity threshold for the neuron being recorded will be more likely to evoke neural activity (Figure 2.8, middle frequency resonance tuning curve). Second, those vibrissae with lower fundamental resonance frequencies will require greater amplification to reach the velocity threshold (Figure 2.8, lower frequency resonance tuning curve). Third, those

* While it remains an open question whether neurons in the vibrissa sensory system are sensitive to "velocity" or to some other factor that co-varies with velocity, we will follow the trend in the literature and will use the term velocity as shorthand for this class of features.

resonances that are expressed above the high frequency threshold may fail to express neural resonance tuning for a given amplitude of stimulation. This is because of a ceiling effect as these motions are already effective in driving robust neural activity and there is little dynamic range in the mean firing rate remaining to allow the expression of resonance-related tuning (Figure 2.8, high frequency resonance tuning curve).

An important implication of this framework is that, for any given amplitude of stimulation, vibrissae with intermediate resonance tuning frequencies may be in the proper range to evoke neural resonance tuning. In our previous experiments, preliminary data suggested that the 1 to 3 arcs demonstrated the highest incidence of observable resonance-related neural tuning and that MUA responses showed the greatest sensitivity to these velocity interactions, i.e., single neurons were less susceptible to ceiling or non-amplification effects (M. Andermann and C. Moore, unpublished observation). A second implication of the apparent saturation of mean firing rate at high velocities of stimulation is that a secondary coding mechanism, beyond resonance or velocity sensitivity — e.g., the fine timing of neural activity — may be necessary to represent frequencies in this range if they are perceived distinctly by the behaving animal (see Section V).

E. Higher Harmonics: Implications for the Vibrissa Resonance Hypothesis

Higher harmonics of vibrissa motion may also generate high frequency/high velocity input to the vibrissa sensory system. In our previous studies,[26,28] we did not systematically characterize higher harmonics for a variety of study-design and methodological reasons. Nevertheless, we observed multiple examples of higher harmonics and many of these evoked neural activity. One example of this kind of response is shown in Figure 2.9.

While the importance of higher harmonics remains to be assessed, several factors suggest that they may not play a major role in perceptually relevant neural transduction. First, because a vibrissa approximates a tapered conical beam, the impact of higher harmonics should be diminished relative to the impact predicted in other structures (e.g., a plucked string).[27,28] Second, the ceiling effect described above for velocity sensitivity at high frequencies could reduce the impact of higher harmonics,

† The fact that these authors did not also observe the frequency tuning that results from vibrissa resonance is likely the result of several factors. First, they sparsely sampled frequency space, such that the distance between any two frequencies sampled (e.g., 131 and 211 Hz) was significantly larger than the entire span of an average vibrissa-resonance tuning curve in this frequency range (Figure 1). Second, these authors stimulated the bases of the vibrissae by resting them against a vibrating bar, a stimulus that likely damped or contaminated the expression of resonance prior to transmission to the follicle. Third, these authors recorded MUA activity, a class of neural responses that typically showed a greater bias towards velocity sensitivity when vibrissa resonance tuning was assessed (26. Andermann, M. L., Ritt, J., Neimark, M. A., and Moore, C. I., Neural correlates of vibrissa resonance; band-pass and somatotopic representation of high frequency stimuli, *Neuron* 42 (3), 451-63, 2004.: see Figure 10). Fourth, they did not analyze their data for the latencies of response post-stimulus that demonstrated the greatest precision of frequency tuning (25–100 msec post-stimulus onset), and, therefore, would not have observed the dynamic evolution of frequency tuning resulting from the vibrissa resonance time constant (see Section IV and Figure 10).

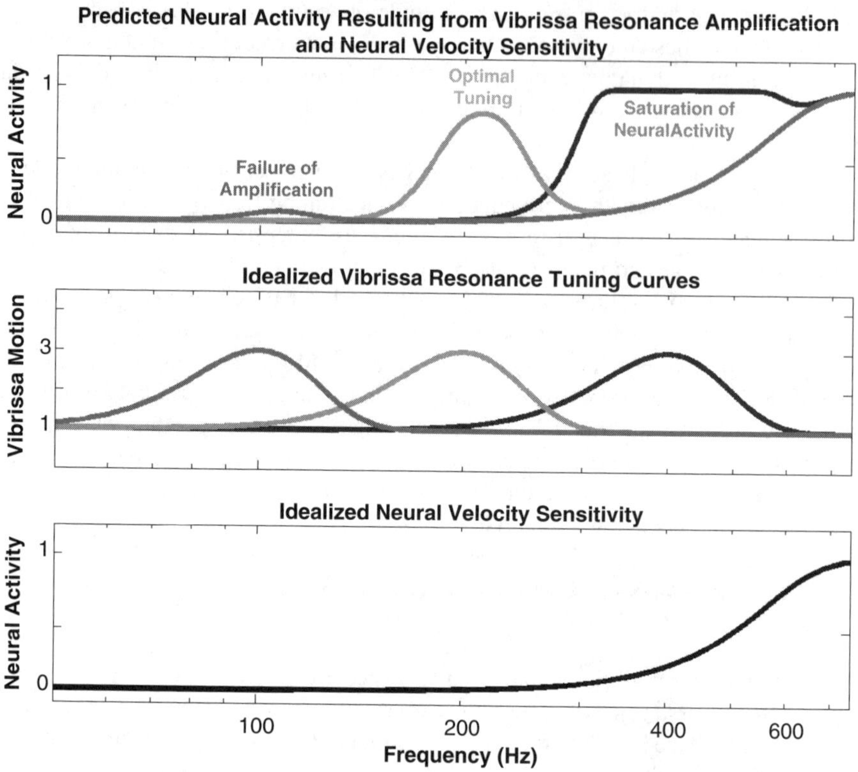

FIGURE 2.8 (See color figure following page 78) Neural velocity sensitivity may impact the expression of vibrissa resonance. *Bottom panel* A model of the neural response to vibrissa stimulation frequency in the absence of resonance amplification. This function was modeled as $\sin^2(pi*f/2000)$, $0 > f > 1000$ Hz, to emulate the neural sensitivity to higher frequency stimulation resulting from velocity sensitivity. Examples of this kind of increase in firing as a thresholded function of vibrissa velocity can be observed in real neural data in Figures 2.4, 2.5, and 2.10 (see also Reference number 63). *Middle panel* Three idealized examples of vibrissa resonance tuning showing a 3:1 gain in motion amplitude at the fundamental resonance frequency and bandwidth proportional to this frequency. *Top panel* The predicted neural response to vibrissa stimulation frequency as a function of resonance amplification of peak motion velocity, and intrinsic velocity sensitivity thresholds. For a given amplitude of stimulation, vibrissa resonance amplification that does not drive a neuron near its velocity threshold may fail to be amplified (purple curve, left resonance peak), while resonance amplification that is significantly above the velocity threshold (shown in the bottom panel) may fail to demonstrate tuning due to an upper limit on the range of possible firing rates for a given neuron (blue curve, right resonance peak). A subset of vibrissa resonance tuning curves near to but not above the intrinsic velocity threshold will, in this model, show optimal frequency tuning. Preliminary data suggest that these effects occur in a subset of trigeminal and cortical neurons, and that, within SI, FSU and multi-unit recordings are more susceptible to these impacts of velocity sensitivity.

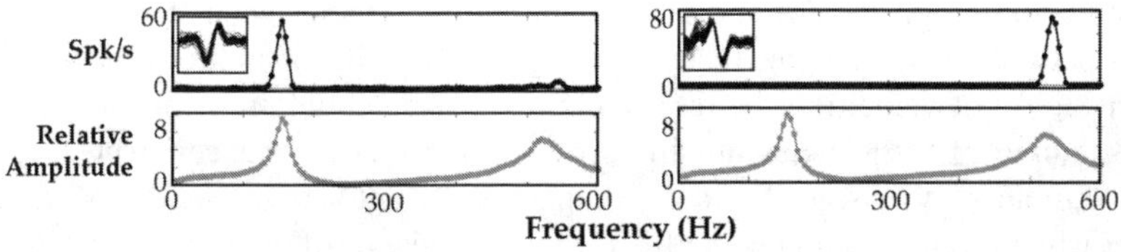

FIGURE 2.9 Higher resonance harmonics can drive evoked neural activity. Recordings were made from two trigeminal single units simultaneously (top panels) while frequency-varying sinusoidal stimuli were applied to their primary vibrissa (bottom panels). One of the single units (left panels) showed a significant increase in mean firing rate in response to the fundamental resonance frequency at ~150 Hz, while the other showed a selective response to the higher harmonic seen at ~520 Hz (right panels).

as higher harmonics are likely to fall above this velocity threshold for a wide range of amplitudes of vibrissa stimulation. Third, many natural surfaces have spatial frequency power spectra that fall off exponentially.[41] This implies that, for natural surfaces (but not iso-amplitude sine waves), high-frequency stimulus components that would drive higher harmonics may not be as prevalent.

Perhaps the most important observation to be made at this stage of investigation of the vibrissa resonance hypothesis is that higher harmonics, if they are expressed and translated into neural activity in perceptually relevant contexts, should enhance the detection of high-frequency stimuli, and may or may not impair their discrimination. Specifically, if higher harmonics provide amplification of subtle high-frequency inputs, they should facilitate the detection of these signals, e.g., the detection of surface roughness. Where higher harmonics pose a potential challenge to the benefits of vibrissa resonance is in the place coding model of discrimination proposed above. Fundamental resonance frequencies and higher harmonics represented in the same position of the somatotopic map could create ambiguity in the interpretation of these signals by a sensing animal. That said, the pattern of positional activation — the specific regions of the map activated by a fundamental and higher harmonics — should still have a unique spatial signature that could be used to decode the frequency applied to the vibrissa. Thus, if higher harmonics are meaningfully expressed in relevant perceptual contexts, we predict that they will facilitate the detection of high frequency stimuli, and may contribute to or undermine the discrimination or identification of high frequency input.

IV. ACTIVE SENSING AND VIBRISSA RESONANCE

A. Modulation of Whisking Velocity

Rats may actively modulate their sensory exploration strategy to enhance or suppress the impact of vibrissa resonance. During whisking, rats typically engage in a series of bouts of whisking with significant variation in the rate of vibrissa motion between bouts.[65] One potential benefit of this variation is that it could provide a broader search space over which a match could be generated between a given vibrissa's frequency tuning preference and the vibrational frequencies generated by the interaction between a spatial frequency and a given velocity of motion. By searching over a variety of velocities, the rat can circumvent potential problems posed by the limited range of frequencies amplified by the vibrissa resonances expressed in a given sample of vibrissae. Further, comparison of activation evoked by faster or slower whisk cycles, combined with knowledge of the position of optimal activation within the pad (e.g., more anterior or posterior on a given sweep), could refine the representation of a textured surface. The suggestion that velocity modulation may assist perception of spatial textures through generation of different frequency inputs is consistent with recent studies of human perception of textured surfaces using a probe: Under these conditions, variation in the velocity of sampling is observed to impact roughness judgments.[66,67]

B. Modulation of Vibrissa Damping

Active sensing may also be engaged at the level of the follicle. Given that vibrissa resonance may not facilitate the perception of contact (or could even impair processing, for example, by introducing "ringing" in the system when precise contact times are desired), an intriguing hypothesis is that a perceiving rodent could modulate the expression of vibrissa resonance by regulating damping in the follicle (e.g., through blood pressure or through muscular tension[68]).

Initial calculations suggest that damping by the follicle would be particularly relevant for relatively shorter vibrissae, particularly microvibrisssae, but may not significantly affect the biomechanics of longer vibrissae. The follicle surrounding the base of a greek arc macrovibrissa, for example, occupies only ~0.5–5% of its total vibrissa length (F. Rice, personal communication). Variations in tension could affect many aspects of transduction (e.g., change the relative sensitivity of classes of receptors), but would not change the propensity of the long vibrissae themselves to resonate. As such, the longer posterior vibrissae may be more important for encoding airborne signals.

V. TEMPORAL CODING AND VIBRISSA RESONANCE

Vibrissa resonance is obviously not required for the vibrissa sensory system to demonstrate temporal coding of high frequency stimuli.[46,69-73] That said, vibrissa resonance has several potentially important implications for the possibility of temporal encoding on broad and finer time scales.

A. Dynamic Evolution of Vibrissa Resonance Tuning: 'Contact' vs. Frequency Coding During Different Epochs of the Response

When sinusoidal input is applied at a vibrissa's fundamental resonance frequency, vibrissae require several cycles of oscillation to develop full resonant amplification (Figures 2.10 and 2.11). Differential frequency amplification in response to sinusoidal stimuli[26] or to vibrissa contact with a complex surface,[42] is more robust 10–100 msec following the onset of sensory stimulation. This property is mirrored in the temporal development of neural frequency tuning (Figure 2.10). As shown in Figure 2.11, the number of tuned neurons and the incidence of significant driven activity both increase for the epoch 25–100 msec post-stimulus onset, as compared to the epoch 0–25 msec post onset. An even higher relative number of RSUs were observed to demonstrate tuning over the epoch from 100–500 msec, although the incidence of neurons driven by any frequency was decreased for this epoch.[26]

These differences between shorter- and longer-latency responses suggest that each temporal window may code for different information. The onset response could encode initial vibrissa contact and/or somatotopic position, while the later sustained/developing response could encode the frequency of vibrissa motion. The map in SI may dynamically evolve over the first 100 msec, transitioning from a map representing space to one representing frequency. This temporal distribution of signals could help resolve an important ambiguity in the place code proposed above, specifically, that a given place in a map also has a somatotopic assignment in addition to a frequency meaning. Several other mechanisms, including the integration across multiple vibrissae in an isofrequency column, and fine temporal cues, could also help resolve this potential interpretive problem.

The time constant that determines the rate of resonant amplification is not constant across vibrissae. It varies inversely with the period of the resonance oscillation. Just as the biomechanical properties of the vibrissae generate a resonance frequency map, they also provide a latency gradient. Longer posterior vibrissae that possess lower frequency resonance tuning require a longer rise time and shorter anterior vibrissae a shorter rise time (Figure 2.11). Further, vibrissae in an isofrequency arc possess similar time constants. The synchrony generated by their simultaneous activation could enhance the neural representation of frequency-specific signals. This shift in onset latency as a function of position in a cortical map has precedent in the primate AI map, where a latency gradient is also observed although in AI it runs approximately perpendicular to the axis of the isofrequency bands.[58]

B. Vibrissa Resonance and Fine Timing of Action Potential Activity

In addition to these larger timescale temporal implications of vibrissa biomechanics, vibrissa resonance may also affect the fine timing of peripheral and cortical neural activity.

1. Vibrissa Resonance and Trigeminal Temporal Coding

Several studies have described the temporal following properties of trigeminal ganglion neurons, suggesting that temporally precise high frequency information is represented in the initial processing of the vibrissa sensory system.[71-74] We also observed several examples of precise neural following in response to low amplitude, high frequency stimuli. This finding, together with an absence of firing away from the resonance peak in mean firing rate tuning curves, led to the observation of tuning of the temporal following properties of trigeminal neurons. Examples of high frequency following and tuning of trigeminal single units are shown in Figure 2.12. In each of these cases, no

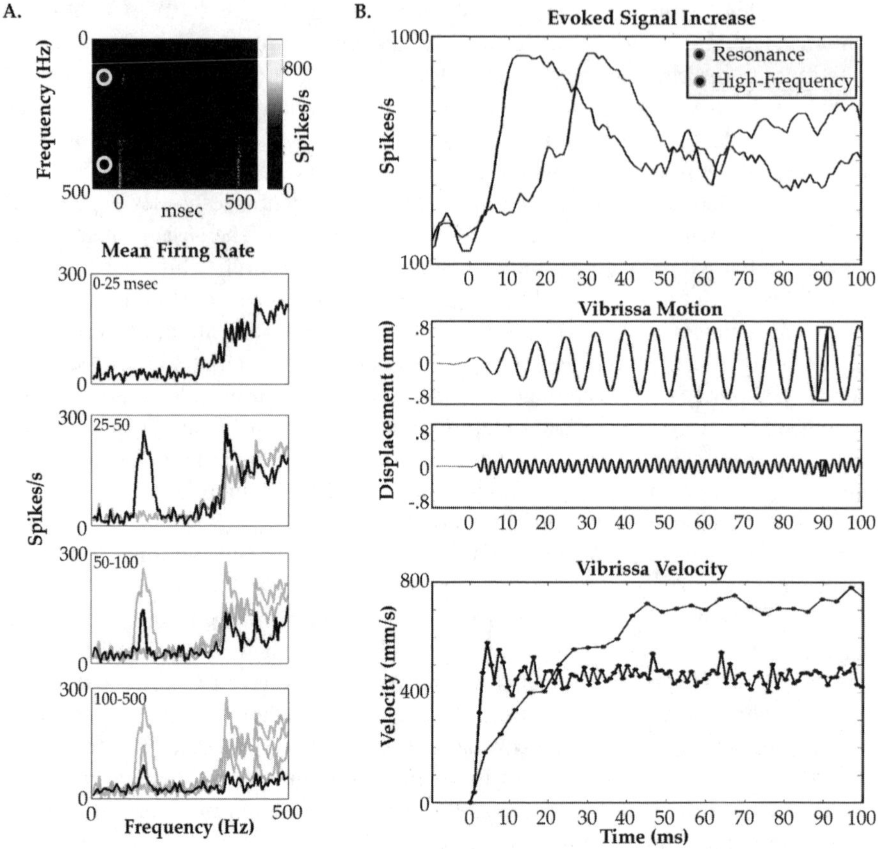

FIGURE 2.10 (See color figure following page 78) An example of the temporal evolution of neural frequency tuning. A. *Top* Peri-stimulus time histograms (PSTHs) are plotted as a function of frequency of stimulation (ordinate) and time (abscissa). Stimuli were applied as 160 μm sinusoids for 500 msec epochs. Resonance tuning can be seen in the selective band of increased firing at ~135 Hz: Intrinsic frequency (velocity) sensitivity can be seen in the increased firing above the threshold of ~ 350 Hz. *Bottom* Neural tuning curves showing mean firing rate for four different epochs post-stimulus onset. Resonance driven activity was not observed in the first epoch (0–25 msec post-stimulus onset) although robust high frequency responses were present. In later epochs, responses above the intrinsic high frequency threshold diminished in relative prominence while resonance driven neural activity increased. **B.** *Top panel* PSTH of activity evoked at the fundamental resonance frequency (red, 135 Hz) and at a frequency above the intrinsic high frequency threshold (black, 460 Hz). The slower rise time of resonance driven neural activity can be appreciated in this PSTH. *Middle panels* Traces of vibrissa motion driven by fundamental resonance frequency and high frequency stimuli. The fundamental resonance frequency driven motion shows a gradual increase in motion amplitude (red trace). *Bottom panel* Plots of the peak velocity of vibrissa motion for the fundamental (red) and high frequency stimuli (black). This time constant for the amplification of vibrissa motion is likely a key factor in the delayed increase in resonance driven activity in this example (see also Figure 2.11).

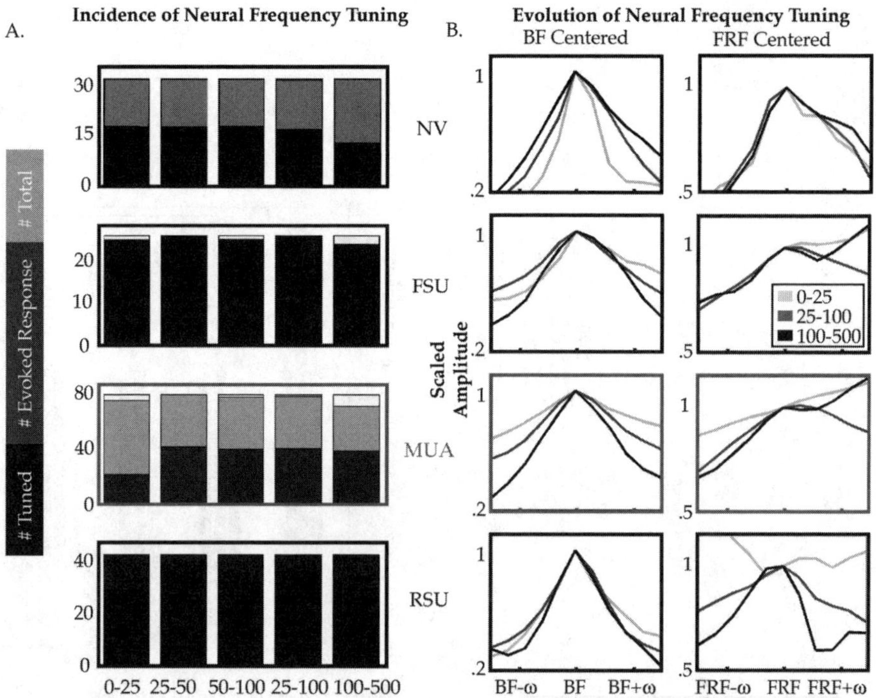

FIGURE 2.11 (See color figure following page 78) Vibrissa resonance tuning is delayed for all Si neural response types, correlated with the time constant of vibrissa resonance amplification. A. Vibrissae were stimulated with a 500 msec sinusoidal train during extracellular recording of trigeminal ganglion neurons (NV, green) or SI barrel recordings, including fast spiking units (FSU, blue), multi-unit activity (MUA, purple) and regular spiking units (RSU, red). Each bar plot shows the probability of observing a frequency tuned neural response (dark shading), a significant evoked increase in mean firing rate without frequency tuning (medium shading), or a neuron not driven by the stimulus (light shading). For all cortical response categories, the incidence of tuned responses and of driven neurons without tuning was increased for epochs >25 msec as compared to the epoch 0–25 msec. B. Average tuning curves are shown for each neural type for responses centered on the best frequency that drove the largest increase in neural firing (BF), or on the vibrissa fundamental resonance frequency (FRF). The apparent velocity sensitivity of FSU and MUA responses, shown as a sensitivity to higher frequency input, can be seen in the FRF-centered tuning curves for the epochs 100–500 msec post-stimulus onset. Similar velocity sensitivity was not observed in any epoch for RSU responses. Each plot is presented relative to the vibrissa frequency tuning bandwidth[26] (ω: see Figure 2.4). (Adapted from Andermann et al., *Neuron* 42, 2004. With permission.) (continued)

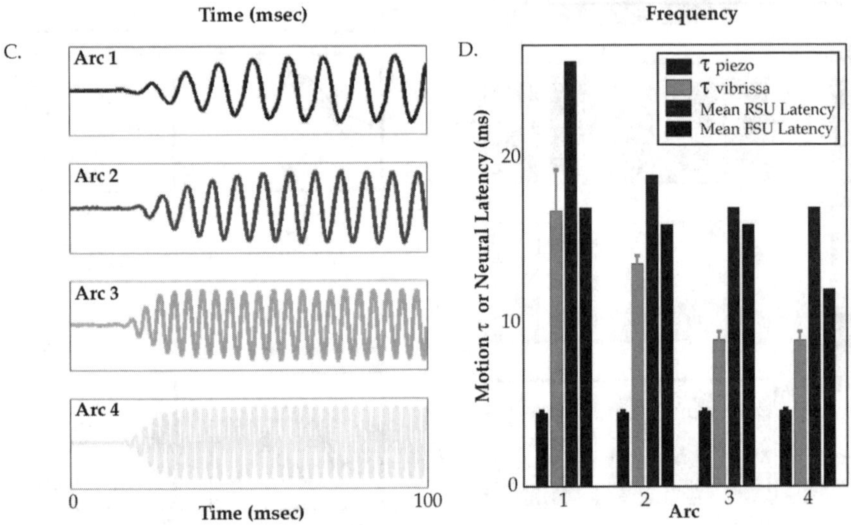

FIGURE 2.11 (CONTINUED) (See color figure following page 78) Vibrissa resonance tuning is delayed for all si neural response types, correlated with the time constant of vibrissa resonance amplification. C. Examples are shown of the evolution of vibrissa resonance when stimuli were applied at the fundamental resonance frequency for vibrissae from the 1—4 arcs. Shorter vibrissae (4 arcs) show a faster rise time than longer vibrissae (1 arc). D. Bar plots showing the time constant for the stimulator (black bars), for vibrissa amplification (gray bars), and for the latency to onset in neural activity (RSU, red bars and FSU, blue bars). The vibrissa time constants and neural latencies shift as a systematic function of the arc of vibrissae stimulated, with longer time constants and neural latencies observed for more posterior vibrissae. These delays provide one cause for the delay in cortical frequency tuning (shown in the above panels and Figure 2.10). Active neural mechanisms including inhibition and thalamocortical depression likely also play a role. The systematic shift in time constants across vibrissae also generates a map of onset timing within SI that may provide relevant coding information for the behaving animal, and that could increase the efficacy of input to an isofrequency column through enhanced neural synchrony.[26] (Adapted from Andermann et al., *Neuron* 42, 2004. With permission.)

sensory activity would be observed without resonance amplification. If temporal coding and not place coding is the relevant mechanism for texture extraction, vibrissa resonance may still play an essential role in the perception of these small amplitude, high frequency stimuli because these inputs would not otherwise be represented by a temporal code, or in any fashion, without this amplification.

In these examples of fine temporal tuning, only a subset of the amplitudes of stimulation that evoked a mean firing rate increase, also showed high frequency following. As shown in Figure 2.12, increased firing was evoked for amplitudes of tip stimulation ranging from ≥ 32 μm, but temporal following at the driving frequency was observed only for amplitudes ≥ 48 μm. The amplitude-dependence of temporal tuning can also be appreciated by examination of the inter-spike interval histograms in Figure 2.12 (see also Figure 2.5). While higher amplitudes of stimulation evoked only intervals at the period of the stimulus, lower amplitudes of stimulation evoked intervals that were more likely to be multiples of the frequency period.

This difference between the amplitudes necessary for mean firing rate increases and those necessary for temporal following parallels differences observed during auditory stimulation in the cat[75] and vibratory stimulation of the monkey skin.[76] An atonal interval exists in tactile and auditory perception between the relatively lower amplitudes of stimulation required for stimulus detection and those required for frequency discrimination.[76,77] Mountcastle and colleagues conjectured that tactile stimulus detection could be accomplished at lower amplitudes of stimulation using a mean firing rate increase. They hypothesized that the larger amplitudes of stimulation required for primate frequency discrimination might, in contrast, reflect the larger amplitude of stimulation necessary to entrain neural firing.[76,77] Whether a similar atonal interval exists for high frequency behavioral tasks performed by the rat vibrissa sensory system remains to be seen.

2. Vibrissa Resonance and the Volley Principle

As suggested throughout this chapter, several of the questions posed in attempting to understand high frequency encoding in the vibrissa sensory system mirror those traditionally posed in study of the auditory system. The central debate that shaped and continues to influence the attempt to understand neural mechanisms of auditory perception has been recapitulated here. Specifically, can high frequency encoding be understood as a place code where the frequency content of a stimulus is determined from a spatial map of frequency established by the biomechanical transduction properties of the peripheral sensors, or, can the fine timing of neural activity, a temporal code, account for pitch perception? Most likely, these mechanisms work in combination with predominance in different ranges of the perceptual spectrum.[40,78]

As mentioned previously, one problem posed by temporal coding of high frequency information is that even the fastest neurons can only fire at rates approaching 1 kHz, due to their refractory period. As a solution to this challenge to temporal coding, Wever proposed the volley principle, suggesting that while any given neuron might not follow every cycle of a stimulus, different subpopulations within a group of neurons could fire in a phase-locked manner to different

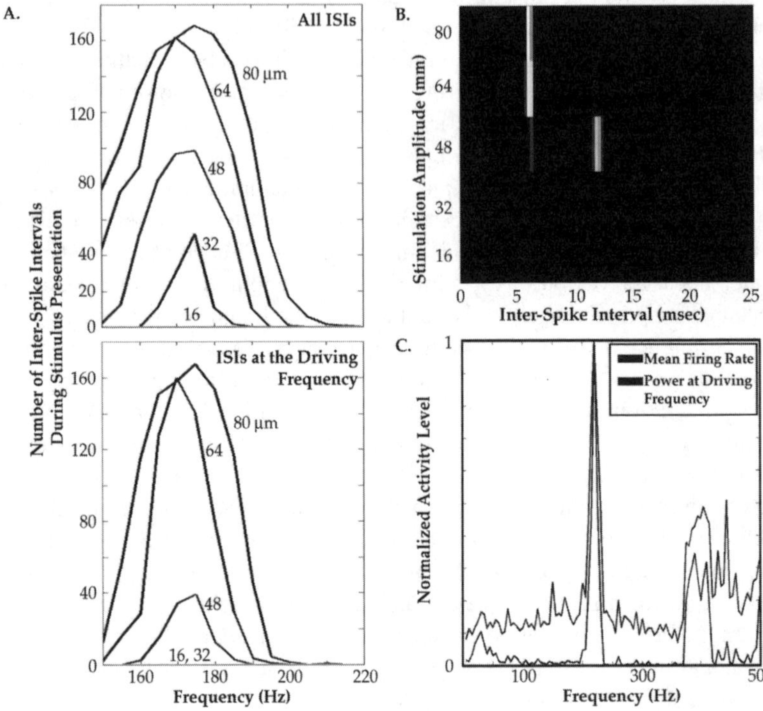

FIGURE 2.12 (See color figure following page 78) Trigeminal ganglion neurons demonstrate neural tuning and an atonal interval in the fine timing of their evoked activity. A. The top graph shows a frequency tuning curve for a trigeminal unit, constructed by counting all evoked inter-spike intervals (ISIs), a measure that is functionally equivalent to the mean firing rate. The lower graph shows the count of ISIs at the driving period, indicating fine temporal following of the neuron. Numbers adjacent to each curve indicate the amplitude of vibrissa stimulation applied. Frequency tuning was observed in both the mean firing rate and in the fine timing of neural evoked activity. An increased mean firing rate was observed for stimuli ≥32 μm, while temporal following at the driving frequency was present only for larger amplitudes of stimulation, ≥48 mm. This finding parallels similar observations made in the primate somatosensory system.[76] B. A graph of the incidence of ISIs at the fundamental resonance frequency, plotted as a function of the amplitude of stimulation (yellow indicates increased incidence) for the example in A. At larger amplitudes of stimulation, only firing at the fundamental resonance frequency was observed, as shown by the exclusive presence of ISIs at ~7 msec at 80 μm stimulation. In contrast, lower amplitudes of stimulation evoked ISIs at multiples of the driving period. C. A plot from a different single-unit trigeminal recording, showing the mean firing rate (red) and power at the driving frequency (blue). As in the example shown in A and B, temporal following provides a more precise tuning function at frequencies surrounding the vibrissa resonance frequency.

individual pulses of the stimulus.[30] In this way, the total output of this neural activity could represent the fine timing of an input, and downstream targets could receive the full frequency representation through the convergence of these precisely timed neural inputs. There were three central tenets to the original proposal. First, that phasic excitation of the system occurred; second, that a given stimulus drove multiple nerve fibers; and third, that subpopulations of the driven nerve fibers were tuned for different aspects of the stimulus presented.[30] The first two tenets of the volley principle are satisfied for encoding of vibrissa-driven responses in the trigeminal nerve. Peripheral neurons respond in a phasic fashion to frequency-varying stimuli, and multiple fibers from a given follicle respond to a given vibrissa.

Wever's third condition for instantiation of the volley principle is also well met and suggests a novel property of the system. Specifically, when a vibrissa is deflected by a high-frequency sinusoid, high-velocity transients are present on the ascending and descending phase of each cycle of stimulation and each of these high velocity transients occurs in an opposite direction. As discussed previously, the vibrissa sensory system is more sensitive to velocity transients than to the peak amplitude of vibrissa motion. As such, each single cycle of a high-frequency stimulus actually contains two events that the vibrissa sensory system will be optimally sensitive to, namely, two high-velocity transients in opposite directions. Both of these stimuli should drive neural activity in subpopulations of trigeminal ganglion neurons tuned for motion in opposite directions, and a given frequency of stimulation applied to the system should lead to a frequency doubling of the response in downstream representations where the inputs from these two neural subpopulations converge.

This framework differs from the proposal made for the volley principle in hearing only in the sense that the vibrissa sensory system is less concerned with the frequency of the peak amplitude of motion of a vibrissa stimulus, and more sensitive to the frequency and direction of high-velocity transients in that stimulus. That said, frequency doubling has also been reported in the auditory nerve in response to low-frequency, high-intensity tones.[75] This proposal is in agreement with a recent study by Keller and colleagues showing that consideration of the output of two trigeminal ganglion neurons with opposite directional tuning properties leads to a more accurate predictive model of the stimulus input.[74]

While we have not examined this prediction extensively, preliminary data suggests that frequency doubling should occur. For example, Figure 2.13 shows an example of a directionally selective trigeminal ganglion neuron that demonstrates 1:1 following at a specific phase (blue trace) that shifts by 180 degrees for this frequency of stimulation (125 Hz) when the initial stimulation direction is reversed (red trace). The summed output of two neurons with these kinds of opposing tuning properties would provide a total of 250 Hz input to a downstream neuron that received input from both channels. The sharp directional tuning of trigeminal neurons obviously has several other consequences for stimulus encoding including enhancing the sensitivity of the system to high frequency changes in the direction of vibrissa motion.

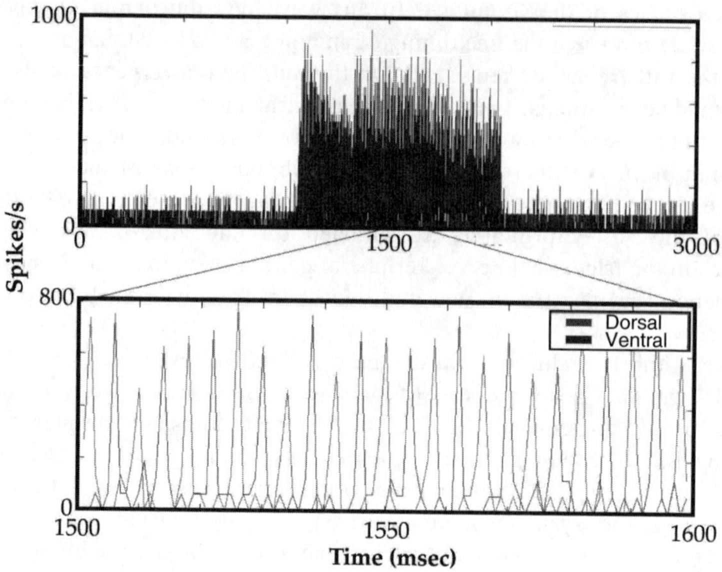

FIGURE 2.13 Velocity sensitivity of the vibrissa sensory system predicts a frequency doubling in response to high-frequency stimulation. The firing pattern of a trigeminal single unit evoked by a 500 msec duration sinusoidal stimulus (125 Hz) is shown. On 2 separate trials, stimulation was initiated in either a dorsal or ventral direction: the red trace is the response to the initial dorsal deflection, the blue trace to the ventral. The responses of this unit are highly directionally tuned for motion in the dorsal direction, leading to precise firing in phase with stimulation in that direction. The previously demonstrated existence of a population of dorsal- and ventral-tuned trigeminal neurons suggests that a high frequency stimulus presented to the vibrissa will ultimately lead to a frequency doubling in the temporal input transmitted to higher neural centers. As such, a volley principle may exist for the transmission of information by the trigeminal nerve. See text for further discussion.

3. Vibrissa Resonance and SI Temporal Coding

Do these high-fidelity temporal signals reach SI? The most commonly reported effect of vibrissa stimulation in the 0–40 Hz range is profound adaptation of the mean firing rate in SI neurons.[50,79,80] However, a distinct high-frequency processing mode may exist in SI that cannot be predicted from the lower-frequency adaptation patterns typically recorded.[81] Specifically, several neural mechanisms, ranging from the more complete adaptation of inhibitory inputs at higher frequencies[60] to the amplification of rapidly presented inputs through excitatory summation,[82] suggest that despite the robust adaptation observed when stimuli are applied in the range from 0–40 Hz, SI neurons are differentially sensitive to high-frequency stimuli.[81] Further, the precision of temporal following in SI neurons has been shown to increase even as the mean firing rate decreases at higher-frequencies.[79,83] This prediction is supported by the high-velocity and high frequency sensitivity of this system as shown in several examples presented here and previously of sensory responses driven by stimuli presented at hundreds of Hz.[26,63] Further, recent studies show that neurons of the ventral posterior medial nucleus of the thalamus, the primary lemniscal input to SI, can represent high frequency information in the hundreds of Hz, suggesting that these signals can be transmitted to SI under at least a subset of contexts.[69,84]

A common measure for assessing the temporal fidelity of neural following of periodic stimuli is vector strength, an indicator of phase-locking.[85] In a subset of multi-unit and single-neuron recordings, we observed increased vector strength at the vibrissa resonance frequency and corresponding tuning of the vector strength (Figure 2.14). Two examples of this tuning are shown for neurons recorded in SI during high-frequency stimulation. In both examples, the mean firing rate response (top panel) was more broadly tuned than the vector strength (2nd panel, blue) around the vibrissa resonance (2nd panel, gray), suggesting an improvement of temporal coding at an amplitude level where saturation in rate coding has occurred. This cortical finding is in agreement with the sharper temporal tuning observed in trigeminal responses (Figure 2.12). Further, there was limited evidence for frequency doubling, or vector strength at half the stimulation period (2nd panel, red), providing support for the description of the volley principle in the vibrissa sensory system described above. In both examples, fast-spiking units were shown. The finding of high-frequency fidelity in the response of fast-spiking units is in agreement with previous studies reporting higher firing rates and more precise following in thin spike neurons[86] and suspected inhibitory neurons.[87]

C. Vibrissa Resonance and Intrinsic Neural Frequency Tuning

Several studies have demonstrated the presence of intrinsically generated high-frequency oscillations (~200–600 Hz) in SI local field potentials that can be entrained to tactile input.[88-90] Synchronization of single neurons to these high-frequency oscillations has been demonstrated for action potential firing[88] and membrane oscillations.[91] The intrinsic oscillatory capability of SI provides support for the suggestion that temporal coding may occur in this system at the level of SI, and that these oscillations may facilitate processing in a specific frequency channel. High frequency neural resonances have also been observed in trigeminal neurons suggesting that active neural amplification or damping could be present throughout the system.[92]

These intrinsic oscillatory patterns of neural activity may be innate or could result from the experience throughout the life of the animal. If experience contributes to establishing these intrinsic rhythms, vibrissa resonance likely plays a key role. Even if vibrissa resonance does not facilitate perception, myriad stimuli will generate vibrissa resonances,[28,38] and neurons throughout the system will receive a significant fraction of their input in the high frequency range dictated by vibrissa resonance. As such, the system may become temporally tuned towards (or, potentially, away from) these high frequency inputs. In support of this suggestion, Rector and colleagues have recently provided preliminary evidence for a spatial map of high-frequency oscillations occurring across SI in response to punctate vibrissa stimulation (400–600 Hz).[90] The range of frequencies represented in this map and the spatial organization of these frequencies across SI generally parallels the map generated by vibrissa resonance.[26]

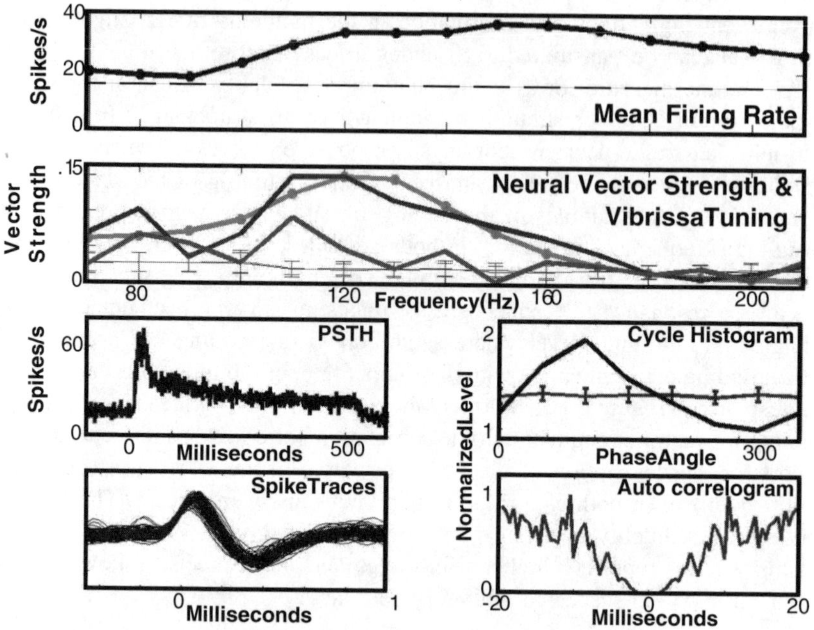

FIGURE 2.14 SI neurons demonstrate frequency tuning in the fine timing of their evoked activity. Two examples of fine temporal following in SI fast-spiking units are shown. *Top panel* The wide top panel of both examples shows the mean firing rate response across a range of frequencies. In both examples, the vector strength shows tuning surrounding a peak at the vibrissa fundamental resonance frequency, as seen by comparison of the blue and gray curves. Consistent with trigeminal recordings, these recordings demonstrate sharper tuning using a measure of temporal fidelity than of mean firing rate. This finding also demonstrates that the increased neural tuning measured with vector strength is not an artifact of an increased firing rate. The example in the lower frame shows no mean firing rate frequency tuning around the FRF (*top panel*), despite precise temporal tuning (vector strength tuning). A similar result is observed in the other example. Estimates of expected vector strength for Poisson spike trains with equal mean firing rates at each frequency are also shown (mean +- std. dev; thin blue line), further validating the presence of significant temporal following. The thick red curve is the vector strength at double the driving frequency, i.e., F(2)/F(0), and the thin red line is the shuffled control for this measurement. These measures provide some evidence for frequency doubling in the convergence of signals within SI, a finding predicted by the velocity sensitivity of the vibrissa sensory system. *Smaller panels* In the smaller panels, cycle histograms (upper right panels within the 4 panel grid) taken at the fundamental resonance frequency are also shown (black line), and demonstrate clear modulation at the cycle length that is not apparent in Poisson control spike trains (blue line, mean +- std. dev.). The autocorrelograms (lower right panels) and spike traces (lower left panels) identify these recordings as coming from single fast spiking unit recordings. The peri-stimulus time histogram (upper left panel: PSTH) was summed over all frequencies of presentation. Both PSTH examples demonstrate sustained firing in response to the entire 500 msec stimulus train. (continued)

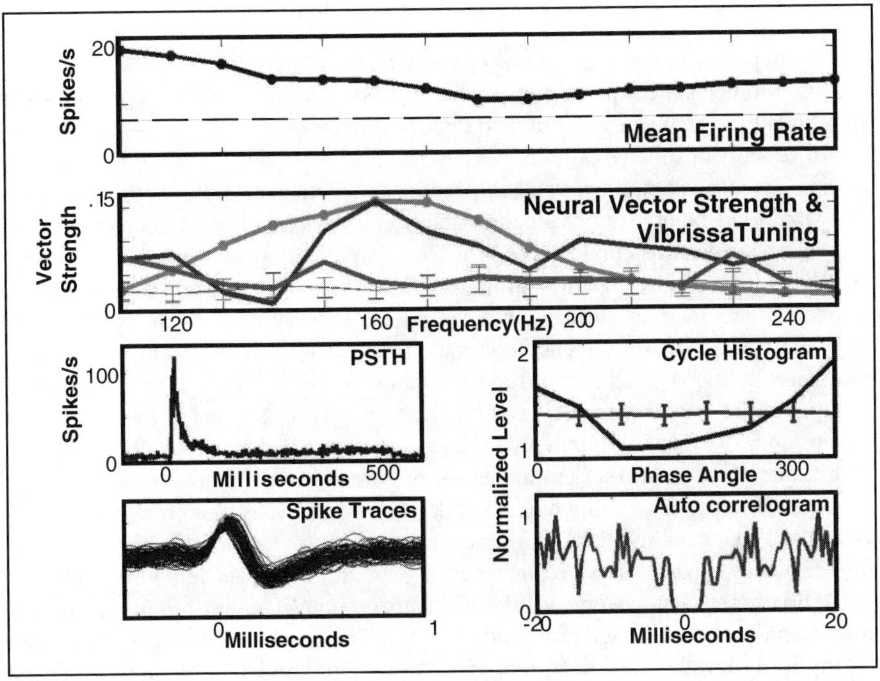

FIGURE 2.14 (CONTINUED) SI neurons demonstrate frequency tuning in the fine timing of their evoked activity. Two examples of fine temporal following in SI fast-spiking units are shown. *Second panel* The lower wide panel shows vibrissa frequency tuning curves in gray. In this panel, the thicker blue line shows the vector strength calculated for the driving frequency. Vector strength at a given frequency indicates the power at that frequency normalized by the power over all frequencies (i.e., mean firing rate), and provides a measure of the phase locking or temporal fidelity of a response. In both examples, the vector strength shows tuning surrounding a peak at the vibrissa fundamental resonance frequency, as seen by comparison of the blue and gray curves. Consistent with trigeminal recordings, these recordings demonstrate sharper tuning using a measure of temporal fidelity than of mean firing rate. This finding also demonstrates that the increased neural tuning measured with vector strength is not an artifact of an increased firing rate. The example in the lower frame shows no mean firing rate frequency tuning around the FRF (*top panel*), despite precise temporal tuning (vector strength tuning). A similar result is observed in the other example. Estimates of expected vector strength for Poisson spike trains with equal mean firing rates at each frequency are also shown (mean +- std. dev; thin blue line), further validating the presence of significant temporal following. The thick red curve is the vector strength at double the driving frequency, i.e., F(2)/F(0), and the thin red line is the shuffled control for this measurement. These measures provide some evidence for frequency doubling in the convergence of signals within SI, a finding predicted by the velocity sensitivity of the vibrissa sensory system. *Smaller panels* In the smaller panels, cycle histograms (upper right panels within the 4 panel grid) taken at the fundamental resonance frequency are also shown (black line), and demonstrate clear modulation at the cycle length that is not apparent in Poisson control spike trains (blue line, mean +- std. dev.). The autocorrelograms (lower right panels) and spike traces (lower left panels) identify these recordings as coming from single fast spiking unit recordings. The peri-stimulus time histogram (upper left panel: PSTH) was summed over all frequencies of presentation. Both PSTH examples demonstrate sustained firing in response to the entire 500 msec stimulus train.

VI. SUMMARY AND HUMAN IMPLICATIONS

In this chapter we proposed a vibrissa resonance hypothesis in which we suggested that the biomechanical properties of the vibrissae facilitate the detection of small amplitude high-frequency stimuli and the discrimination of frequency-specific stimuli. In support of this hypothesis, we described recent evidence demonstrating that vibrissa resonance amplifies small amplitude high-frequency inputs such that signals that would not otherwise drive neural activity in peripheral or SI neurons are able to evoke significant neural responses. This amplification of sensory input could greatly facilitate high-frequency stimulus detection. Further, the presence of vibrissa resonance generates frequency tuning in somatosensory neurons and a frequency map and system of isofrequency columns in SI (and, presumably, in other central somatotopic maps). This neural representation suggests that a place code may be engaged in the vibrissa sensory system. The velocity sensitivity of the system and the presence of higher harmonics may affect the quality and precision of transduction within this place code: further studies are required to assess both factors.

In addition to discussing enhanced detection achieved through signal amplification and place coding through the resonance tuning map, we also described two different ways that vibrissa resonance may create or impact temporal coding in the vibrissa sensory system. Vibrissa resonance amplification requires ~ tens of milliseconds to evolve, and as a product of this delay, relatively greater frequency-tuning is observed in later temporal epochs. This finding led to the hypothesis that initial neural activity, occurring in the epoch ~0–25 msec after stimulus onset, may encode vibrissa position and/or the initial contact of a vibrissa, while later sustained activity may represent frequency information. We also presented several examples of fine temporal following of sensory input on a millisecond time scale in trigeminal and cortical neurons and presented evidence for the hypothesis that frequency doubling should occur through a volley principle present in the vibrissa sensory system.

The existing evidence for the vibrissa resonance hypothesis is correlative. Under controlled stimulus presentation conditions, vibrissae generate signals that are well positioned to amplify and selectively filter frequency information. Further, in the anesthetized animal, these signals are transduced into complementary patterns of neural activity. While this kind of characterization has been the first step in the elucidation of several fundamental principles of sensory encoding (e.g., orientation tuning, tonotopic mapping), the crucial test of whether vibrissa resonance is relevant to perception, and, in turn, whether temporal coding and/or rate coding mechanisms contribute to the transmission of vibrissa resonance signals, will require targeted behavioral studies with simultaneous neural monitoring. Systematic alteration of the resonance properties of the vibrissae (e.g., by the partial or patterned trimming of the vibrissae of rats trained to perform texture perception tasks) may also provide evidence for or against this hypothesis.

The predominance of the resonance effects described here suggest that even if resonance does not enhance perception, the behaving animal must compensate for the impact that resonance has on incoming sensory information. As discussed in Section IV, active manipulation of whisking velocity and damping in the follicle could serve this end, and as discussed in Section V, intrinsic neural frequency tuning resulting from plasticity in specific somatotopic positions could reduce or amplify resonance-driven neural signals. Dynamic regulation of frequency dependent processing, on the millisecond to second timescale likely plays a role.[69,81,84,93]

An important question is what the study of vibrissa biomechanics may teach us about the neural mechanisms of human perception. While humans obviously do not possess vibrissae, there are several potential connections between the principles discussed here and our understanding of human tactile perception. One important parallel is the study of how biomechanical properties of the human body impact the perception of vibratory stimuli.[33,34,94] A further straightforward connection is in the consideration of sensory substitution or amplification devices. Cochlea-like devices applied to the skin can successfully relay auditory stimuli.[10,33,95,96] As Kirman and Keidel both describe, the historical attempt to engineer a device that could optimally represent speech using tactile perception converged on a place-coding mechanism and a series of vibrational transducers.[10,33,95,96] This kind of approach in many ways parallels the vibrissa pattern that evolved on the rodent face.

A great deal of modern human tactile perception is mediated through haptic probes — for example, the perception of the texture of a writing surface through contact of a pen or pencil tip. Lederman, Klatzky and colleagues have found that perception of surface roughness with this kind of probe is mediated at least in part through vibratory mechanisms.[66,67,97] As such, the resonance properties of these kinds of devices in common use (e.g., writing implements), and how they interact with damping due to the biomechanical properties of the hand, should be a central design consideration. Beyond common applications, several specialized work environments have led to the development of haptic experts who routinely conduct high-resolution perception through tools. Geologists provide one example as they analyze sub-millimeter surface features using sharp-tipped probes (e.g., hypodermic needles). This kind of mechanical analysis is now being developed for automated sample measurement, for example, in unmanned NASA geological expeditions.[41] Similarly, haptic vibration feedback is currently being considered for improving tele-surgery and surgical training.[98] These examples highlight the need for humans in the modern environment to solve the problem that the vibrissa sensory system is already optimized to solve, the representation of the world by temporally varying input translated through a non-innervated sensor. A better understanding of the active strategies and mechanical filtering employed in the model vibrissa sensory system may provide not only a better understanding of fundamental principles of neural representation, but also insight into important aspects of human perception.

ACKNOWLEDGMENTS

The authors would like to acknowledge Maria Neimark and Dr. Jason Ritt for collaboration on development of many of the ideas presented here, and Cheryl Cheney for assistance in preparation of the manuscript.

REFERENCES

1. Woolsey, T. A. and Van der Loos, H., The structural organization of layer IV in the somatosensory region (SI) of mouse cerebral cortex. The description of a cortical field composed of discrete cytoarchitectonic units, *Brain Res* 17 (2), 205-42, 1970.
2. Narins, P. M., Lewis, E. R., Jarvis, J. J., and O'Riain, J., The use of seismic signals by fossorial southern African mammals: a neuroethological gold mine, *Brain Res Bull* 44 (5), 641-6, 1997.
3. O'Connell-Rodwell, C. E., Arnason, B. T., and Hart, L. A., Seismic properties of Asian elephant (Elephas maximus) vocalizations and locomotion, *J Acoust Soc Am* 108 (6), 3066-72, 2000.
4. Johnson, K. O. and Hsiao, S. S., Evaluation of the relative roles of slowly and rapidly adapting afferent fibers in roughness perception, *Can J Physiol Pharmacol* 72 (5), 488-97, 1994.
5. Taylor, M. M. and Lederman, S. J., Tactile roughness of grooved surfaces: A model and the effect of friction., *Percep. Psychophys.* 17, 23-36, 1975.
6. Hollins, M., Bensmaia, S. J., and Roy, E. A., Vibrotaction and texture perception, *Behav Brain Res* 135 (1-2), 51-6, 2002.
7. Gamzu, E. and Ahissar, E., Importance of temporal cues for tactile spatial- frequency discrimination, *J Neurosci* 21 (18), 7416-27, 2001.
8. Cascio, C. J. and Sathian, K., Temporal cues contribute to tactile perception of roughness, *J Neurosci* 21 (14), 5289-96, 2001.
9. Sherrick, C. E. and Rogers, R., Apparent haptic movement, *Percep. Psychophys.* 1, 175-180, 1966.
10. Kirman, J. H., Tactile apparent movement: the effects of number of stimulators, *J Exp Psychol* 103 (6), 1175-80, 1974.
11. Gardner, E. P. and Sklar, B. F., Discrimination of the direction of motion on the human hand: a psychophysical study of stimulation parameters, *J Neurophysiol* 71 (6), 2414-29, 1994.
12. Rinberg, D. and Davidowitz, H., Do cockroaches 'know' about fluid dynamics?, *Nature* 405 (6788), 756, 2000.
13. Rinberg, D. and Davidowitz, H., Wind spectra and the response of the cercal system in the cockroach, *J Comp Physiol A Neuroethol Sens Neural Behav Physiol* 189 (12), 867-76, 2003.
14. Shaw, S., Detection of Airborne Sound by a Cockroach 'Vibration Detector': A Possible Missing Link in Insect Auditory Evolution, *J Exp Biol* 193 (1), 13-47, 1994.
15. Levanen, S. and Hamdorf, D., Feeling vibrations: enhanced tactile sensitivity in congenitally deaf humans, *Neurosci Lett* 301 (1), 75-7, 2001.
16. Brecht, M., Preilowski, B., and Merzenich, M. M., Functional architecture of the mystacial vibrissae, *Behav Brain Res* 84 (1-2), 81-97, 1997.
17. Hartmann, M. J., Active sensing capabilities of the rat whisker system, *Auton Robots* 11, 249-254, 2001.
18. Hutson, K. A. and Masterton, R. B., The sensory contribution of a single vibrissa's cortical barrel, *J Neurophysiol* 56 (4), 1196-223, 1986.
19. Dehnhardt, G., Mauck, B., Hanke, W., and Bleckmann, H., Hydrodynamic trail-following in harbor seals (Phoca vitulina), *Science* 293 (5527), 102-4, 2001.
20. van Netten, S. M. and Khanna, S. M., Mechanical demodulation of hydrodynamic stimuli performed by the lateral line organ, *Prog Brain Res* 97, 45-51, 1993.
21. Carvell, G. E. and Simons, D. J., Biometric Analyses of Vibrissal Tactile Discrimination in the Rat, *J. Neurosci.* 10 (8), 2638-2648, 1990.

22. Carvell, G. E. and Simons, D. J., Task- and subject-related differences in sensorimotor behavior during active touch, *Somatosens. Mot. Res.* 12 (1), 1-9, 1995.
23. Carvell, G. E., Miller, S. A., and Simons, D. J., The relationship of vibrissal motor cortex unit activity to whisking in the awake rat, *Somatosens Mot Res* 13 (2), 115-27, 1996.
24. Guic-Robles, E., Valdivieso, C., and Guajardo, G., Rats can learn a roughness discrimination using only their vibrissal system, *Behav. Brain Res.* 31, 285-289, 1989.
25. Guic-Robles, E., Jenkins, W. M., and Bravo, H., Vibrissal roughness discrimination is barrelcortex-dependent, *Behav. Brain Res.* 48, 145-152, 1992.
26. Andermann, M. L., Ritt, J., Neimark, M. A., and Moore, C. I., Neural correlates of vibrissa resonance; band-pass and somatotopic representation of high frequency stimuli, *Neuron* 42 (3), 451-63, 2004.
27. Mehta, S. and Kleinfeld, D., Frisking the whiskers: patterned sensory input in the rat vibrissa system, *Neuron* 41, 181-184, 2004.
28. Neimark, M. A., Andermann, M. L., Hopfield, J. J., and Moore, C. I., Vibrissa resonance as a transduction mechanism for tactile encoding, *J Neurosci* 23 (16), 6499-509, 2003.
29. Fend, M., Bovet, S., Yokoi, H., and Pfeifer, R., An Active Artificial Whisker Array for Texture Discrimination, in *IEEE/RSJ Int. Conf. on IROS*, Las Vegas, 2003.
30. Wever, E. G., *Theory of Hearing* Wiley, New York, 1949.
31. Shimozawa, T. and Kanou, M., Varieties of filiform hairs: range fractionation by sensory afferents and cecal interneurons of a cricket, *J Comp Physiol* 155, 485-493, 1984.
32. Kondoh, Y., Arima, T., Okuma, J., and Hasegawa, Y., Filter characteristics of cercal afferents in the cockroach, *J Comp Physiol [A]* 169 (6), 653-62, 1991.
33. Keidel, W., The sensory detection of vibrations, in *Foundations of Sensory Science*, Dawson, W. W. and Enoch, J. M. Springer-Verlag, Berlin, 1984, pp. 465-512.
34. Sherrick, C. E., Jr., Variables affecting sensitivity of the human skin to mechanical vibration, *J Exp Psychol* 45 (5), 273-82, 1953.
35. Vega-Bermudez, F. and Johnson, K. O., Fingertip skin conformance accounts, in part, for differences in tactile spatial acuity in young subjects, but not for the decline in spatial acuity with aging, *Percept Psychophys* 66 (1), 60-7, 2004.
36. Helmholtz, H. L. F., *On the Sensations of Tone as a Physiological Basis for the Theory of Music* Longmans, Green and Co., London, 1885.
37. Neimark, M. A., Anderman, M. A., Hopfield, J. J., and Moore, C. I., A mechanical model of vibrissae as filters of textural information, in *Barrels XIV*, Carlsbad, CA, 2001.
38. Hartmann, M. J., Johnson, N. J., Towal, R. B., and Assad, C., Mechanical characteristics of rat vibrissae: resonant frequencies and damping in isolated whiskers and in the awake behaving animal, *J Neurosci* 23 (16), 6510-9, 2003.
39. deCharms, R. C. and Merzenich, M. M., Primary cortical representation of sounds by the coordination of action-potential timing, *Nature* 381 (6583), 610-3, 1996.
40. Cariani, P., Temporal coding of periodicity pitch in the auditory system: an overview, *Neural Plast* 6 (4), 147-72, 1999.
41. Costa, M. A., *Fractal Description of Rough Surfaces for Haptic Display*, Stanford University, 2000.
42. Ritt, J., Andermann, M. A., and Moore, C. I., Tactile natural scenes in the rat vibrissa system, in *Computational and Systems Neuroscience*, Cold Spring Harbor, N.Y., 2004.
43. Ritt, J., Andermann, M. A., and Moore, C. I., Signal transduction of periodic and aperiodic surfaces by rat vibrissae, in *Soc. Neurosci.*, New Orleans, 2003.

44. Heffner, R. S., Heffner, H. E., Contos, C., and Kearns, D., Hearing in prairie dogs: transition between surface and subterranean rodents, *Hear Res* 73 (2), 185-9, 1994.
45. Popelar, J., Groh, D., Mazelova, J., and Syka, J., Cochlear function in young and adult Fischer 344 rats, *Hear Res* 186 (1-2), 75-84, 2003.
46. Zucker, E. and Welker, W. I., Coding of somatic sensory input by vibrissae neurons in the rat's trigeminal ganglion, *Brain Res* 12 (1), 138-56, 1969.
47. Lichtenstein, S. H., Carvell, G. E., and Simons, D. J., Responses of rat trigeminal ganglion neurons to movements of vibrissae in different directions, *Somatosens Mot Res* 7 (1), 47-65, 1990.
48. Lee, S. H. and Simons, D. J., Angular tuning and velocity sensitivity in different neuron classes within layer 4 of rat barrel cortex, *J Neurophysiol* 91 (1), 223-9, 2004.
49. Simons, D. J. and Carvell, G. E., Thalamocortical response transformation in the rat vibrissa/barrel system, *J Neurophysiol* 61 (2), 311-30, 1989.
50. Chung, S., Li, X., and Nelson, S. B., Short-term depression at thalamocortical synapses contributes to rapid adaptation of cortical sensory responses *in vivo, Neuron* 34 (3), 437-46, 2002.
51. Kyriazi, H. T., Carvell, G. E., Brumberg, J. C., and Simons, D. J., Quantitative effects of GABA and bicuculline methiodide on receptive field properties of neurons in real and simulated whisker barrels, *J Neurophysiol* 75 (2), 547-60, 1996.
52. Galarreta, M. and Hestrin, S., Frequency-dependent synaptic depression and the balance of excitation and inhibition in the neocortex, *Nat Neurosci* 1 (7), 587-94, 1998.
53. Moore, C. I. and Nelson, S. B., Spatio-temporal subthreshold receptive fields in the vibrissa representation of rat primary somatosensory cortex, *J Neurophysiol* 80 (6), 2882-92, 1998.
54. Carvell, G. E. and Simons, D. J., Membrane potential changes in rat SmI cortical neurons evoked by controlled stimulation of mystacial vibrissae, *Brain Res* 448 (1), 186-91, 1988.
55. Zhu, J. J. and Connors, B. W., Intrinsic firing patterns and whisker-evoked synaptic responses of neurons in the rat barrel cortex, *J Neurophysiol* 81 (3), 1171-83, 1999.
56. Harvey, M. A., Bermejo, R., and Zeigler, H. P., Discriminative whisking in the head-fixed rat: optoelectronic monitoring during tactile detection and discrimination tasks, *Somatosens Mot Res* 18 (3), 211-22, 2001.
57. Abeles, M. and Goldstein, M. H., Jr., Functional architecture in cat primary auditory cortex: columnar organization and organization according to depth, *J Neurophysiol* 33 (1), 172-87, 1970.
58. Cheung, S. W., Bedenbaugh, P. H., Nagarajan, S. S., and Schreiner, C. E., Functional organization of squirrel monkey primary auditory cortex: responses to pure tones, *J Neurophysiol* 85 (4), 1732-49, 2001.
59. Hahn, J. F., Stimulus-response relationships in first-order sensory fibres from cat vibrissae, *J Physiol* 213 (1), 215-26, 1971.
60. Hellweg, F. C., Schultz, W., and Creutzfeldt, O. D., Extracellular and intracellular recordings from cat's cortical whisker projection area: thalamocortical response transformation, *J Neurophysiol* 40 (3), 463-79, 1977.
61. Simons, D. J., Temporal and spatial integration in the rat SI vibrissa cortex, *J Neurophysiol* 54 (3), 615-35, 1985.
62. Pinto, D. J., Brumberg, J. C., and Simons, D. J., Circuit dynamics and coding strategies in rodent somatosensory cortex, *J Neurophysiol* 83 (3), 1158-66, 2000.

63. Arabzadeh, E., Petersen, R. S., and Diamond, M. E., Encoding of whisker vibration by rat barrel cortex neurons: implications for texture discrimination, *J Neurosci* 23 (27), 9146-54, 2003.
64. Temereanca, S. and Simons, D. J., Local field potentials and the encoding of whisker deflections by population firing synchrony in thalamic barreloids, *J Neurophysiol* 89 (4), 2137-45, 2003.
65. Berg, R. W. and Kleinfeld, D., Rhythmic whisking by rat: retraction as well as protraction of the vibrissae is under active muscular control, *J Neurophysiol* 89 (1), 104-17, 2003.
66. Klatzky, R. L. and Lederman, S. J., Tactile roughness perception with a rigid link interposed between skin and surface, *Percept Psychophys* 61 (4), 591-607, 1999.
67. Klatzky, R. L., Lederman, S. J., Hamilton, C., Grindley, M., and Swendsen, R. H., Feeling textures through a probe: effects of probe and surface geometry and exploratory factors, *Percept Psychophys* 65 (4), 613-31, 2003.
68. Yohro, T., Structure of the sinus hair follicle in the big-clawed shrew, Sorex unguiculatus, *J Morphol* 153 (2), 333-53, 1977.
69. Deschenes, M., Timofeeva, E., and Lavallee, P., The relay of high frequency sensory signals in the whisker-to-barreloid pathway, *J Neurosci* 23 (17), 6778-87, 2003.
70. Detsch, O., Kochs, E., Siemers, M., Bromm, B., and Vahle-Hinz, C., Increased responsiveness of cortical neurons in contrast to thalamic neurons during isoflurane-induced EEG bursts in rats, *Neurosci Lett* 317 (1), 9-12, 2002.
71. Nier, K., Effects of different ionic components on afferent responses of vibrissa receptors, *Pflugers Arch* 347 (1), 27-38, 1974.
72. Gibson, J. M. and Welker, W. I., Quantitative studies of stimulus coding in first-order vibrissa afferents of rats. 2. Adaptation and coding of stimulus parameters, *Somatosens Res* 1 (2), 95-117, 1983.
73. Gottschaldt, K. M. and Vahle-Hinz, C., Merkel cell receptors: structure and transducer function, *Science* 214 (4517), 183-6, 1981.
74. Jones, L. M., Depireux, D. A., Simons, D. J., and Keller, A., Robust temporal coding in the trigeminal system, *Science* 304 (5679), 1986-9, 2004.
75. Johnson, D. H., The relationship between spike rate and synchrony in responses of auditory-nerve fibers to single tones, *J Acoust Soc Am* 68 (4), 1115-22, 1980.
76. LaMotte, R. H. and Mountcastle, V. B., Capacities of humans and monkeys to discriminate vibratory stimuli of different frequency and amplitude: a correlation between neural events and psychological measurements, *J Neurophysiol* 38 (3), 539-59., 1975.
77. Hamalainen, H., Jarvilehto, T., and Soininen, K., Vibrotactile atonal interval correlated with activity in peripheral mechanoreceptive units innervating the human hand, *Brain Res* 333 (2), 311-24, 1985.
78. Green, B. G., Vibrotactile temporal summation: effect of frequency, *Sens Processes* 1 (2), 138-49, 1976.
79. Garabedian, C. E., Jones, S. R., Merzenich, M. M., Dale, A., and Moore, C. I., Band-pass response properties of rat SI neurons, *J Neurophysiol.* 90, 1379–1391, 2003.
80. Simons, D. J., Response properties of vibrissa units in rat SI somatosensory neocortex, *J Neurophysiol* 41 (3), 798-820, 1978.
81. Moore, C. I., Frequency-dependent processing in the vibrissa sensory system, *J Neurophysiol* 91 (6), 2390-9, 2004.
82. Laaris, N., Carlson, G. C., and Keller, A., Thalamic-evoked synaptic interactions in barrel cortex revealed by optical imaging, *J Neurosci* 20 (4), 1529-37, 2000.

83. Khatri, V., D.J. Simons, Firing synchrony between layer 4 barrel cells *in vivo.* Program No. 58.15, *2003 Abstract Viewer/Itinerary Planner*, 2003.
84. Castro-Alamancos, M. A., Different temporal processing of sensory inputs in the rat thalamus during quiescent and information processing states *in vivo, J Physiol* 539 (Pt 2), 567-78, 2002.
85. Eggermont, J. J., Temporal modulation transfer functions in cat primary auditory cortex: separating stimulus effects from neural mechanisms, *J Neurophysiol* 87 (1), 305-21, 2002.
86. Mountcastle, V. B., W.H. Talbot, H. Sakata, J. Hyvarinen, Cortical neuronal mechanisms in flutter-vibration studied in unanesthetized monkeys. Neuronal periodicity and frequency discrimination, *J Neurophysiol* 32 (3), 452-484, 1969.
87. Swadlow, H. A., Efferent neurons and suspected interneurons in S-1 vibrissa cortex of the awake rabbit: receptive fields and axonal properties, *J Neurophysiol* 62 (1), 288-308, 1989.
88. Kandel, A. and Buzsaki, G., Cellular-synaptic generation of sleep spindles, spike-and-wave discharges, and evoked thalamocortical responses in the neocortex of the rat, *J Neurosci* 17 (17), 6783-97, 1997.
89. Barth, D. S., Submillisecond synchronization of fast electrical oscillations in neocortex, *J Neurosci* 23 (6), 2502-10, 2003.
90. Dansereau, R. P., Rojas, M. J., Carter, K. M., and Rector, D. M., Frequency Dependent Topography of 400-600 Hz Oscillations in the Rat Whisker Barrel, in *Soc Neurosci Abstracts (Online)*, 2003.
91. Jones, M. S., MacDonald, K. D., Choi, B., Dudek, F. E., and Barth, D. S., Intracellular correlates of fast (>200 Hz) electrical oscillations in rat somatosensory cortex, *J Neurophysiol* 84 (3), 1505-18, 2000.
92. Puil, E., Gimbarzevsky, B., and Miura, R. M., Quantification of membrane properties of trigeminal root ganglion neurons in guinea pigs, *J Neurophysiol* 55 (5), 995-1016, 1986.
93. Moore, C. I., Nelson, S. B., and Sur, M., Dynamics of neuronal processing in rat somatosensory cortex, *Trends Neurosci* 22 (11), 513-20, 1999.
94. Phillips, J. R. and Johnson, K. O., Tactile spatial resolution. III. A continuum mechanics model of skin predicting mechanoreceptor responses to bars, edges, and gratings, *J Neurophysiol* 46 (6), 1204-25, 1981.
95. von Bekesy, G., *Experiments in Hearing* McGraw Hill, New York, 1960.
96. von Bekesy, G., *Sensory Inhibition*, Princeton Univ. Press, Princeton, NJ, 1967.
97. Brisben, A. J., Hsiao, S. S., and Johnson, K. O., Detection of vibration transmitted through an object grasped in the hand, *J Neurophysiol* 81 (4), 1548-58, 1999.
98. Okamura, A. M., Dennerlein, J. T., and Howe, R. D., Vibration feedback models for virtual environments, in *Proc. of the IEEE Int. Conf. on Robotics and Automation*, 1998, pp. 674-679.

3 Spatial and Temporal Rules Underlying Rat Barrel Cortex Plasticity

Mathew E. Diamond

CONTENTS

I. Introduction 61
  A. Summary 61
  B. Rat Vibrissal Sensory Pathway as a General Model 62
II. Spatial Rules Underlying Rat Barrel Cortex Plasticity 63
  A. Modular Learning in the Rat Whisker System 63
  B. Modular Learning in the Human Tactile System 65
  C. Physiological Account for Modular Learning 66
  D. Population Coding in Rat Barrel Cortex 67
  E. Decoding the Activity of Rat Barrel Cortex Populations 68
  F. Concluding Observations Concerning Spatial Plasticity Rules 69
III. Temporal Rules Underlying Rat Barrel Cortex Plasticity 70
  A. Fluctuations in Cortical Excitability 70
  B Timing-Based Plasticity of Intracortical Connections 71
  C. Timing-Based Plasticity of Sensory Responses 74
  D. Possible Mechanisms for Rapid Fluctuations in Plasticity 74
  E. Locus of Modification 75
  F. Concluding Observations Concerning Temporal Plasticity Rules 76
Acknowledgments 76
References 76

I. INTRODUCTION

A. Summary

We outline a systematic approach to investigating the role of modular map-like cortical organization in the processing of sensory information. We survey current evidence concerning the functional significance of cortical maps and modules, arguing that the topographic framework of primary sensory cortex renders the encoding and storage of sensory information efficient, fast, and reliable. The above set of experiments constitute what we term the spatial rules for barrel cortex plasticity. Recent experiments indicate that plasticity also is governed by temporal rules. Neuronal populations in sensory cortex exhibit fluctuations in excitability, characterized by bursts of spikes synchronized across barrel cortex. By comparing the plasticity induced by sensory inputs delivered during bursts compared to those delivered between bursts, we discovered that the strength of co-activity between columns in the barrel cortex can be modified by sensory input patterns during discrete, intermittent intervals time-locked to bursts. These spatial and temporal rules provide a context for understanding the barrel cortex in an awake exploring rat.

B. Rat Vibrissal Sensory Pathway as a General Model

We will review a series of experiments carried out in laboratory rats and, in a few cases, in human subjects. Rats have superb tactile capacities achieved by using their facial whiskers. Indeed, as nocturnal animals, albino rats depend on their vibrissal sensory system for object localization, judgment of the roughness or texture of surfaces, and the size and shape of small objects. It is of particular interest that humans and rats have roughly equivalent capacities in texture discrimination,[1] the haptic task that can be most directly compared between species. In short, the whisker sensory system is efficient and highly evolved. We believe that insights into its functioning will generalize to other specialized systems like the primate visual system.

Tactile sensation in the rat whisker system is an active process. Rather than passively allowing objects to encounter the whisker, the rat sweeps its whiskers backwards and forwards through the air, seeking out things.[2] The rat creates its own tactile stimuli and has the continuous task of determining the meaning or identity of objects (or other rats), the existence of which might be recorded as nothing more than the mechanical energy carried by the whisker shaft.

As the whisker is a specialized hair, the transduction process occurs within the whisker follicle. The neural pathway begins at the primary afferent terminations resting on the whisker shaft. Whisker deflections or vibrations cause a stretching of the membrane of the primary afferent fiber, depolarizing the terminal and inducing a train of action potentials that stream along the afferent nerve. Each follicle contains a bundle of about 200–300 sensory fibers.

Of particular convenience for the neuroscientist is the arrangement of the large whiskers of the snout into a horizontal and vertical grid, the whisker rows and arcs, respectively. Nearly identical in each individual rat, the particular arrangement allows the experimenter to quickly identify a particular whisker as, say C1 or E3.

Sensory signals travel along the afferent nerve, past the trigeminal ganglion, to the trigeminal nuclei in the brain stem. Here the first synapse is located. The axons of second-order neurons cross the brain midline in the medial lemniscus and travel to the thalamic somatosensory nuclei – ventral posterior medial (VPM) and posterior medial (PoM) – where the second synapse is located. Thalamic neurons project to the primary somatosensory cortex, conveying information mainly to layer IV targets but not excluding sites in other cortical layers. Our focus here will be limited to the cortical stage of the sensory pathway.

The whisker area of rodent somatosensory cortex is one of the most robust examples of mammalian columnar organization. In 1970 Woolsey and Van der Loos discovered that in the mouse, discrete clusters of small cells could be delineated in a tangential section through layer IV, and that the spatial arrangement of clusters replicated the spatial arrangement of the large facial whiskers on the opposite side of the snout.[3] From here was born the theory of the one-to-one mapping between sensory receptor cluster (the whisker follicle) and its corresponding cortical column. Shortly thereafter, a similar structure was found in rats.[4] In the succeeding decades, various laboratories carried out a large number of physiological experiments to determine whether the anatomical one-to-one projection extended to a functional one-to-one projection.[5–7]

Does the special columnar organization of rodent somatosensory cortex limit its appropriateness as a general experimental model for sensory cortex? Another example of the spectacular columnar pattern of organization is the somatosensory cortex of the star-nosed mole where each segment of each nose ray possesses a discrete cortical territory.[8] But it is not only in the cases of dramatic anatomical specialization that columns exist as functional units. Our argument is that the barrel cortex is but one expression of the universal columnar organizational unit of somatosensory cortex. We regard columnar organization as a general principle characterizing somatosensory cortex in all mammals and we would suggest that many of the principles of information processing and plasticity in rat somatosensory columns are likely to apply to other mammals.

II. SPATIAL RULES UNDERLYING RAT BARREL CORTEX PLASTICITY

A. Modular Learning in the Rat Whisker System

Does tactile learning take place in a modular framework? The question also can be phrased: Is the signal from one whisker processed mainly in its single, homologous columnar module, or does a much wider ensemble support the processing of a single whisker? Undoubtedly, the rat's natural behavior entails using many whiskers together. Arguing from this that single columns must not be functional units, would miss the point, for even a set of 30 columns could be simultaneously active while each column's circuitry processed information associated with a single whisker. The relevant question then, is whether one column is capable of supporting by itself, the information transformation necessary to some meaningful behavior. That is, are all the columns of the barrel map bound together as a functional unit or can different areas of a map participate independently in sensory processing?

To look for the role of columns in cortical plasticity we designed an experiment based upon single-whisker learning. We chose the gap-cross behavior because it suited our purpose since it has been known from its first publication[9] that a rat with a single whisker can learn and perform the task. In our experiments,[10] on one side of the snout all whiskers except one were clipped to the base. On the opposite side of the face, no whiskers were left intact. The training was conducted under dim red

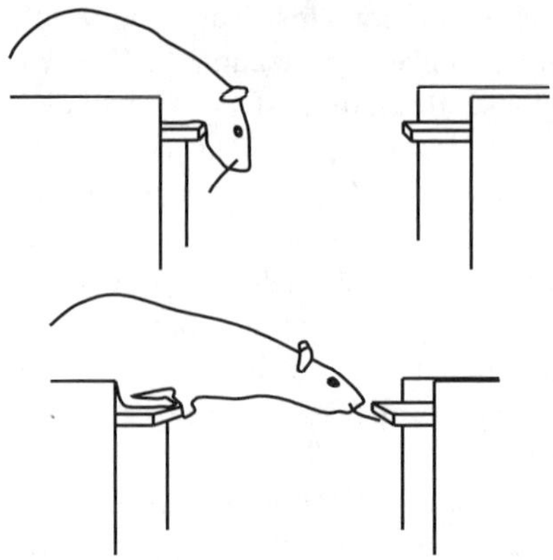

FIGURE 3.1 Drawings made from video frames to show two phases of gap crossing. In the upper illustration, the rat finds the front edge of the start platform. In the lower illustration, the rat perches and leans forward, sweeping its single intact whisker in front of the snout to detect the opposite platform.

light (visible to the experimenter but not to the albino rat) so that the animal was forced to rely on its whisker.

Rats learned to use this single whisker to detect the presence of the opposite platform before jumping across to obtain its chocolate cereal reward (Figure 3.1). During the initial training sessions, the gaps were small permitting the rat to become familiar with the paradigm while facing little risk. Over the course of 5–7 d (about 20 trials/d), the gap distance was increased until the goal could be detected only if the rat extended the tip of its remaining whisker about 1–2 cm in front of its nose. To make certain that rats truly detected and identified the goal platform (rather than settling into the stereotyped behavior of positioning itself and jumping regardless of whisker contact), the goal platform had an unpredictable location and was even occasionally removed.

After performance on the task had reached a fixed criterion, the intact whisker was clipped and a prosthetic whisker was attached to the stub of the whisker that had just been trained, or to the stub of a different whisker (one clipped at the start of training). The prosthetic whisker was simply a natural whisker taken from the collection of whiskers previously harvested at the initial clipping. The prosthetic was carefully attached to the selected whisker stub using a combination of super glue and hot glue. The rat's performance using the prosthetic was taken as an index of how proficiently the rat could transfer what had been learned with the trained whisker to the sensory signal arising from a new whisker. For each rat, the relative locations of the trained and prosthetic whisker were carefully controlled. If the trained whisker was, for example, D2 and the prosthetic attached to the D3 stub, the rat would be labeled as row1/arc0, this being the distance in follicle coordinates between the two sites. A range of different relative positions was used. To make certain that the prosthetic attachment procedure itself did not interfere with gap-crossing behavior, one group of rats had the prosthetic whisker glued back onto the same site from which the trained whisker had just been clipped. These rats (members of the group called Same) crossed the gap immediately, meaning that the prosthesis did not pose problems.

Two predictions may be contrasted. If learning involved globally distributed networks, then the task would be immediately reacquired even if the prosthetic whisker were implanted far from its original site. On the other hand, if learning involved only a localized module corresponding to the trained whisker, then reacquisition time would be very high even if the prosthetic whisker were adjacent to the trained whisker.

Our results indicate an intermediate degree of localization. The cortical network that participates in learning the task (or has access to the learned information) appears not to be uniformly distributed, as would be predicted by the hypothesis of global distribution. Nor is the relevant cortical network restricted to a single module. Rats in the Same group were the only ones able to utilize the prosthetic whisker's sensory signal immediately. Rats which had the prosthetic whisker attached to an adjacent neighbor of the trained whisker (i.e., row1/arc0 or row0/arc1) required a significant number of trials to relearn the task. Rats which had the prosthetic whisker attached to a site more distant from the trained whisker (i.e., row2/arc0), required an even larger number of trials to relearn the task. All retraining values were compared to those obtained from naïve rats, a group which was originally trained, at smaller gap distances, with all whiskers clipped such that they were familiar with gap crossing, but had never used a whisker to guide their behavior. These naïve rats were then retrained at the wider gap distance with a prosthetic whisker. This comparison revealed that when the prosthetic whisker was attached to any site beyond the immediate neighbors of the trained whisker, performance during retraining was not significantly better than for a naïve rat. Thus, the benefit of the original training was not shared beyond the immediate neighbors.

The topographic rule held up across the midline. Rats could rapidly transfer learning to whiskers symmetrically opposite the trained whiskers (e.g., left C_3 to right C_3), but required additional training before successfully transferring to non-symmetric opposite whiskers.[11] Since vibrissal information cannot be transferred between the two sides of the brain at any subcortical level of the sensory pathway, the corpus callosum connection linking homotopic sites in somatosensory cortex is the best candidate as the neural substrate for this transfer of learning.

While these data should not be taken to mean that all aspects of learning occur in the topographic framework (strategy, motivation, and context are certainly not represented in a topographic map), they do suggest that cortical topography acts as a bottleneck for learned information. Otherwise, learned information would have transferred to all whisker sites, neighboring and distant.

B. Modular Learning in the Human Tactile System

The finding that tactile learning in rats is constrained by a topographic framework motivated us to ask whether the human tactile system might operate according to the same principles. There were good reasons for thinking that topography might be irrelevant in humans. Possessing few cortical tactile processing regions, the rat whisker system might be forced to rely on the somatosensory primary area (the SI barrel field) – and its strict inherent topography – not just for online processing but also for storing learned information. In contrast, the primate tactile processing stream includes SI, SII, a nearby retroinsular field, posterior parietal fields (areas 5 and 7), and several fields in frontal cortex. Thus, it was conceivable that storage of learned information within higher, non-topographic areas could nullify spatial effects.

To find out whether learned information is stored within a map-like framework, we asked to what extent discriminations learned with one finger are accessible when people are tested using different fingers.Though learning of different sorts of tactile stimuli was investigated, here we report the findings only on vibratory stimuli. In each experiment, subjects learned to use one trained fingertip to discriminate between two vibrations of different frequency. Immediately after reaching criterion performance with the trained fingertip (this required, on average, about 260 training trials), subjects were tested with the same fingertip, its first and second neighbors on the same hand as well as the three corresponding fingers on the contralateral hand. Under these conditions, the subjects' learned ability to discriminate between vibrations did not transfer to any other fingertips.[12,13]

C. Physiological Account for Modular Learning

Satisfied that to a first approximation, tactile learning in humans and rats is constrained by the same topographic principle, we now return to the barrel cortex to discuss the physiological organization of sensory cortex and its relation to behavior. The method was to insert 100-electrode arrays into barrel cortex of urethane-anesthetized rats to measure the cortical representation of individual whiskers.[5,10] From the resulting response maps, we calculated the overlap (dot product summated across all channels) of the representations of separate whiskers as a function of the distance between the two whiskers.

Across all rats and all whisker positions, the overlap between cortical activity patterns was compared to gap cross reacquisition speed (number of trials required) for the same relative whisker positions. The correlation coefficient between the two data sets was 0.98. In other words, knowing the extent to which any two whiskers engage a common set of cortical barrel columns allows one to predict with confidence the speed of transfer of learning between the same two whiskers.[10]

The behavioral and physiological data fit together to give the following simple model. Neural modifications in the barrel cortex during gap cross training are localized to the set of barrel columns engaged by the trained whisker. After attachment of a prosthetic whisker, a second set of cortical columns is engaged. The transfer of previous learning depends on the degree to which the second set of cortical columns participated in the original learning. If several columns are common to both the original and the second sets, then the synaptic modifications are immediately utilized and little relearning is required. If the second group of columns does not overlap with the original group (i.e., the prosthetic whisker is distant from the trained whisker), then the previous synaptic modifications are not available when the prosthetic whisker is used.

If one accepts the equivalence between the experiments with gap crossing in rats and vibration learning in human subjects, then we can generalize the idea that during perceptual learning, a critical component of the relevant sensory information is stored within a restricted region of the topographic map by the same neuronal population that processes the sensory signal during training.

D. Population Coding in Rat Barrel Cortex

Maps and modules thus provide a structural framework that constrains the spatial distribution of tactile processing and learning. What features of neuronal activity within the responding neuronal population carry information about the stimulus?

To address this question, it was essential to reverse the stimulus-to-response paradigm that had yielded earlier observations. Rather than determining the response that a given whisker stimulus evokes in a set of cortical neurons, we asked: if a given set of neurons generates a particular response (one or more action potentials in some temporal order), what information have the neurons transmitted about a given set of stimuli? Posed this way, the problem might more closely resemble the challenge faced by the rat's nervous system. For the rat to make decisions based upon ongoing neural activity, averaging responses across trials for a set of stimuli, is not a realistic processing strategy. The stimuli used in the dataset were single whisker deflections (the stimulus set was comprised of nine whiskers), and we determined in what way the resulting spike train on each trial reported the identity of the deflected whisker. The design of the experiment and the analysis can be thought of as a simplified physiological analogue of gap crossing. During the behavioral task the rat must continuously evaluate whether its barrel cortex activity has reported contact of the spared whisker with the opposite platform.

One of the preferred strategies for determining the precise meaning of neuronal activity – for quantifying what neurons report about sensory events – is to employ the Mutual Information Theory of Shannon,[14] or one of its several variants and derivatives. Information analyses measure how well an ideal observer of neuronal responses can discriminate between the stimuli that evoked them. For a given set of stimuli, the information transmitted about the stimuli increases when the neural response to each stimulus, on a typical trial, is increasingly unique. An observer of neuronal activity could decode the stimulus from the selected neurons' activity. Information analyses give an objective scale to how neurons represent stimuli. Informative responses carry more bits of information permitting the observer to greatly narrow down the estimate of which stimulus occurred. We were particularly interested in three questions. (1) What is the role of spike timing in information encoding – are stimuli encoded entirely by the number of spikes occurring over long time intervals or does the millisecond precision of spike times convey additional information? (2) What is the role of spike correlations – does each spike encode the stimulus independently, or do correlated spike patterns convey additional information? (3) How does the columnar organization of the barrel cortex influence the decoding mechanisms available to downstream regions receiving the barrel cortex output?

For the collection of physiological data, spike trains of small sets of single-units in barrel cortex were simultaneously recorded while whiskers were stimulated one at a time.[6,15,17] On a given trial, the spike trains of up to three recorded neurons were binned with a certain precision, which was varied in the range from 2.5 to 10 msec. For example, using a resolution of 10 msec, across 40 msec a neuron's evoked response yielded a 4-letter binary word, such as <1 0 0 1> or <0 1 0 0> on two different trials. On these same trials, the spike counts (where the entire 40-msec response period is considered as one bin) would be <2> and <1>.

Taking the timing of spikes into consideration makes the cortical population code potentially very rich and complex. The application of a stimulus in fact activates many cortical neurons, each of which could emit several spikes. The situation is complicated by the fact that those spikes are often correlated, and correlations may be a key ingredient to the neuronal code. We addressed this problem by taking advantage of a recently developed information theoretic approach – the series expansion.[6] The series expansion reveals the information transmitted by the spike trains of single neurons and small neuronal populations.

We first examined whether the timing of spikes conveys additional information not available in the spike counts alone and found that, for single cells, spike timing at 2.5 msec precision transmits 44% information more than spike counts alone. A similar result was found for neuronal pairs.

To what extent does the information in spike timing arise from correlations between spikes? When we examined pairs of neurons located in column D2, we found that individual spikes were far more informative about the stimulus set than were correlated spike patterns. Results were very similar for neuron pairs recorded in the same or in adjacent barrel columns. For example, for neuron pairs in barrel columns D1 and D3, 79% of the total information was carried by individual spikes, and 21% by correlated spike patterns. Nearly all the information in correlations was due to within-cell spike patterns; cross-cell spike patterns had a negligible effect. Thus the main information-bearing unit is the individual spike.

E. Decoding the Activity of Rat Barrel Cortex Populations

The results summarized above suggest that a downstream neuron-reading barrel cortex output can capture almost all the available information about the sensory stimulus (at least for the stimulus set analyzed here) by integrating spikes individually and not registering the correlations among them. Although the absence of information in cross-neuron correlations simplifies the decoding problem, to make full use of the available information, the target neurons still must conserve the label of each incoming spike – the identity of the neuron from which the spike arises. To use the terminology above, the total information carried by two neurons with simultaneous spike trains <1 0 0 1> and <0 1 0 0> depends on the spike trains being integrated separately. However, this would entail complicated biophysical mechanisms at the level of the dendritic tree of the target neurons. A simpler alternative is the pooling together by the postsynaptic neuron of the spikes coming from all neurons. However, if the neurons converging on a target report different aspects of the external world (that is, if their receptive fields are different), then the information loss will be drastic: dissimilar messages, averaged together, conserve little of the original information.

A dilemma thus appears to be posed by two incompatible strategies – labeling input neurons yields an enormous information capacity but inherent decoding complexity, whereas pooling input neurons could lead to substantial information loss but an inherent decoding simplicity. We proposed that columnar organization might contribute toward resolving this dilemma. Columns facilitate the pooling of neurons with similar receptive fields and the separation of neurons with dissimilar receptive fields. Moreover, columnar bundling seems to be respected in the output of sensory cortex. Many of the output projections preserve some degree of columnar organization, such that target neurons indeed receive convergent input from afferent fibers arising in the same column.We tested the proposal of informational efficiency by quantifying how much the information about stimulus location was diminished due to pooling.[16] Cross column pooling led to large information losses, in the range of 25%–55%. Pooling neurons within the same column, on the other hand, caused a loss of only about 5% of the information. These results suggest that to decode stimulus position from the discharge of barrel cortex populations, target neurons could simply pool the activity arising from neurons of the same column, while maintaining the activity arising from neurons of separate columns at least partially segregated. In this way, afferent spikes would need to be labeled only according to columnar identity.

Additional studies investigated the representation of sinusoidal whisker vibrations by barrel cortex neurons.[17,18] Such vibrations are a simplified model of the complex vibrations induced by sweeping across textures. While vibration-coding mechanisms are not the main point of this chapter, the relation between columnar organization and stimulus decoding is relevant and can be summarized as follows. For decoding vibration parameters, just as was true for decoding stimulus sites, pooling of the spike output of neurons from the same column had only a negligible effect on the information available to an ideal decoder. However, because barrel cortex neurons encode vibrations in the same manner regardless of the column in which they reside, pooling of different column spikes led to the loss of just 7% of the total labeled spike information. This suggests a decoding scheme whereby target neurons could capture all available information simply by summating the signals from separate barrel cortex neurons. Whisker identity then is represented in a topographic framework whereby which neuron is active and which is not carries considerable information. Vibration is represented in our dataset in a nontopographic framework whereby differences in the activity of different columns are not informative.

F. Concluding Observations Concerning Spatial Plasticity Rules

The investigations reviewed above show that the cortical processing of sensory stimuli couldn't be fully understood unless wedded to the concepts of maps and columns. Once these organizational principles are taken into consideration, it becomes evident that individual columns (or small sets of columns) can have a crucial role in perception. We could find no evidence that essential sensory information is carried by synergistic interactions between neurons. The column's contribution then is to assemble individual spikes emitted by the member neurons.

Because we attempted to study the rat and human tactile systems under the simplest possible experimental conditions, it is too early to know exactly how observations we made will generalize to more complex situations. How does the rat barrel cortex represent complex and dynamic stimuli such as textured surfaces? How does it represent the size and shape of objects? How does the human cortex process stimuli that move across multiple digits? It is likely that the principles we have emphasized will need to be updated as new data accumulate. For now, our findings constitute elements of a beginning hypothesis to be challenged against more complex conditions.

III. TEMPORAL RULES UNDERLYING RAT BARREL CORTEX PLASTICITY

A. Fluctuations in Cortical Excitability

Neuronal populations in sensory cortex exhibit fluctuations in excitability. This simple and well known observation suggested that a second set of rules – temporal rules – might be important in guiding barrel cortex plasticity. We, therefore, designed an experiment to test the hypothesis that the variations in the state of cortical networks coincide with peaks and troughs in cortical modifiability.[19]

Cyclical oscillations in neuronal firing coordinated across widespread regions of cortex, are a characteristic common to many different brain states. They are present in anesthetized, in sleeping, and in alert animals under a variety of behavioral conditions.[19] In general, during sleep and anesthesia states, the oscillation amplitude is more profound and the rhythm is slower (0.5–3 Hz). The EEG correlates of these states are high voltage, low frequency waves. In somatosensory cortex of rats, urethane anesthesia produces activity resembling slow-wave sleep during which single-unit bursts are synchronized across distant cortical sites. During alert waking states, the oscillation amplitude is often shallow and the rhythm is faster. Here the EEG is characterized by low-voltage, high-frequency waves. Individual neurons in different brain areas have a different pattern of activity depending on the sleep stage and the phase of the EEG wave cycle.

In a recent report[20] we used 100-microelectrode arrays to characterize the behavior of neuronal populations in the barrel cortex (the cortical whisker representation) of lightly anesthetized rats. The spontaneous firing pattern of neurons consisted of clusters of spikes (bursts) separated by periods of low-spike density and the burst and inter-burst intervals were correlated across the entire barrel field. The key observation was that measured by the magnitude and spatial extent of response, neurons showed the greatest responsiveness to single-whisker deflections when stimuli were delivered during bursts and the least responsiveness when stimuli were delivered during the interval between bursts. Thus, during bursts there seemed to be the greatest opportunity for individual neurons to integrate inputs from multiple whiskers creating conditions favoring timing-dependent forms of Hebbian synaptic plasticity. Based on these observations, we hypothesized that bursts may modulate plasticity enabling cortex to be modified by patterns of sensory input preferentially at discrete intervals aligned with the bursts.

B Timing-Based Plasticity of Intracortical Connections

Testing the above hypothesis requires comparing the cortical modification induced by sensory inputs delivered when the cortex is at the peak of the oscillation as opposed to when it is at the trough of the oscillation. As an input pattern, we selected paired whisker stimulation (simultaneous deflection of two adjacent whiskers) because this induces receptive field plasticity and changes in the effective connectivity between cortical barrel columns in awake freely moving rats.[21] To test whether cortical modifiability is determined by the precise timing of sensory inputs in relation to the instantaneous cortical state, we repeatedly deflected one whisker pair during bursts while deflecting a second whisker pair during inter-burst intervals across a 50-minute conditioning procedure

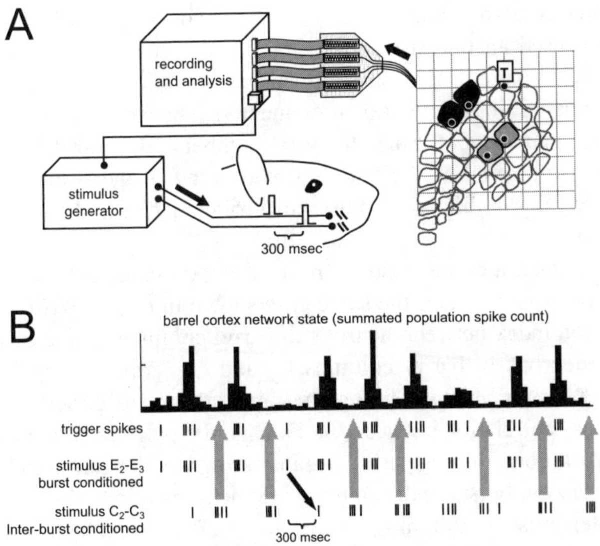

FIGURE 3.2 Experimental setup and conditioning procedure. A. The plot at lower-right indicates the 10 × 10 electrode array position relative to the cortical barrel field for this experiment (R20). Spikes from channel n. 76, located in column δ, were led to a stimulus generator. This produced pulses that were delivered to piezoelectric wafers situated in contact with two pairs of whiskers. Cortical columns matching the stimulus sites (C_2-C_3 and to E_2-E_3) are shaded in gray and black, respectively. Site of the trigger electrode indicated by T. B. Relation of sensory conditioning stimuli to cortical bursts. The histogram shows the cortical bursting pattern by summating spikes at all electrodes except those activated by the specific whiskers of interest. The fact that spikes are temporally clustered in evident. Spikes from the trigger electrode (T in part A) are shown below the population record and it is clear that activity at this electrode was aligned with cortical population activity. Below the trigger spikes, the times of E_2-E_3 paired stimuli are shown and below those, the times of C_2-C_3 stimuli. Note that the E_2-E_3 stimuli usually occurred during bursts while the C_2-C_3 stimuli tended to occur during inter-burst intervals (upward arrows).

(Figure 3.2). More specifically, activity at one electrode in the 100-electrode grid (labeled T for trigger, Figure 3.2A) acted as a monitor of cortical state. Each spike at this electrode was used as a trigger for two pairs of whisker stimuli (Figure 3.2B). One whisker pair was stimulated 1 msec after each trigger spike so the paired sensory inputs were integrated when the cortical network was excitable (burst-conditioned whisker pair). The second whisker pair was stimulated 300 msec after each trigger spike so the paired sensory inputs were integrated when the cortical network was likely to be less excitable (inter-burst-conditioned whisker pair). We predicted that the barrel columns of the first whisker pair would become more strongly linked having received the conditioning stimuli when plasticity was facilitated. On the other hand, the barrel columns of the second whisker pair would show no change in linkage having received the same conditioning stimuli when plasticity was minimal.

The degree of plasticity induced by sensory inputs was determined by the cortical network state at the precise time of stimulus delivery. Our evidence concerning the temporal rules for barrel cortex plasticity is given in Figure 3.3 (State-dependent changes in inter-columnar connectivity). Over the course of the 50-min-period burst conditioning caused an increase in the functional connectivity between the barrel columns receiving paired input. In contrast, inter-burst conditioning caused no significant modification in the functional connectivity between the cortical barrel columns receiving paired input. Since the total number of stimuli delivered to the two pairs of whiskers as well as the temporal patterning of the stimuli was equal, the different degree of modification must be attributed to rapid modulations in the modifiability of cortex.

The figure illustrates the results from an experiment that was typical of the complete set of experiments,[19] the same case shown in Figure 3.2. Part A shows the cross-correlation index between neurons in barrel columns E_2 and E_3 (black points) and between neurons in barrel columns C_2 and C_3 (gray points). The index was measured over 1-min blocks. During the preconditioning period (0–10 min), the strength of cross correlation between the E_2-E_3 and C_2-C_3 barrel columns was equal. At t = 10 min, the conditioning period began, during which neurons in barrel columns E_2 and E_3 underwent burst conditioning, while neurons in barrel-columns C_2 and C_3 underwent inter-burst conditioning (see Figure 3.2B). For both stimulated pairs, the amount of correlated activity increased immediately as would be expected due to the paired whisker stimulation. At t = 30 min, the correlation index for the E_2-E_3 pair rose dramatically while the correlation index for the C_2-C_3 pair remained at a constant level. By the conclusion of the 50 min conditioning paradigm (t = 60 min), the stimulus induced correlation was approximately twice as great for the burst conditioned neurons (barrel columns E_2-E_3) as for the inter-burst conditioned neurons (barrel columns C_2-C_3).

Conditioning terminated at t = 60 min. The cross-correlation index for the inter-burst-conditioned neurons immediately returned to the preconditioning level. But the cross correlation index for the burst conditioned neurons remained elevated. The modification in effective connectivity for the burst conditioned neurons was still present at t = 85 min, 25 min after the conclusion of the conditioning protocol.

Changes in intracortical connectivity for this experiment are summarized in Figure 3.3B. Cross correlation histograms were constructed separately from five 10-min intervals labeled a–e in Figure 3.3A. Before conditioning (interval a) and shortly after the onset of conditioning (interval b), the cross correlation histograms for the burst-conditioned neurons, the inter-burst conditioned neurons were equivalent. By the final 10 min of the conditioning period (interval c), the cross-correlation histograms for the burst-conditioned neurons had grown larger, while that for the inter-burst-conditioned neurons was practically unchanged. The heightened synchrony

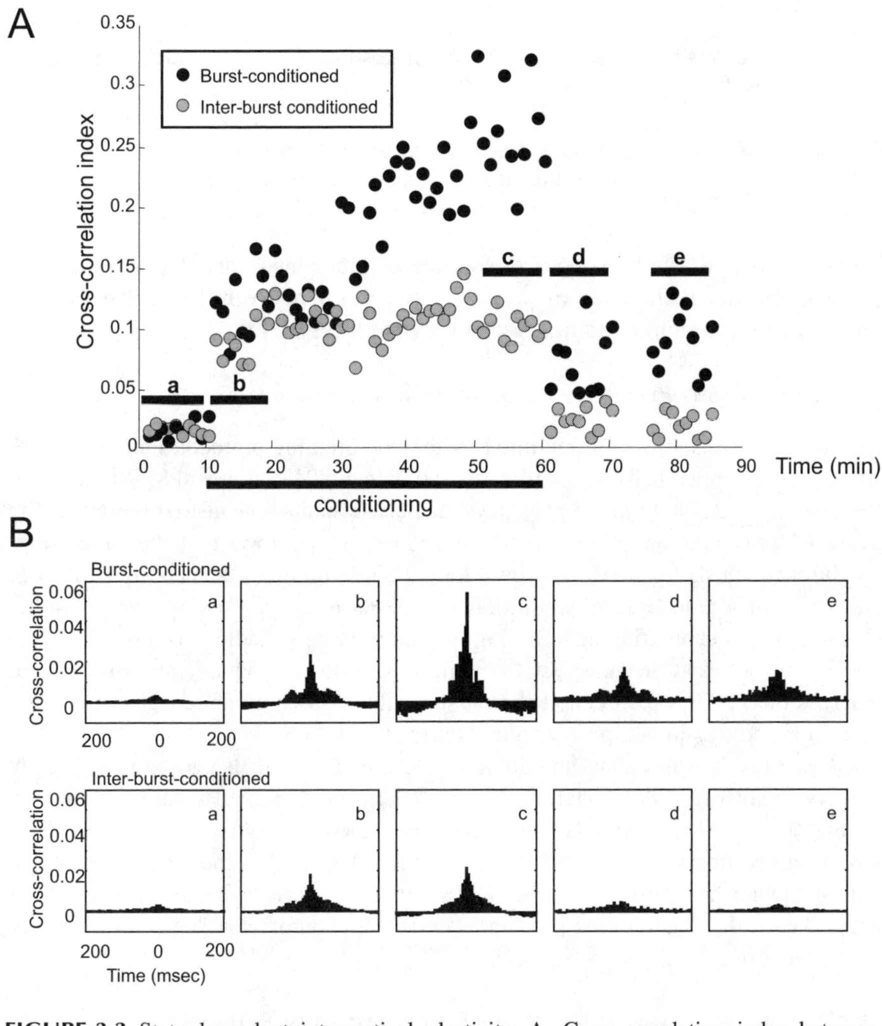

FIGURE 3.3 State-dependent intracortical plasticity. A. Cross-correlation index between burst-conditioned and inter-burst-conditioned cortical columns. B. Cross correlograms generated during the time intervals labeled a–e in part A. (continued)

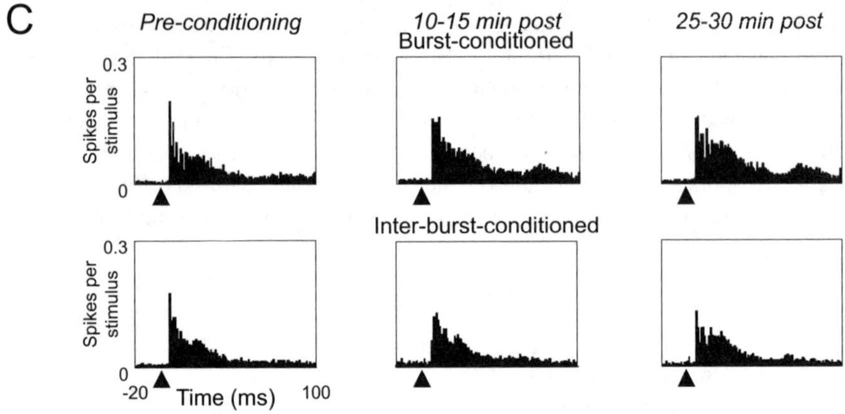

FIGURE 3.3 (CONTINUED) State-dependent intracortical plasticity. C. Sensory responses before and after conditioning in the burst-conditioned and inter-burst-conditioned cortical columns.

between spikes of the burst-conditioned neurons remained after the conclusion of conditioning (intervals d and e), while the synchrony between the inter-burst-conditioned neurons returned to the preconditioning level.

C. Timing-Based Plasticity of Sensory Responses

Sensory responses also were modified by the conditioning protocol. Paired whisker stimuli were applied before conditioning and then at 70–75 min and 85–90 min. For the same experiment, Figure 3.3C shows the peri-stimulus time histograms (PSTH's) evoked by stimulation of the burst-conditioned (upper row) and the inter-burst-conditioned whisker pairs (lower row). Each PSTH combines the activity from both electrodes of a pair. Before conditioning (left side), the PSTH's appeared similar. Compared to preconditioning values after conditioning (middle and right side), the PSTH's recorded at the inter-burst-conditioned cortical columns were unchanged, whereas the PSTH's recorded at the burst-conditioned columns revealed increased sensory responses in late post-stimulus intervals.

These results show that the cortex in anesthetized animals can be modified by sensory input patterns only during discrete intermittent intervals. Stimuli delivered during bursts – short intervals of elevated spontaneous activity – were effective in modifying cortical connectivity, whereas stimuli delivered in the intervals between bursts, though matching in number and temporal pattern, were ineffective. In all experiments, the conditioning paradigm produced a comparable degree of plasticity in the burst-triggered channels.

D. Possible Mechanisms for Rapid Fluctuations in Plasticity

Our preferred explanation is that neuromodulatory substances, in particular Acetylcholine (ACh), are released in sensory cortex during the peak of the cortical burst/pause rhythm[22–24] and that sensory inputs arriving during that phase can induce modifications in cortical circuitry. If the bursting of basal forebrain neurons causes a transient elevation in cortical ACh concentration, this could complement the increased synaptic reliability present during cortical bursts.[25] The picture that is suggested then, is that under urethane anesthesia, cortical bursts are linked to basal forebrain firing and the transient release of ACh combines with the ongoing cortical burst to enable plasticity. Another candidate as a modulatory input, whose activity is linked to barrel cortex bursts, is the intralaminar thalamic nuclei.[26]

E. Locus of Modification

In awake, attentive animals, training on sensory discrimination tasks or periods of altered sensory experience cause changes in somatosensory cortical-receptive fields and cortical maps. Modification of effective intracortical connectivity has been proposed as one of the candidate mechanisms to account for the plasticity of the sensory cortex.[27,28] The clearest evidence for the hypothesis comes from an experiment in alert monkeys. A high level of contingency between the firing of two neurons in auditory cortex (spontaneous activity of one neuron triggered an immediate sensory input to the second neuron) altered the connectivity between the pair of neurons as measured by cross correlation.[29]

Our experiments showed heightened correlated activity between burst-conditioned columns both during spontaneous activity, revealed by simple cross correlograms, and during whisker stimulation, revealed by JPSTH's (not illustrated here[19]). The augmented stimulus-related correlations could reflect either strengthened inter-columnar connectivity or else a more synchronized level of common ascending input to the two columns; we cannot specify the locus of this effect. In other sensory-conditioning paradigms, however, cortex is more plastic than its inputs. For example, in primates, synchronous multi-digit stimulation causes increased correlation between the neurons with paired inputs in the somatosensory cortex but not in the corresponding nucleus of the thalamus.[28] From the circuitry of the rat somatosensory system, stimulus-induced common input seems unlikely to account for the increased cross correlation in spontaneous activity after the conditioning period. Changes in common input synchrony specific to the paired cortical columns during spontaneous activity could occur only if the two thalamic barreloids projecting to two cortical columns were active in a more synchronous manner implying some communication between the barreloids. However, neurons in barreloids have no direct interconnection. It is simpler to suggest that repetitive costimulation of two whiskers timed to coincide with transient increases in cortical plasticity caused the burst-conditioned, cortical barrel columns to become more strongly connected, presumably through Hebbian mechanisms.

Our dataset came from infragranular layer recordings. In pilot experiments, we did not detect changes in inter-columnar connectivity when recordings were made from layer IV. This, too, is consistent with the proposal that plasticity is expressed largely in lateral intracortical connections. Layer IV neurons are only weakly connected to neighboring columns whereas the neurons we sampled are part of a dense network of lateral connections.[30] Our experiments add one more piece of evidence to support the idea that horizontal connections above and below layer IV are particularly modifiable.

F. Concluding Observations Concerning Temporal Plasticity Rules

What the present study may add to the understanding of cortical plasticity is that even in the anesthetized brain, sensory cortex can be modifiable as long as the sensory input pattern is timed to occur coincident with peaks in intrinsic fluctuations in excitability. The possibility that these findings can be of general relevance is supported by the fact that cortical fluctuations occur in every brain state. Their detailed nature depends on the behavioral conditions. We suggest that in every state, the timing of the peaks and troughs of these variations may modulate the ability of ascending inputs to modify the cortex. In rats, intermittent fluctuations occur in relation to natural behaviors such as whisking and sniffing. We speculate that sensory information is collected and stored in the cortex at discrete time intervals dictated by such fluctuations. The earlier work concerning spatial rules suggests that when information is stored, it resides within a topographic framework – within a small set of cortical modules matching the input whiskers. Further confirmation of these ideas will require the study of unanesthetized animals.

ACKNOWLEDGMENTS

I thank Justin Harris, Rasmus Petersen, Stefano Panzeri, Irina Harris, Ehsan Arabzadeh, and Irina Erchova for carrying out the experiments that this review is based on. I am grateful to Miguel Maravall, and Moritz von Heimendahl for stimulating discussions related to these projects. The research discussed here has been supported by European Community IST-2000-28127, Telethon Foundation GGP02459, J.S. McDonnell Foundation 20002035, Consiglio Nazionale delle Ricerche 02.00536.ST97, Regione Friuli Venezia Giulia, Human Frontiers Science Programme grant RGP43/2004, Wellcome Trust 066372/Z/01/Z, and The Royal Society.

REFERENCES

1. Carvell, G.E., and Simons, D.J., Biometric analyses of vibrissal tactile discrimination in the rat, *J. Neurosci.*, 10, 2638, 1990.
2. Welker, W.I. Analysis of sniffing of the the albino rat, *Behaviour,* 12, 223, 1964.
3. Woolsey, T.A. and Van der Loos H., The structural organization of layer IV in the somatosensory region (SI) of mouse cerebral cortex. The description of a cortical field composed of discrete cytoarchitectonic units, *Brain Res.*, 17, 205, 1970.
4. Welker, C., Microelectrode delineation of the fine grain somatotopic organization of SmI cerebral neocortex in albino rat, *Brain Res.*, 26, 259, 1971.
5. Petersen, R.S. and Diamond, M.E., Spatial and temporal distribution of neural activity in rat barrel cortex and the coding of stimulus location, *J. Neurosci.*, 20, 6135, 2000.

6. Petersen, R.S., Panzeri, S., and Diamond, M.E., Population coding of stimulus location in rat somatosensory cortex, *Neuron*, 32, 503, 2001.
7. Petersen, R.S., Panzeri, S., and Diamond, M.E., Population coding in somatosensory cortex, *Curr. Opin. Neurobiol.*, 12, 441, 2002.
8. Catania, K.C., A nose that looks like a hand and acts like an eye: the unusual mechanosensory system of the star-nosed mole, *J. Comp. Physiol. A.*, 185, 367, 1999.
9. Hutson, K.A. and Masterton, R.B., The sensory contribution of a single vibrissa's cortical barrel, *J. Neurophysiol.*, 1196, 1986.
10. Harris, J.A., Petersen, R.S., and Diamond, M.E., Distribution of tactile learning and its neural basis, *Proc. Natl. Acad. Sci. USA.*, 96, 7587, 1999.
11. Harris, J.A. and Diamond, M.E., Ipsilateral and contralateral transfer of tactile learning, *Neuroreport*, 11, 263, 2000.
12. Harris, J.A., Harris, I.M., and Diamond, M.E., The topography of tactile learning in humans, *J. Neurosci.*, 21, 1056, 2001.
13. Harris, J.A., Petersen, R.S., and Diamond M.E., The cortical distribution of sensory memories, *Neuron*, 30, 315, 2001.
14. Shannon, C., A mathematical theory of communication, *Bell Sys. Tech., J.*, 27, 379, 1948.
15. Panzeri, S., Petersen, R.S., Schultz, S., Lebedev, M., and Diamond, M.E., The role of spike timing in the coding of stimulus location in rat somatosensory cortex, *Neuron*, 29, 769, 2001.
16. Panzeri, S., Petroni, F., Petersen, R.S., and Diamond, M.E., Decoding neuronal population activity in rat somatosensory cortex: role of columnar organization, *Cerebral Cortex*, 13, 45, 2002.
17. Arabzadeh, E., Petersen, R.S., and Diamond, M.E., Encoding of whisker vibration by rat barrel cortex neurons: implications for texture discrimination, *J. Neurosci.*, 23, 9146, 2003.
18. Arabzadeh, E., Panzeri, S., and Diamond, M.E., Whisker vibration information carried by rat barrel cortex neurons, *J. Neurosci.*, in press, 2004.
19. Erchova, I.A. and Diamond M.E. (2004) Rapid fluctuations in rat barrel cortex plasticity, *J. Neurosci.*, in press, 2004.
20. Erchova, I., Lebedev, M.A., and Diamond, M.E., Somatosensory cortical neuronal population activity across states of anesthesia. *Eur. J. Neurosci.*, 15, 744, 2002.
21. Lebedev, M.A., Mirabella, G., Erchova, I., and Diamond, M.E., Experience-dependent plasticity of rat barrel cortex redistribution of activity across barrel-columns, *Cerebral Cortex*, 10, 23, 2000.
22. Detari, L., Rasmusson, D.D., and Semba, K., The role of basal forebrain neurons in tonic and phasic activation of the cerebral cortex, *Prog. Neurobiol.*, 58, 249, 1999.
23. Duque, A., Balatoni, B,, Detari, L., and Zaborszky, L., EEG correlation of the discharge properties of identified neurons in the basal forebrain, *J. Neurophysiol.*, 84, 1627, 2000.
24. Zaborszky, L. and Duque, A., Sleep-wake mechanisms and basal forebrain circuitry, *Front Biosci.*, 8, 1146, 2003.
25. Lisman, J., Bursts as a unit of neural information: making unreliable synapses reliable, *Trends Neurosci.*, 20, 38, 1997.
26. Fox, K. and Armstrong-James, M., The role of the anterior intralaminar nuclei and N-methyl D-aspartate receptors in the generation of spontaneous bursts in rat neocortical neurones, *Exp. Brain Res.*, 63, 505, 1986.

27. Armstrong-James, M.A., Diamond, M.E., and Ebner, F.F., An innocuous bias in whisker use in adult rats modifies receptive fields of barrel cortex neurons, *J. Neurosci.,* 14, 6978, 1994.
28. Wang, X., Merzenich, M.M., Sameshima, K., and Jenkins, W.M., Remodelling of hand representation in adult cortex determined by timing of tactile stimulation, *Nature*, 378, 71, 1995.
29. Ahissar, E., Vaadia, E., Ahissar, M., Bergman, H., Arieli, A., and Abeles, M., Dependence of cortical plasticity on correlated activity of single neurons and on behavioral context, *Science*, 257, 1412, 1992.
30. Kim, U. and Ebner, F.F. (1999) Barrels and septa: separate circuits in rat barrels field cortex, *J. Comp. Neurol.*, 408: 489-505.

4 Probing the Cortical Evidence of Somatosensory Discrimination

*Ranulfo Romo, Adrián Hernández,
Antonio Zainos, Luis Lemus,
Victor de Lafuente, and Rogelio Luna*

CONTENTS

I. Overview 79
II. Introduction 79
III. Psychophysics in the Flutter Discrimination Task 80
IV. Neural Coding of Vibrotactile Stimuli in S1 83
V. Neuronal Correlates of Flutter Discrimination in S1 86
VI. Artificial Induction of Activity in S1 Underlying Flutter Discrimination 89
VII. General Comments 92
Acknowledgments 94
References 94

I. OVERVIEW

We focus on studies that address the question of which components of the evoked neuronal activity in the somatosensory cortex represent the stimulus features and probe whether these neuronal representations are essential to somatosensation.

II. INTRODUCTION

An important problem in sensory physiology is the isolation of the neural codes related to perception. The underlying belief is that unraveling the coding of sensory stimuli from the periphery to early stages of cortical processing is key to understanding the beginning of perceptual processes. We could trace the beginning of the study of neural coding to Adrian who in 1928, recorded action potentials from isolated axons of peripheral nerves and observed discharges evoked by mechanical stimuli applied to the skin.[1] It was implicit in that observation that the discharges associated with the stimuli were transmitting information to the central nervous system regarding the physical attributes of the stimulus and that this could be considered as the material for sensation and perception in highly evolved brains. Four decades later, Mountcastle and colleagues began to study the responses of cutaneous primary afferents evoked by well-controlled mechanical stimuli applied to the skin.[33,38] Their key conceptual advance was to closely coordinate two types of experiments that had been divorced from previous studies. The first experiment quantified the relationship between the actual value of a stimulus and its subjective value as perceived and reported by humans — classical psychophysics. The second experiment recorded the discharges of cutaneous afferent fibers in anesthetized monkeys aroused by the same stimulus sets used in psychophysical experiments — cutting edge neurophysiology.[12] The goal of the experiments was to identify the functional dependence between stimulus and evoked discharges (i.e., the neural code of the stimulus). The experiments were seeking an explanation for the psychophysical relationship between stimulus and sensation. Indeed, close correspondences were found.

This novel approach, which evolved to experiments combining psychophysics and neurophysiology in behaving monkeys, paved the way for new questions that are now more closely related to linking neural coding to perception. However, in order to address these questions clearly, experimental methods should conform to two essential conditions. The first condition is the sensory stimulus must be under precise quantitative control. The second condition is the subject's psychophysical responses should be well controlled and quantitatively measured. Experimental methods conforming to these two essential conditions have been used to investigate nontrivial cognitive tasks — sensory discrimination — using highly simplified stimuli so the neural codes for simple stimuli should be identified in early stages of cortical processing and compared with the psychophysical responses.[8,12,16,35] However, one of the main challenges of this experimental method is that even the simplest cognitive task engages many cortical areas. Each one might represent sensory information in a different way or combine it with other types of stored signals representing past experience of future actions.[8,16] We reviewed results obtained from combined psychophysical and neurophysiological experiments aimed at understanding neural coding in the somatosensory cortex. First, we scrutinized the optimal conditions for frequency discrimination in the sense of flutter.[8,16] Then we studied the neuronal responses of primary somatosensory cortex (S1) and sought to determine which component(s) of the neural signals related to the psychophysical performance.[9,30] Finally, we probed whether the evoked neuronal responses in S1 account for the discrimination responses.[24,26] This was done by artificially activating neuronal clusters of S1 that responded to the vibrotactile stimuli.

III. PSYCHOPHYSICS IN THE FLUTTER DISCRIMINATION TASK

Mountcastle and colleagues made a number of important observations in a sensory submodality called the sense of flutter.[17,33] Flutter is felt by touching an object that

Color Figures

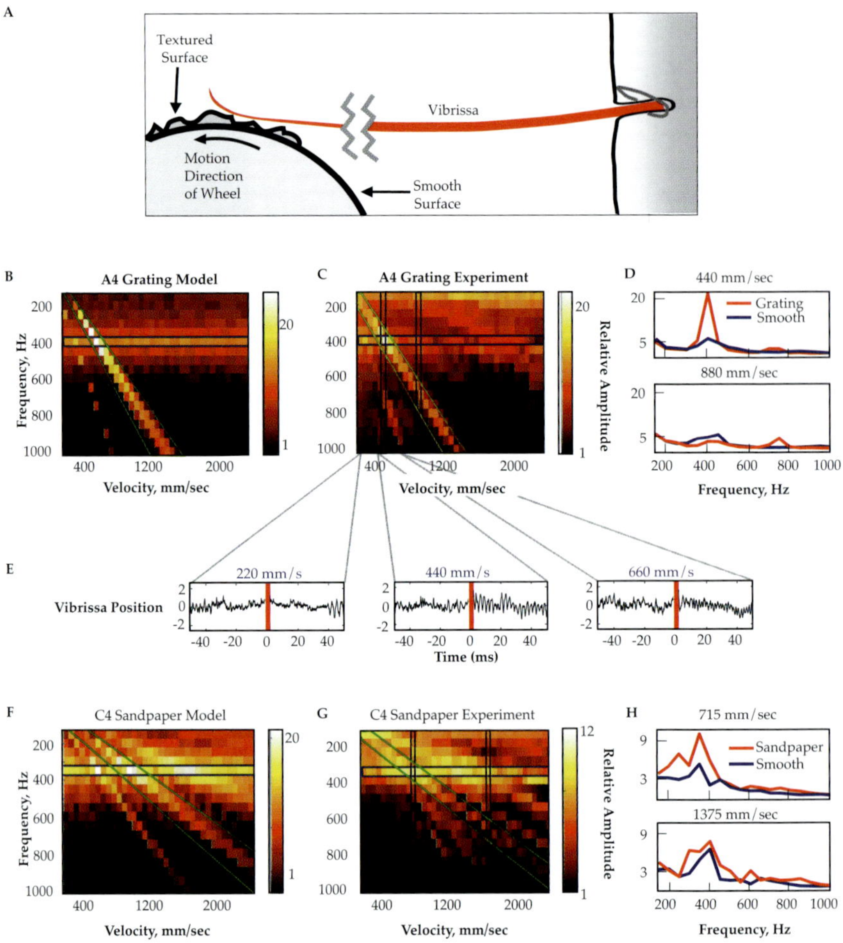

A
Textured Surface
Vibrissa
Motion Direction of Wheel
Smooth Surface
B
A4 Grating Model
Frequency, Hz
Velocity, mm/sec
C
A4 Grating Experiment
Relative Amplitude
D
440 mm/sec
Grating
Smooth
880 mm/sec
Frequency, Hz
E
Vibrissa Position
220 mm/s
440 mm/s
660 mm/s
Time (ms)
F
C4 Sandpaper Model
G
C4 Sandpaper Experiment
H
715 mm/sec
Sandpaper
Smooth
1375 mm/sec

FIGURE 2.2 Vibrissae resonate when driven by complex natural stimuli. A. Half of a smooth wheel was covered with a textured surface, either a grating or sandpaper, and was moved at different velocities while contacting a vibrissa. This motion in turn drove vibrations in the vibrissa that were monitored with an optical sensor. B. and C. Plots comparing the power spectra of vibrissa oscillations driven by a grating at different speeds of wheel motion. Color scale indicates the relative amplitude of vibrissa motion. In C, increasing wheel speed caused an increased rate of vibrissa vibration as predicted by a one-to-one translation of the predominant frequencies of the grating as a function of wheel velocity (increased diagonal band of activation bounded by green lines). This signal was amplified when the grating drove the vibrissa at its fundamental resonance frequency (~400 Hz, horizontal band of activation bounded by black lines). A model of the vibrissa as a thin elastic beam[28] predicted this pattern of resonance amplification (B). D. The relative amplification of the vibrissa was shown for the grating surface and for a smooth surface for two speeds of wheel motion. The grating surface evoked peak amplification of vibrissa motion, although a small increase was observed at the vibrissa fundamental resonance frequency at ~400 Hz in both instances. E. Traces of vibrissa oscillation are shown for three distinct wheel speeds, with the red bar indicating the point at which the textured region of the wheel surface came into contact with the vibrissa. These traces show the amplification of vibrissa motion observed when the grating was moved at 440 mm/sec. F. and G. Data plotted as in B. and C. for the vibrissa response. When 80-grit sandpaper was used as the textured stimulus. Note that both the smooth and textured surfaces drove increased vibrissa resonance at ~375 Hz, and that the textured stimulus was again more effective at driving motion at the fundamental resonance frequency when the wheel speed generated an optimal driving frequency. See Reference number 28 for further details and Figure 2.5 for an example of neural frequency tuning evoked under parallel stimulus conditions. (Adapted from Neimark et al., *J. Neurosci.* 23, 2003. With permission).

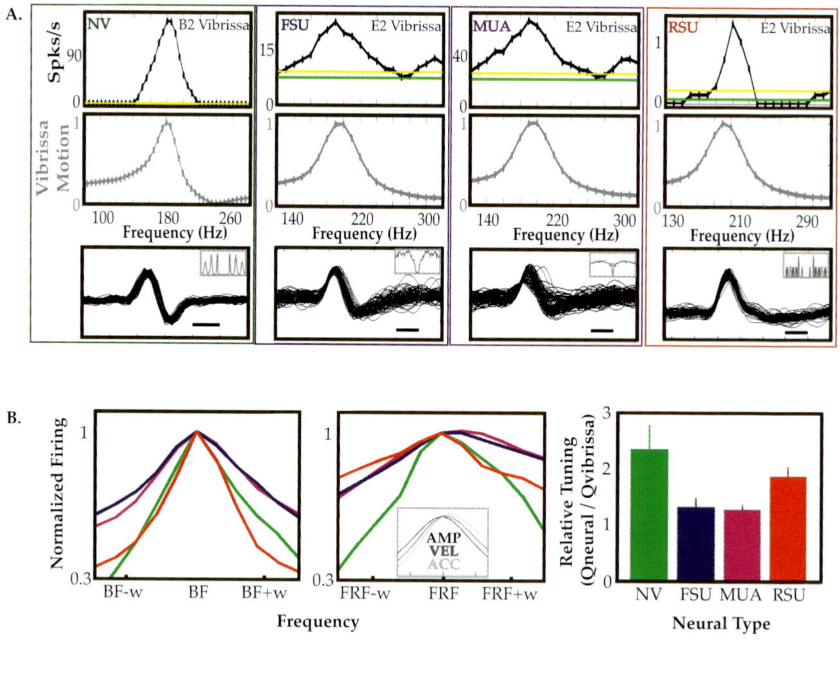
A.
NV
B2 Vibrissa
FSU
E2 Vibrissa
MUA
E2 Vibrissa
RSU
E2 Vibrissa
Spks/s
Vibrissa Motion
Frequency (Hz)
B.
Normalized Firing
BF-w
BF
BF+w
FRF-w
FRF
FRF+w
AMP
VEL
ACC
Frequency
Relative Tuning (Qneural / Qvibrissa)
NV
FSU
MUA
RSU
Neural Type

FIGURE 2.4 Vibrissa resonance tuning is translated into neural frequency tuning in somatosensory peripheral and cortical neurons. A. Vibrissa resonance tuning curves (gray lines, middle row of boxes) and corresponding neural frequency tuning curves (black lines, top row) are shown for peripheral and cortical recordings. The bottom row shows the corresponding spike traces for trigeminal (NV), fast spiking unit (FSU), regular spiking unit (RSU) and multi-unit activity (MUA). Green horizontal lines indicate the spontaneous firing rate; yellow lines indicate the threshold for significant evoked activity. Note that off-resonance stimuli were unable to evoke a significant increase in neural activity, demonstrating the potential importance of resonance for the amplification of sensory information. B. *Left and Center Boxes* Average neural tuning curves are shown for all four types of neural recording. In the graph on the left, average neural activity was normalized to peak firing rate and centered on the best frequency (BF), the frequency that drove the greatest increase in mean firing rate. On the right, average neural activity was normalized to peak firing rate and centered on the fundamental resonance frequency (FRF), the frequency that drove the greatest increase in the amplitude of vibrissa motion. All four classes of neural recording showed frequency tuning. *Right* The quality of neural frequency tuning (Qneural) normalized by the quality of vibrissa frequency tuning (Qvibrissa) for all four neural recording types. As seen in the BF- and FRF-centered average tuning curves, RSUs (red curve) and trigeminal neurons (green) demonstrated more refined tuning than FSUs (blue) and MUA (purple) for both averaging approaches.[26] See Figure 2.11. (Adapted from Andermann et al., *Neuron* 42, 2004. With permission.)

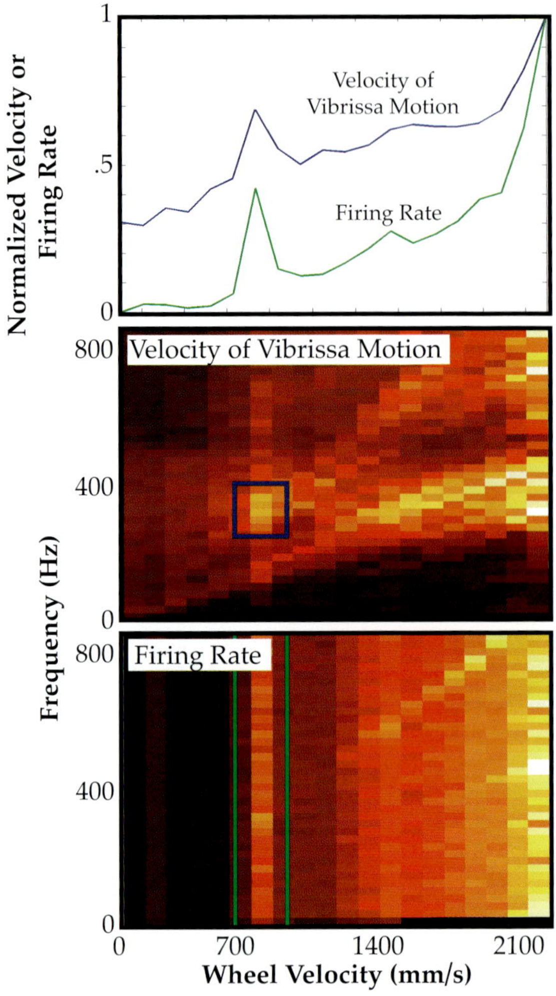

FIGURE 2.5 Vibrissa resonance evokes increased neural activity when natural complex (sandpaper) stimuli are applied. A. Multi-unit activity was recorded in the trigeminal ganglion while a stimulus wheel covered in 80-grit sandpaper was rolled against the primary vibrissa (see Figure 2.2 for a parallel example). As the wheel velocity increased, so did the vibrissa oscillation velocity (blue line). Vibrissa resonance amplification can be observed in the spike in vibrissa velocity (P(f)*f, *top*) at a wheel speed of 800 mm/s. Neural activity also showed a spike in mean firing rate at this velocity (green line). Neural activity also demonstrated a thresholded sensitivity to the increasing velocity of vibrissa oscillation at higher frequencies (≥ a wheel speed of 2000 mm/sec; see also Figure 2.8). B. Power spectra showing increased velocity of vibrissa motion or increased amplitude of neural activity (*bottom*) as a function of oscillation frequency and wheel speed. In the top panel, the peak in velocity signal at ~350 Hz (global increase in power) reflects the increased velocity of vibrissa motion generated when the wheel speed drove the predominant spatial frequency present in the texture (shown in the diagonal bands) at the vibrissa resonance (blue box). The increased mean firing rate in the associated neural response is indicated by the vertical band of increased activity observed at a wheel speed of 800 mm/s in the bottom panel. Note also that a peak is present in MUA power spectrum at the vibrissa resonance (~350 Hz), indicating fine temporal fidelity of spiking activity in response to a complex stimulus presentation (see also Figures 2.12–2.14).

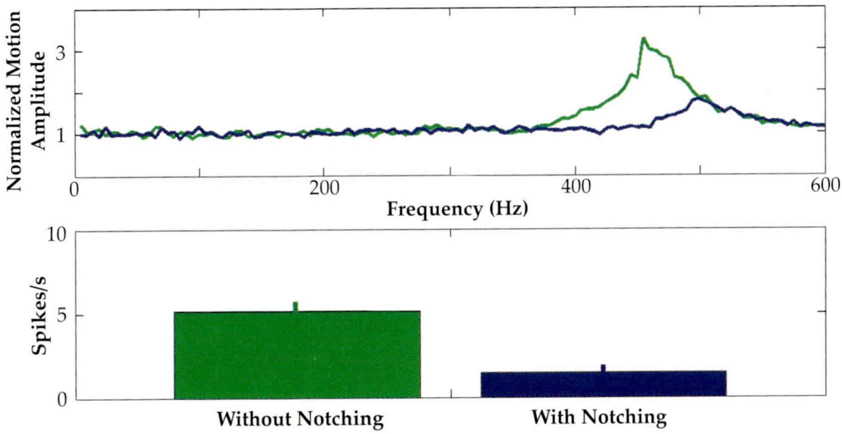

FIGURE 2.6 Vibrissa resonance evokes increased neural activity when synthesized complex stimuli are applied. A. White noise stimuli constructed as the sum of phase-shifted sinusoids from 0–600 Hz were presented through a piezoelectric stimulator to the vibrissa. A notched stimulus was also created in which the fundamental resonance and surrounding frequencies (400–500 Hz) were removed from the stimulus and the power adjusted across remaining frequencies. Vibrissa oscillations showed a resonance amplification at ~450 Hz when white noise stimuli were applied (green line) that is not present when notched stimuli were applied (blue line). B. These complex stimuli were presented while recording from a trigeminal ganglion single unit. Average neural activity was summed over the stimulation period (500 msec). As predicted by the differential increase in vibrissa motion, greater mean firing rate was evoked by the non-notched (green bar) than the notched stimulus (blue bar) (N = 37 trials, mean ± SE).

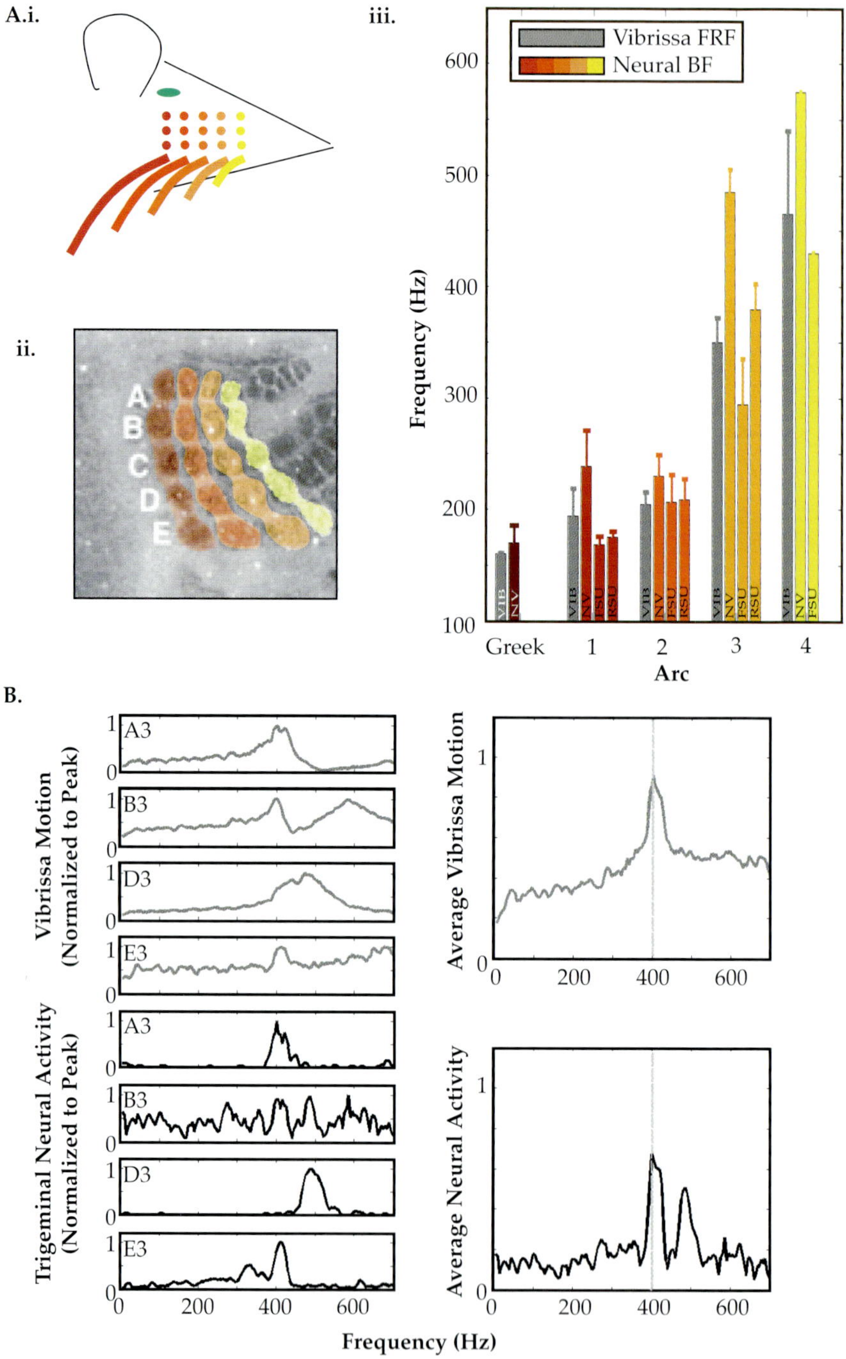
A.i.
iii.
ii.
A
B
C
D
E
Vibrissa FRF
Neural BF
Frequency (Hz)
Greek
1
2
3
4
Arc
VIB
NV
FSU
RSU
B.
A3
B3
D3
E3
Vibrissa Motion (Normalized to Peak)
Trigeminal Neural Activity (Normalized to Peak)
Average Vibrissa Motion
Average Neural Activity
Frequency (Hz)

FIGURE 2.7 Vibrissa resonance creates a somatotopic frequency map and isofrequency columns in Si. A. i. A cartoon of the rat face, showing decreasing vibrissa length in more anterior vibrissae. ii. The similar lengths of vibrissae within an arc predict the existence of isofrequency columns, spanning multiple vibrissa representations. This prediction is shown on a cytochrome oxidase stain of the SI barrel map (anatomy from http://www.neurobio.pitt.edu/barrels). iii. Vibrissa fundamental resonance frequencies (FRF: gray bars) and the neural best frequency (BF: colored bars) increased as a function of arc position of the stimulated vibrissa.[26] B. *Left* Examples of four trigeminal single unit recordings obtained during primary vibrissa stimulation. Recordings were made from the same arc of vibrissae from one animal. Vibrissa frequency tuning curves are shown in the upper panels (gray), and neural frequency tuning curves in the lower panels (black). *Right* When responses were normalized and summed across the arc, a peak in vibrissa amplitude and neural activity was observed at ~400 Hz. This example highlights the coding of isofrequency information within an arc of vibrissae.

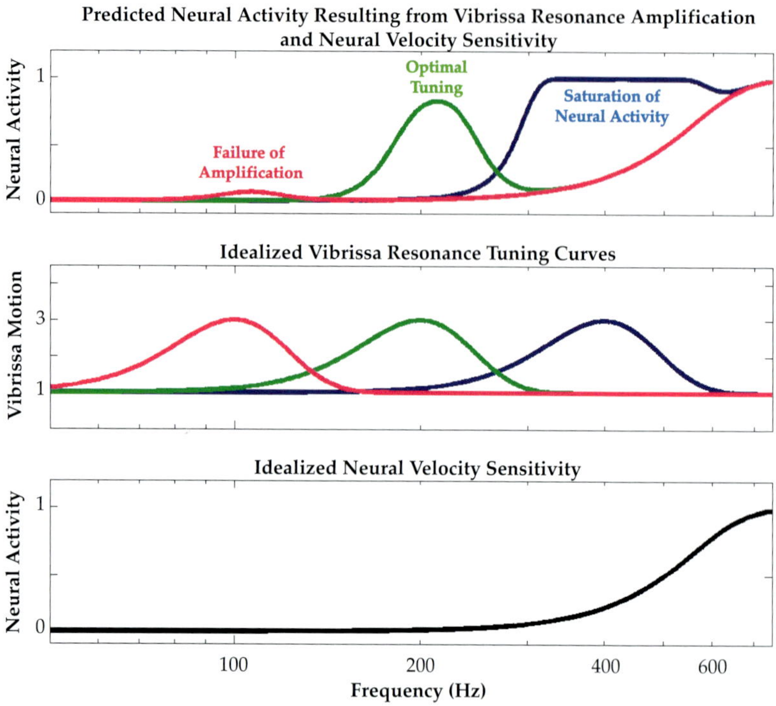

FIGURE 2.8 Neural velocity sensitivity may impact the expression of vibrissa resonance. *Bottom panel* A model of the neural response to vibrissa stimulation frequency in the absence of resonance amplification. This function was modeled as $\sin^2(pi*f/2000)$, $0 > f > 1000$ Hz, to emulate the neural sensitivity to higher frequency stimulation resulting from velocity sensitivity. Examples of this kind of increase in firing as a thresholded function of vibrissa velocity can be observed in real neural data in Figures 2.4, 2.5, and 2.10 (see also Reference number 63). *Middle panel* Three idealized examples of vibrissa resonance tuning showing a 3:1 gain in motion amplitude at the fundamental resonance frequency and bandwidth proportional to this frequency. *Top panel* The predicted neural response to vibrissa stimulation frequency as a function of resonance amplification of peak motion velocity, and intrinsic velocity sensitivity thresholds. For a given amplitude of stimulation, vibrissa resonance amplification that does not drive a neuron near its velocity threshold may fail to be amplified (purple curve, left resonance peak), while resonance amplification that is significantly above the velocity threshold (shown in the bottom panel) may fail to demonstrate tuning due to an upper limit on the range of possible firing rates for a given neuron (blue curve, right resonance peak). A subset of vibrissa resonance tuning curves near to but not above the intrinsic velocity threshold will, in this model, show optimal frequency tuning. Preliminary data suggest that these effects occur in a subset of trigeminal and cortical neurons, and that, within SI, FSU and multi-unit recordings are more susceptible to these impacts of velocity sensitivity.

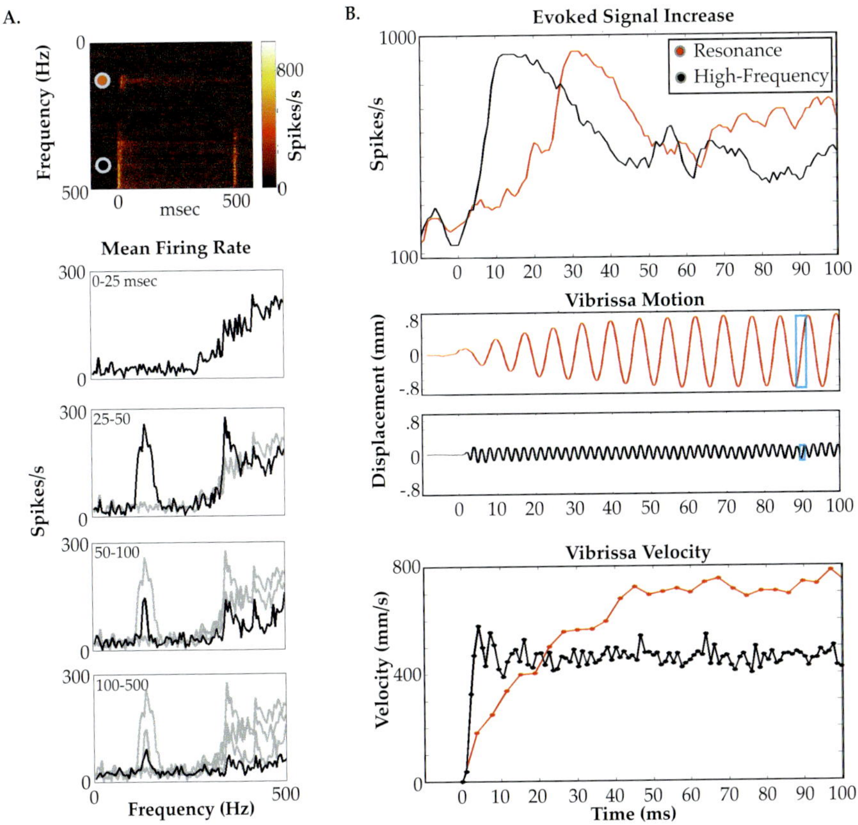

FIGURE 2.10] An example of the temporal evolution of neural frequency tuning. A. *Top* Peri-stimulus time histograms (PSTHs) are plotted as a function of frequency of stimulation (ordinate) and time (abscissa). Stimuli were applied as 160 μm sinusoids for 500 msec epochs. Resonance tuning can be seen in the selective band of increased firing at ~135 Hz: Intrinsic frequency (velocity) sensitivity can be seen in the increased firing above the threshold of ~ 350 Hz. *Bottom* Neural tuning curves showing mean firing rate for four different epochs post-stimulus onset. Resonance driven activity was not observed in the first epoch (0–25 msec post-stimulus onset) although robust high frequency responses were present. In later epochs, responses above the intrinsic high frequency threshold diminished in relative prominence while resonance driven neural activity increased. **B.** *Top panel* PSTH of activity evoked at the fundamental resonance frequency (red, 135 Hz) and at a frequency above the intrinsic high frequency threshold (black, 460 Hz). The slower rise time of resonance driven neural activity can be appreciated in this PSTH. *Middle panels* Traces of vibrissa motion driven by fundamental resonance frequency and high frequency stimuli. The fundamental resonance frequency driven motion shows a gradual increase in motion amplitude (red trace). *Bottom panel* Plots of the peak velocity of vibrissa motion for the fundamental (red) and high frequency stimuli (black). This time constant for the amplification of vibrissa motion is likely a key factor in the delayed increase in resonance driven activity in this example (see also Figure 2.11).

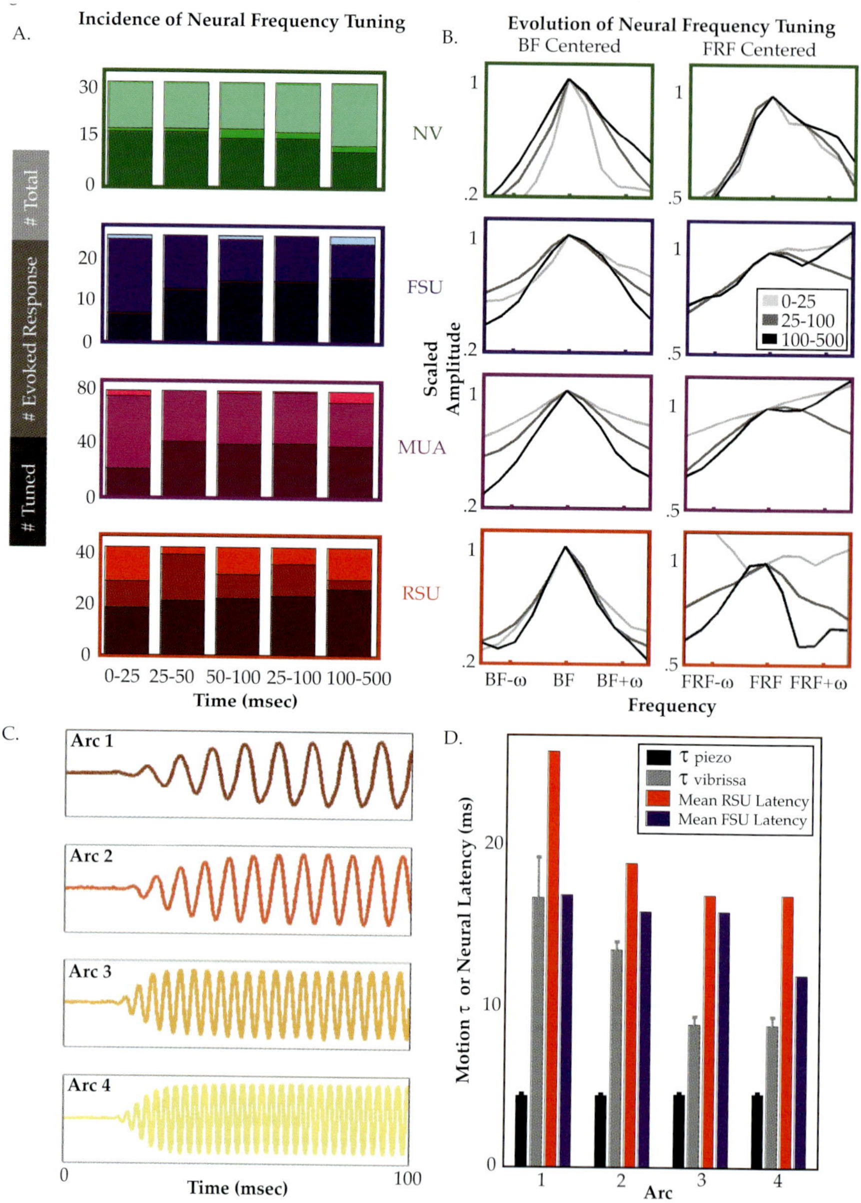
A.
Incidence of Neural Frequency Tuning
Total
Evoked Response
Tuned
NV
FSU
MUA
RSU
0-25 25-50 50-100 25-100 100-500
Time (msec)
B.
Evolution of Neural Frequency Tuning
BF Centered
FRF Centered
Scaled Amplitude
0-25
25-100
100-500
BF-ω BF BF+ω
FRF-ω FRF FRF+ω
Frequency
C.
Arc 1
Arc 2
Arc 3
Arc 4
Time (msec)
D.
τ piezo
τ vibrissa
Mean RSU Latency
Mean FSU Latency
Motion τ or Neural Latency (ms)
Arc

FIGURE 2.11 Vibrissa resonance tuning is delayed for all Si neural response types, correlated with the time constant of vibrissa resonance amplification. A. Vibrissae were stimulated with a 500 msec sinusoidal train during extracellular recording of trigeminal ganglion neurons (NV, green) or SI barrel recordings, including fast spiking units (FSU, blue), multi-unit activity (MUA, purple) and regular spiking units (RSU, red). Each bar plot shows the probability of observing a frequency tuned neural response (dark shading), a significant evoked increase in mean firing rate without frequency tuning (medium shading), or a neuron not driven by the stimulus (light shading). For all cortical response categories, the incidence of tuned responses and of driven neurons without tuning was increased for epochs >25 msec as compared to the epoch 0–25 msec. B. Average tuning curves are shown for each neural type for responses centered on the best frequency that drove the largest increase in neural firing (BF), or on the vibrissa fundamental resonance frequency (FRF). The apparent velocity sensitivity of FSU and MUA responses, shown as a sensitivity to higher frequency input, can be seen in the FRF-centered tuning curves for the epochs 100–500 msec post-stimulus onset. Similar velocity sensitivity was not observed in any epoch for RSU responses. Each plot is presented relative to the vibrissa frequency tuning bandwidth[26] C. Examples are shown of the evolution of vibrissa resonance when stimuli were applied at the fundamental resonance frequency for vibrissae from the 1—4 arcs. Shorter vibrissae (4 arcs) show a faster rise time than longer vibrissae (1 arc). D. Bar plots showing the time constant for the stimulator (black bars), for vibrissa amplification (gray bars), and for the latency to onset in neural activity (RSU, red bars and FSU, blue bars). The vibrissa time constants and neural latencies shift as a systematic function of the arc of vibrissae stimulated, with longer time constants and neural latencies observed for more posterior vibrissae. These delays provide one cause for the delay in cortical frequency tuning (shown in the above panels and Figure 2.10). Active neural mechanisms including inhibition and thalamocortical depression likely also play a role. The systematic shift in time constants across vibrissae also generates a map of onset timing within SI that may provide relevant coding information for the behaving animal, and that could increase the efficacy of input to an isofrequency column through enhanced neural synchrony.[26] (Adapted from Andermann et al., *Neuron* 42, 2004. With permission.)

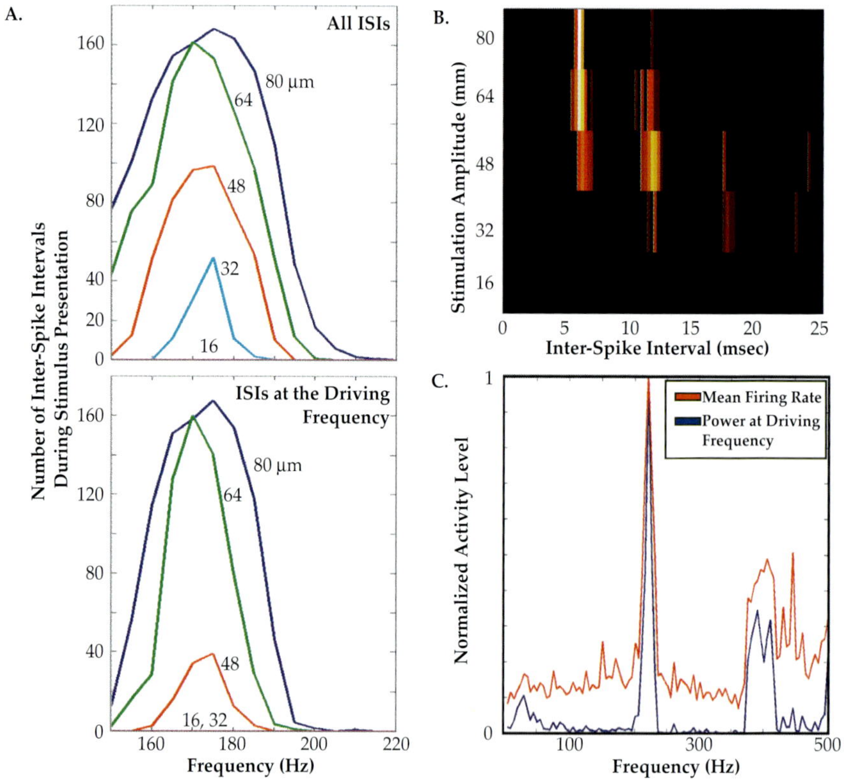

FIGURE 2.12 Trigeminal ganglion neurons demonstrate neural tuning and an atonal interval in the fine timing of their evoked activity. A. The top graph shows a frequency tuning curve for a trigeminal unit, constructed by counting all evoked inter-spike intervals (ISIs), a measure that is functionally equivalent to the mean firing rate. The lower graph shows the count of ISIs at the driving period, indicating fine temporal following of the neuron. Numbers adjacent to each curve indicate the amplitude of vibrissa stimulation applied. Frequency tuning was observed in both the mean firing rate and in the fine timing of neural evoked activity. An increased mean firing rate was observed for stimuli ≥32 μm, while temporal following at the driving frequency was present only for larger amplitudes of stimulation, ≥48 mm. This finding parallels similar observations made in the primate somatosensory system.[76] B. A graph of the incidence of ISIs at the fundamental resonance frequency, plotted as a function of the amplitude of stimulation (yellow indicates increased incidence) for the example in A. At larger amplitudes of stimulation, only firing at the fundamental resonance frequency was observed, as shown by the exclusive presence of ISIs at ~7 msec at 80 μm stimulation. In contrast, lower amplitudes of stimulation evoked ISIs at multiples of the driving period. C. A plot from a different single-unit trigeminal recording, showing the mean firing rate (red) and power at the driving frequency (blue). As in the example shown in A and B, temporal following provides a more precise tuning function at frequencies surrounding the vibrissa resonance frequency.

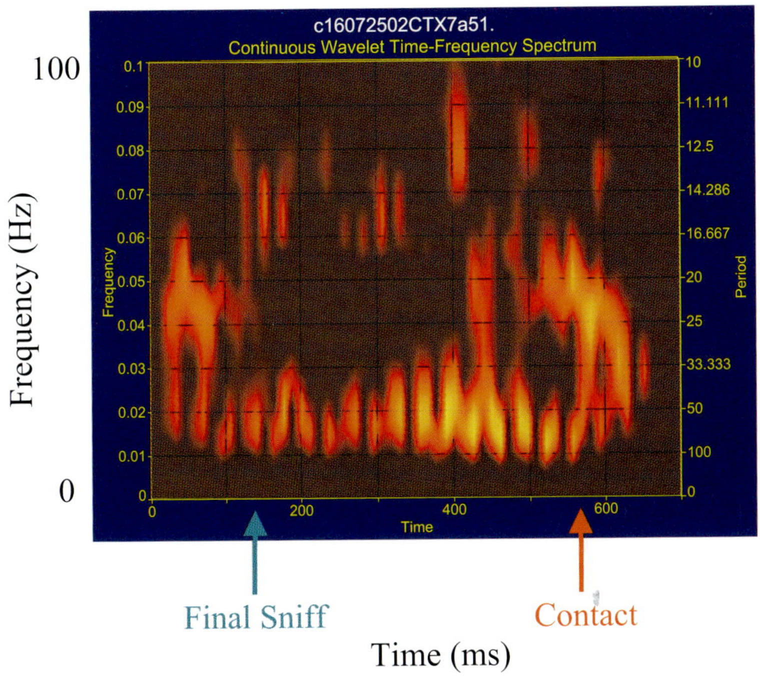

FIGURE 7.6 Sub- and suprathreshold neural activity in the vicinity of one M1 electrode during one trial of skilled reaching. The x-axis shows task-time, the y-axis shows frequency from 0 to 100 Hz, and pixel color represents amount of energy, with hotter colors indicating higher amounts of energy. See text for further description.

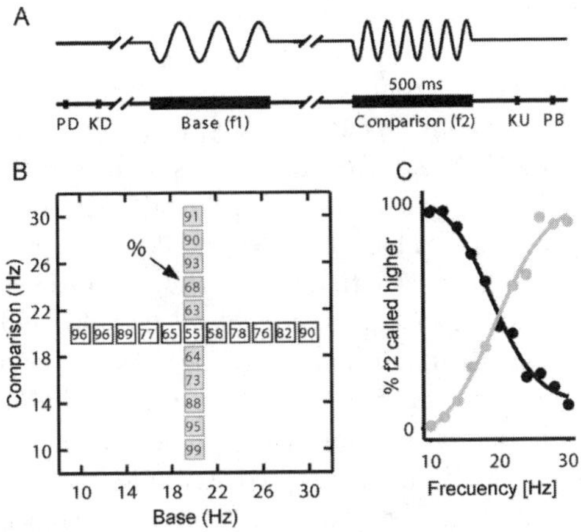

FIGURE 4.1 Discrimination task. A. Sequence of events during discrimination trials. The mechanical probe is lowered, indenting the glabrous skin of one digit of the hand (PD); the monkey places his free hand on an immovable key (KD); the probe oscillates vertically, at the base frequency. After a delay, a second mechanical vibration is delivered at the comparison frequency. The monkey releases the key (KU) and presses one of two push buttons (PB) to indicate whether the comparison frequency was higher or lower than the base. B. Stimulus set used to estimate psychometric thresholds. Each box indicates a base frequency/comparison frequency stimulus pair used. The number inside the box indicates overall percent correct trials for that base/comparison pair. C. Psychometric curves computed from B.[24,26]

vibrates at frequencies between 5 and 50 Hz. They showed that flutter is primarily transmitted by quickly-adapting (QA) cutaneous mechanoreceptors,[17,33] and found that humans and monkeys have similar abilities for detecting and discriminating the frequencies of mechanical sinusoids delivered to the hands.[12,16,33] They also tried to determine how the neural activity triggered by flutter stimuli is related to psychophysical performance.[16,33] In the discrimination task they designed,[16] animals had to indicate whether the frequency of a comparison stimulus was lower or higher than the frequency of a base stimulus presented earlier (Figure 4.1). In principle, the task can be conceptualized as a chain of neural operations or cognitive steps: encoding the first stimulus frequency, maintaining it in working memory, encoding the second frequency and comparing it to the memory trace left by the first stimulus, and communicating the result of the comparison to the motor apparatus. The flutter task offers a number of advantages as a model for sensory processing in the brain.[28,29] Not only do humans and monkeys perform similarly, but the items to be compared are spread out across time and always activate the same well defined population of primary receptors.[17,19,33,34] Spatial variations are basically taken out of the picture. However, for the flutter task to be a useful model, it is essential that it generate a reliable sequence of cognitive events like the one just mentioned. How do we know that this is so?

A crucial step is to scrutinize the psychophysics. In the original paradigm, the base frequency did not vary from trial to trial during a run.[12,16] When we re-examined the flutter discrimination task, we found the paradigm to be ambiguous: when the base stimulus frequency is kept constant, the task can be solved either by comparing the two stimuli or by categorizing the second stimulus as high or low, ignoring the base stimulus.[6] What were the monkeys actually doing? When the base frequency was held constant during long blocks of trials, as done originally, the measured discrimination limens (DLs) and Weber fractions were similar to those reported before. If the monkeys had been discriminating the differences in frequency between the two stimuli, they would also have been able to do so when the frequency of the base stimulus changed from trial to trial. However, when this was not the case, their performance dropped to chance levels. It seemed that the monkeys were paying attention only to the second stimulus, categorizing it as low or high with respect to an internal reference, perhaps the base frequency used during training. To test this possibility, the base stimulus was removed and single stimuli were delivered in each trial. In this new condition the monkeys were rewarded for correctly categorizing stimulus frequency as lower or higher than an arbitrary reference (20, 30, or 40 Hz) kept constant during a block of trials; the monkeys had to determine this reference by trial and error. The monkeys learned this task very quickly and the psychometric curves[6] measured in this condition were practically identical to those measured during the classical discrimination task.[13]

For true discrimination, the monkeys were retrained using multiple pairs of base and comparison frequencies. The key was to vary the base frequency in each trial but in such a way that each frequency could be followed either by a higher or a lower comparison (Figure 4.1C) which would force the subjects to compare.[6] After retraining, which took a few months, performance in this situation was similar to that in the classical discrimination task. From these results it seems almost certain that the animals truly learned to discriminate between frequencies on a trial-by-trial basis.

We learned that although monkeys may indeed learn to discriminate, they can also develop alternate strategies to solve a task, as suggested earlier based on theoretical arguments (Johnson, 1980). In particular, in the classical flutter discrimination paradigm monkeys tend not compare the two stimuli at every trial. Instead they classify the second stimulus, possibly setting the limits of each category during the first few trials in a run. Whenever animals are assumed to discriminate, this problem should not be underestimated, regardless of sensory modality (Johnson, 1980; Vogels et al., 1990; Vázquez et al., 2000).

This appears as a simple observation, but it may reflect fundamentally different mechanisms at work. Consider a task that involves variations in a single feature across trials. To identify or classify a current sensory stimulus, it must be compared to a reference stored in long-term memory. But it is not clear how the comparison process can be studied in this situation. How is information stored in long-term memory read out and made comparable to current sensory events? Where is this information stored and how does it differ from the original sensory-evoked activity? In contrast, in discrimination tasks in which two stimuli are presented sequentially in each trial, the comparison is made against the short-term memory trace left by the first stimulus. This means that if we can identify the neural correlate of the working memory component, it might be possible to study the comparison- or decision-making mechanisms that underlie task performance. Indeed, such neural correlates have been recently reported.

IV. NEURAL CODING OF VIBROTACTILE STIMULI IN S1

Shortly after their work on cutaneous afferent fibers,[18] Mountcastle and colleagues studied the responses of S1 neurons. Two decades later, S1 neurons were recorded again. This time in behaving monkeys trained to detect and discriminate the frequencies of flutter stimuli.[16] The results support previous findings. First, it was found that QA neurons of S1, like their afferent fibers, fire periodically in phase with mechanical oscillations. Second, their firing rates seemed to change little in the flutter range (this conclusion was based on data from 17 neurons). Third, psychophysical performance matched the inferred performance based on the discriminability of the periodic inter-spike intervals.[16] It followed that as proposed before, stimulus frequency could not be encoded by S1 firing rates. Stimulus frequency had to be encoded temporally, in the serial order of evoked spikes.[16,18,33]

In support of this proposal, using flutter stimuli, Merzenich and colleagues compared psychophysical data from monkeys to S1 recordings in separate experiments from the same animals.[21] The comparison was consistent with a temporal coding mechanism and firing rates were not seen to vary with stimulus frequency (however, the range of frequencies tested was quite narrow, and animals were anaesthetised). Merzenich and colleagues made another important observation, that spike timing associated with the sine wave was more precise in trained animals compared to untrained monkeys.[21] Thus, based on these results, a psychophysical observer should exploit the periodic spike timing evoked in the QA neurons of S1 cortex for sensory discrimination.

Arguments in favour of the proposal reviewed above could be strengthened if a large number of neurons were studied and if neurons were studied in behaving animals during the flutter discrimination task. To this end, we trained monkeys to discriminate between flutter-stimulus frequencies and recorded many neurons with QA properties in areas 3b and 1 of S1.[9,30] Each recorded neuron with QA properties was studied during the discrimination task. There were three major results. First, the majority of neurons from S1 were phase-locked to the input stimulus frequency (Figure 4.2A and 4.2B). However, almost a third of QA neurons modulated their firing rates as a function of the stimulus frequency (Figure 4.3A and 4.3B.[30]). The second important finding was that QA neurons that modulate their firing rates were affected by the task condition that is, they increased their transmitted information about the stimulus frequency during task performance.[30] Third, only those neurons that varied their firing rates as a function of the stimulus frequency were affected in error trials.[30]

These findings question a unique role of periodic spike timing in discrimination of flutter stimuli and suggest that a firing rate code cannot be discarded.[30] But apart from this, what do these findings suggest? They suggest the presence of two sub-populations of QA neurons in S1 which behave differently in response to a periodic mechanical stimulus.[9,30] These two sub-populations might be arranged in hierarchical fashion. QA neurons that respond periodically, might be closer to the input stimulus and those that modulate their firing rates might integrate the responses of the periodic neurons and transform them into a rate code.[9] This last order neurons of the QA

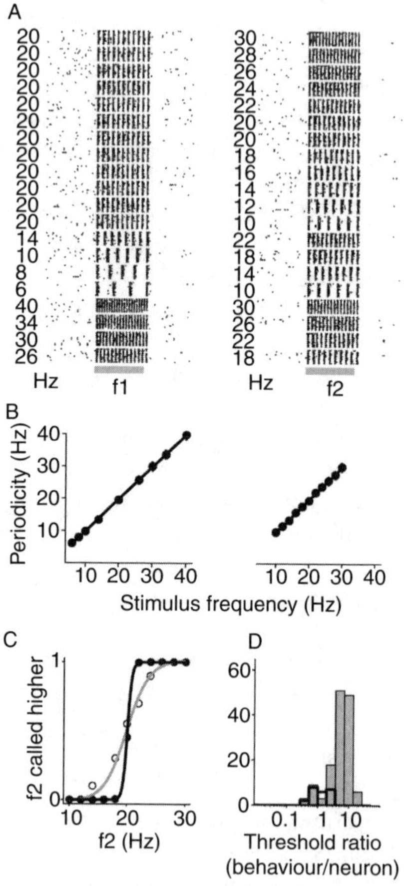

FIGURE 4.2 Periodic responses of a QA neuron of area 1 during the discrimination task. **A.** Rater plots. Each row of ticks represents a trial and each tick represents an action potential. Trials were randomly delivered. Gray horizontal lines indicate the first (f1) and the second (f2) stimulus. B. Periodicity (± SD) as a function of the first (f1) and second stimulus (f2) frequencies. C. Relationship between psychometric and neurometric discrimination functions. This is plotted as the probability that the second stimulus is judged higher than the first. Data and sigmoidal fits (χ^2 test, $p < 0.001$) for eleven pairs of stimulus frequencies in which the base frequency was 20 Hz. Gray and black lines represent psychometric and neurometric functions, respectively. D. Threshold ratios (psychometric/neurometric thresholds) calculated from neurons with periodicity (gray bars). Open bars represent the threshold ratios between psychometric and neurometric thresholds calculated from a small number of neurons with modulations in their firing rate.[9]

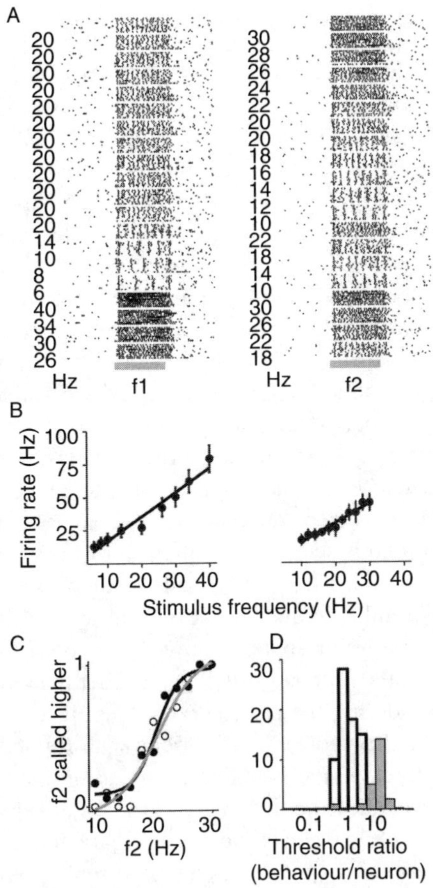

FIGURE 4.3 Firing rate modulation of a QA neuron of area 3b during the discrimination task. Same format as Figure 4.2. A. Rater plots. B. Mean firing rate (± SD) as a function of the stimulus frequency. C. Relationship between psychometric and neurometric discrimination functions. D. Threshold ratios calculated between psychometric and neurometric thresholds for each neuron which varied the firing rate as a function of the stimulus frequency (open bars). Gray bars represent the threshold ratios between psychometric and neurometric thresholds calculated from a small number of neurons that show periodicity.[9]

circuit could distribute the neural representation to those structures anatomically linked to S1, in order to solve the sensory discrimination task. However, further studies are needed to see whether this is so.

V. NEURONAL CORRELATES OF FLUTTER DISCRIMINATION IN S1

A more direct test for the role of periodicity in flutter discrimination is measuring the discrimination capabilities of these subtypes of QA neurons associated with the psychophysical performance (Figure 4.2C, 4.3C, and 4.4C). A second test is to prove whether the evoked neural activity during discrimination in S1 cortex is sufficient for sensory performance. Finally, to test whether the temporal order of the spikes is important for sensory discrimination. These are incisive tests to validate the meaning of the neural encoding of the flutter stimuli in S1 cortex.

The vibrotactile discrimination task requires the comparison of the second stimulus frequency against the first.[9] As indicated above, we found two types of responses in QA neurons of S1 cortex: one that is periodically entrained by the stimulus frequency, and a second that, although not periodically entrained, has average firing rates during the stimulus period that are modulated as a function of the stimulus frequency.[9,30] To investigate which of these two responses is associated with the psychophysical performance, we determined the probability that an observer (a cortical region central to S1 cortex) could distinguish the difference between the two stimuli.[9] This could be based on a comparison of the neuronal response distributions of the second stimulus frequency made against the neuronal response distributions of the first stimulus frequency. According to this, the observer could use a simple rule: if the number of spikes during the second stimulus is higher than during the first stimulus, then second stimulus is higher than the first. The same rule can be used when considering the periodicity values: if the periodicity (estimated as the frequency with greatest power in a Fourier transform of the spiking responses) during the second stimulus period is higher than during the first stimulus, then the second stimulus is higher than the first. The effect is equivalent to determining the area under the curve ROC (receiver operating characteristic[7]) generated by the neuronal response distributions for each pair of stimulus frequencies, using both periodicity and firing rate values.[8] The areas under each of these two ROC curves are an estimate of the proportion of correct trials that an optimal observer would obtain by comparing the numbers of spikes or periodicity. In pairs of stimulus frequencies where the neuronal response distributions during the second stimulus are much higher than the neuronal distributions of the first stimulus, ROC values are close to 1. If the neuronal response distributions during the stimulus are much lower than the neuronal response distributions of the first stimulus, ROC values are close to 0; for overlapping distributions, intermediate ROC values are found. The ROC values were used to compute neurometric functions. Psychophysical and neuronal discrimination thresholds are calculated as half the difference between the stimulus frequency identified as higher than the standard in 75% of trials and that frequency identified as higher in 25% of the trials. These are read directly from the logistic functions expressed in terms of Hz. Using this analysis, we are now in the position to address the question of which of the two representations is meaningful for frequency discrimination.

Neurometric functions based on periodicity or firing rate of single S1 neurons were directly compared to the psychometric thresholds.[9] The results of this analysis show that neurometric threshold values based on periodicity are far lower than psychometric thresholds (Figure 4.2C and 4.2D). This is not the case when neurometric thresholds based on firing rate are compared to the psychometric thresholds (Figure 4.3C and 4.3D). They are very close to the psychometric thresholds. The goal of computing neurometric functions was not only to reveal the relationship between the neuronal responses of S1 to the mechanical stimulus, but also to discern whether these neural signals account for the psychometric behavior. However, what is then the functional meaning of the periodic neural signal in S1? One possible role is that they simply represent the temporal structure of the stimulus and that monkeys do not use this exquisite representation for frequency discrimination. This would be the case if, for example, discrimination were based on the mean number of spikes (or bursts) fired by the population of QA neurons as a function of the stimulus frequency. Consistent with this idea, we found[9] that QA neurons in S1 whose firing rates are modulated by the stimulus frequencies, and their neurometric thresholds based on firing rates, are closely similar to the monkey's psychophysical thresholds (Figure 4.2C and 4.2D). However, these correlations do not prove they are sufficient for discrimination.[24]

One experiment which could give an insight about the functional meaning of the periodic spike structure of the evoked activity in S1, is testing whether monkeys could discriminate between the two stimuli when periodicity is broken. If monkeys fail to discriminate between the in mean frequency of two stimuli, this would strengthen the proposal that discrimination of flutter stimuli depends on the periodic structure of the spike trains evoked in S1. However, monkeys were able to extract the mean frequency from the nonperiodic signals and the psychophysical measures were almost identical with the periodic stimuli.[26]

We then studied QA neurons in each of two conditions: while monkeys discriminated between periodic stimuli, and while monkeys discriminated aperiodic stimuli.[7] Due to the aperiodic stimulus design, even highly stimulus-entrained neurons do not carry information about stimulus frequency in their periodicity. Clearly, neurometric thresholds based on the firing rate were again closely associated with the psychometric thresholds (Figure 4.4C and 4.4D). As in the periodic condition, a psychophysical observer could exploit the firing rate for frequency discrimination of aperiodic stimuli. These results suggest that an observer could solve this task with a precision similar to that of the monkey based only on the firing rate produced during the stimulus periods.

In summary, firing rates that vary as functions of stimulus frequency are seen in multiple areas activated during the task, in particular in S1, and there is evidence that these rate variations have a significant impact on behavior. Clearly, the brain must be able to extract at least some information from the precise timing of S1 spikes evoked during the task for instance, humans can easily distinguish periodic stimuli from aperiodic. However, we found no indication that the high periodicity found in S1 contributes to frequency discrimination although this possibility is hard to rule out entirely.

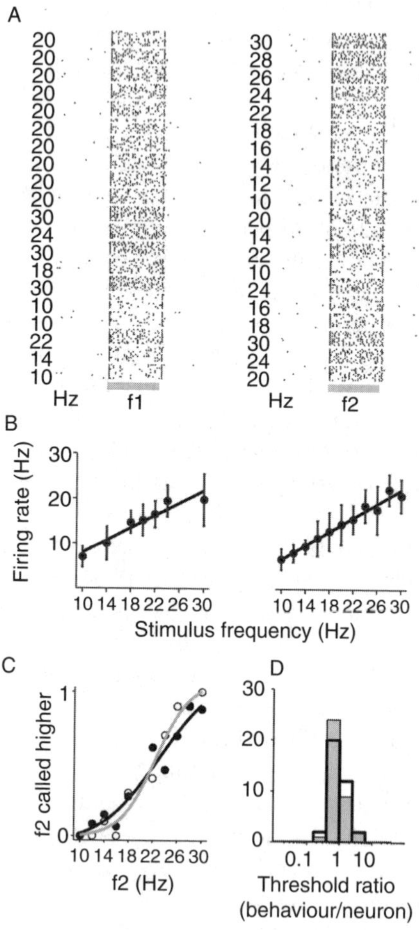

FIGURE 4.4 Firing rate modulation of a QA neuron of area 1 during the discrimination of aperiodic stimuli. Same format as Figure 4.2, but both base (f1) and comparison (f2) frequencies (mean frequencies) lack periodicity. A. Rater plots. B. Mean firing rate (± SD) as a function of the stimulus frequency. C. Relationship between psychometric and neurometric discrimination functions. D. Threshold ratios calculated between psychometric and neurometric thresholds for each neuron during the discrimination of period stimulus frequencies (open bars). Black bars represent threshold ratios between psychometric and neurometric thresholds during the discrimination of aperiodic frequencies.[9]

VI. ARTIFICIAL INDUCTION OF ACTIVITY IN S1 UNDERLYING FLUTTER DISCRIMINATION

How can we be sure that the activity recorded in S1 is actually related to perception and behavior? Intracortical microstimulation is a powerful technique capable of establishing a causal link — not just a correlation — between the activity of localized neuronal populations and specific cognitive functions.[2,6,24,26,31] For flutter discrimination, this approach has provided the most compelling evidence that all the cognitive processes of the task may be triggered directly by the QA circuit in S1 and has also allowed us to explore questions about the neural code for flutter stimuli.[24,26] Importantly, S1 is organized in modules of neurons sharing the same receptive field and mechanorceptor submodality.[14,20,32] The experiments described below, were aimed to drive a column(s) of S1 — mostly of the QA type — in a way that matched the dynamic responses recorded when mechanical stimuli were applied to a patch of skin of the fingertips.

Initially, the idea was to manipulate the comparison stimulus only.[26] The monkeys first learned to discriminate the frequencies of two sinusoidal vibrations delivered successively to the fingertips. Once they mastered the task, neurophysiological recordings were made in area 3b of S1 which allowed the identification of clusters of QA neurons. A microstimulation current was spread around a certain cortical area, activating many neighbouring units. Thus, a key for the success of microstimulation experiments is that the microelectrode must be located in the midst of a functionally homogeneous cluster of neurons. Fortunately, area 3b is indeed organized into modules of units with similar properties or columns.[32] Having identified a set of QA neurons, the comparison stimulus was substituted with microstimulation in one half of the trials (Figure 4.5). Artificial stimuli consisted of periodic current bursts delivered at the same comparison frequencies as mechanical stimuli. Microstimulation sites in S1 were selected to have QA neurons with receptive fields on the fingertip at the location of the mechanical stimulating probe. Remarkably, the monkeys discriminated the mechanical (base) and electrical (comparison) signals with performance profiles indistinguishable from those obtained with mechanical stimuli only (Figure 5A), so the artificially induced sensations probably resembled natural flutter quite closely.[26]

Going back to the question of whether periodicity is crucial for frequency discrimination, we applied aperiodic microstimulation patterns that mimicked the random trains of mechanical pulses discussed earlier (Figure 4.5B). The same mean frequencies were used in this condition — 20 Hz still corresponded to 10 current bursts delivered in 500 ms — but the bursts were separated by random time intervals. Everything else proceeded as before, with mechanical and stimulation trials interleaved, as indicated in Figure 4.5B. From the very first trials, the animals were able to discriminate both mechanical and electrical aperiodic signals (Figure 4.5B), with practically the same performance level reached with mechanical periodic vibrations.[26]

An interesting effect was observed as the intensity of the microstimulation current was varied. At very low intensities, artificial stimuli were not detected; the monkeys simply kept waiting. At intermediate intensities the monkeys detected the stimuli — they reacted and pushed a button — but their psychophysical behaviour

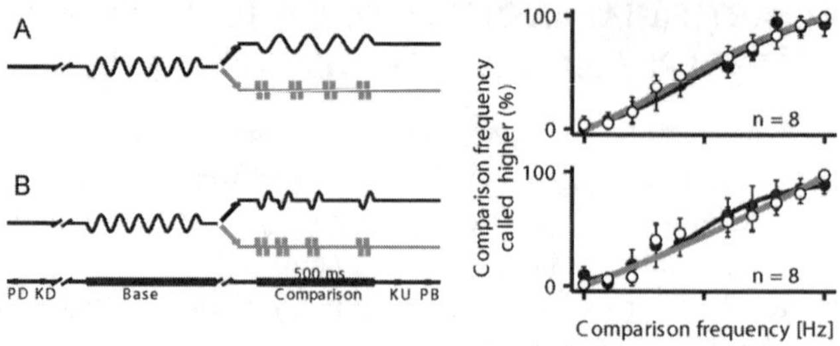

FIGURE 4.5 Psychophysical performance in frequency discrimination with natural mechanical stimuli and with artificial electrical stimuli injected into clusters of quickly adapting (QA) neurons of area 3b. The diagrams on the left show two types of trials that were interleaved during the experiments. In half of the trials, the monkeys compared two mechanical vibrations delivered on the skin. In the other half, one or both stimuli could be replaced by electrical frequencies microinjected into clusters of QA neurons of area 3b. The curves on the right show the animals' performance in the different situations illustrated on the left. Filled and open circles indicate mechanical and electrical stimuli, respectively. In A and B, the y-axis corresponds to the percentage of times the monkeys called the second stimulus frequency (x-axis) higher than the first (20 Hz). A. Psychophysical performance using periodic stimuli. The comparison stimulus could be either mechanical or electrical frequencies. B. Psychophysical performance when the comparison stimulus could be either aperiodic, mechanical, or electrical stimulus frequencies. PD indentation of the glabrous skin by the mechanical probe. KD detection of indentation. KU detection of the end of the comparison stimulus. PB the monkeys presses one of two push buttons to indicate whether the second stimulus frequency was higher or lower than the first.[26]

was at chance levels, as if they could determine the presence but not the frequency of the artificial stimuli. At higher intensities they discriminated normally.[26] These transitions parallel those observed when the amplitude of mechanical vibrations was gradually increased; in particular, there is an atonal interval, in which stimuli can be detected but their frequencies cannot be ascertained.[12,16] In an extra control experiment we investigated the effet of stimulus amplitude, which could potentially act as an alternate discrimination signals.[26] Four frequency pairs and all stimuli were mechanical and the other half of the comparison stimulus was electrical. In both cases, the amplitude of the comparison could take one of three values, a standard amplitude A, 0.6 A, or 1.4 A. The changes in amplitude were in terms of percentage, of the same magnitude (40%) as the differences between base and comparison frequencies. The results showed that performance was not affected by the large variations in amplitude. Had the monkeys been guided by the amplitude changes, one of the three combinations for each frequency pair would have fallen to < 25% correct, because performance was normally > 75% correct. Amplitude corrections, like those mechanical stimuli, were also applied to electrical currents in ~60% of all runs; they had no marked impact on performance.[26]

Because of the design of the paradigm, comparison of the second stimulus is made against a memory trace of the first one.[8] Having shown that the monkeys could use an artificial stimulus during the comparison, we wondered whether they would be able to memorize and use an electrical stimulus delivered during the base period.[24]

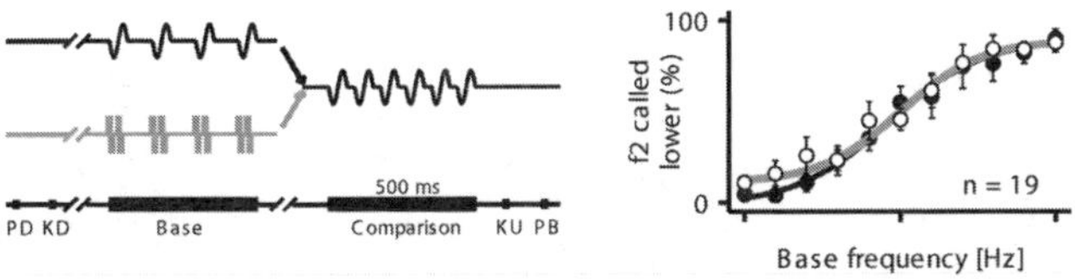

FIGURE 4.6 Psychophysical performance when the base stimulus could be either periodic, mechanical, or electrical stimulus frequencies injected into clusters of quickly-adapting QA neurons in area 3b. The same protocols and labels as in Figure 4.5, but here the base mechanical stimulus frequency was substituted by electrical stimulus frequencies. The y-axis corresponds to the percentage of times the monkeys called the comparison stimulus (20 Hz) lower than base stimuli at the frequency specified in the x-axis.[24]

In this case, in one half of the trials, the base stimulus consisted of electrical microstimulation at a frequency equal to f1 (Figure 4.6), with the electric current again being injected into QA neurons. The frequency pairs and event sequence during the task were the same as in previous experiments with natural stimuli. We stress this because careful design of the stimulus sets was particularly crucial here, in order to ensure that the monkeys paid attention to the base stimulus and stored it in working memory.[22] The monkeys' psychophysical behaviour was again indistinguishable from that observed with natural stimuli only (Figure 4.6), showing that the signals evoked by mechanical and artificial stimuli could be stored and recalled with approximately the same fidelity.[24] Finally, we also investigated whether monkeys could perform the entire task on the basis of purely artificial stimuli. In most sessions in which the two mechanical stimuli were replaced by microstimulation patterns, monkeys were able to reach discrimination levels close to those measured with mechanical stimuli delivered to the fingertips. This demonstrates that activation of QA neurons is sufficient to drive all the cognitive processes involved in the task with little degradation in performance.[24]

A couple of additional observations derived from these experiments are also noteworthy. First, early experiments with primary afferents had demonstrated that the flutter sensation is specifically transmitted by QA fibers.[19,34] But this was more difficult to test at the level of S1.[26] When microstimulation was applied to clusters of neurons identified as having slowly adapting (SA) properties (Figure 4.7A, the monkeys could barely discriminate if at all.[24] As the electrode was advanced to the border between SA and QA clusters, performance became somewhat better (Figure 4.7B) and reached its usual degree of accuracy when QA properties became most evident in the recordings (Figure 4.7C[24]). Hence, QA and SA units are still functionally segregated in the primary cortex consistent with previous observations.[29]

In some sessions, we were able to introduce three microelectrodes into a cluster of QA neurons of area 3b that shared the same receptive field.[24] We knew that the most anterior microelectrode was placed in the superficial layers, because another microelectrode was placed in front of it and recorded units in primary motor cortex that were driven by spontaneous or passive movements of the fingers and lacked cutaneous receptive fields. The most posterior microelectrode was placed, we believe, in the lower layers, and the microelectrode between these two in the middle

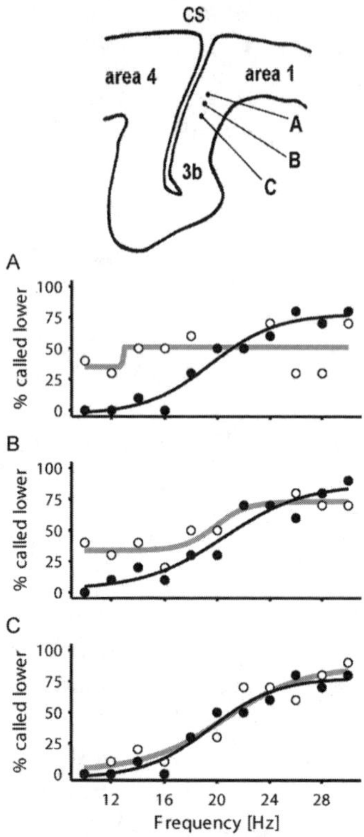

FIGURE 4.7 Psychophysical performance elicited by microstimulating at the base stimulus frequency in three different sites of area 3b. Same protocol and labels as in Figure 4.6. A-C. Data collected in three separate runs during the same electrode penetration A. Psychophysical performance when microstimulation was applied in the center of cluster of slowly adapting (SA) neurons. B. Psychophysical performance when microstimulation was applied in the border between QA and slowly adapting (SA) neurons. C. Psychophysical performance when microstimulation was applied in a cluster of QA neurons.[24]

layers. In separate runs, the frequency pairs and event sequence were the same in both mechanical and microestimulation trials, except that in the microstimulation trials the first mechanical stimuli were substituted with train of electrical current pulses delivered at the frequency of the mechanical stimulus they were substituting. Figure 4.8 shows that discrimination is triggered by microstimulating each of the three different clusters. Thus, activation of any part of the cluster of neurons with similar functional properties is sufficient to initiate discrimination in this task.[24]

VII. GENERAL COMMENTS

The results obtained in these sets of experiments suggest that QA neurons from S1 represent the stimulus frequencies both in the periodic spike intervals and in the firing rate,[9,30] and that activation of the QA circuit is sufficient to activate the entire chain of discrimination processes of this task.[24,26]

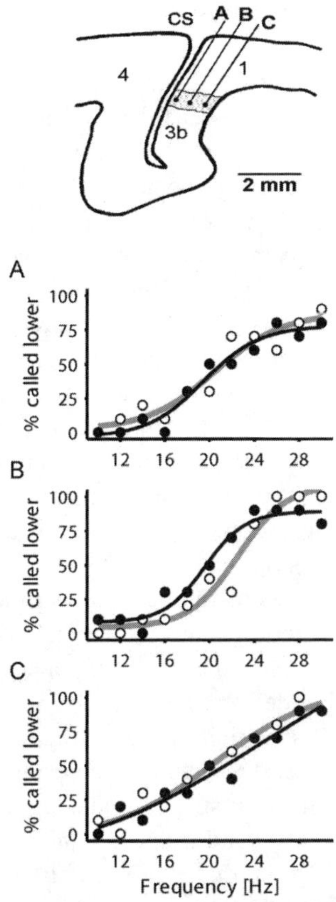

FIGURE 4.8 Psychophysical performance elicited by microstimulating at the base stimulus frequency in three independent microelectrodes in three different sites of a cluster of QA neurons of area 3b. Protocols and labels as for Figure 4.6. CS, central sulcus.[24]

The conclusion previously found in the literature,[16] that frequency discrimination is based on periodicity, came from the observation that a small number of studied QA neurons from S1 reproduce in their activity the periodicity of the mechanical stimulus frequency, and also from the fact that these neurons did not have average firing rates that were modulated by the stimulus frequency.[16] However, the study that reached this conclusion only determined the relationship between the neuronal responses to the mechanical stimulus frequencies; no attempt was made to quantify the neurometric thresholds based on periodicity and to compare these to the psychophysical thresholds. We observed that neurometric thresholds using the periodicity values are far lower than the psychometric thresholds.[9] What is then the functional meaning of this neural signal? One possible role is that this simply represents the temporal structure of the stimulus and that monkeys do not use this exquisite representation for frequency discrimination. Consistent with this interpretation, we found QA neurons in S1 whose firing rates are modulated by the stimulus frequencies,[30] and their neurometric thresholds based on this measure are similar to the monkey's psychophysical thresholds.[9]

These results also suggest that QA neurons of S1, which are classified according to their capacity to react to a slight mechanical indentation applied to the center of their receptive fields, may in fact be composed of two subpopulations, each of which behaves differently in response to a periodic mechanical stimulus. These two subpopulations might be organized in hierarchical fashion: QA neurons that respond periodically might be closer to the input stimulus, and those that modulate their firing rate might integrate the responses of the periodic neurons and transform them into a rate code. Such last order neurons of the QA circuit could distribute the neural representation of the stimulus to those structures anatomically linked to S1,[4,5] to solve the sensory discrimination task. Relevant to this interpretation is the fact that neurons in S2 respond by encoding the stimulus frequencies in their firing rates. This encoding correlates closely with the subject's discrimination performance.[25,27,30] Ascending the cortical hierarchy, neurons in the prefrontal cortex and in the premotor cortices respond by encoding the entire sequence of the discrimination task and correlate closely with the decision motor report.[3,10,22,23]

The results obtained in the microstimulation experiments show that the relationship between the neuronal responses and the animal's behavior in the flutter discrimination task are not simple coincidences.[24,26] Monkeys were able to discriminate the stimulus frequencies either delivered to the fingertips or artificially injected into a cluster of QA neurons. The specificity of QA stimulation for frequency discrimination is suggested by the fact that SA stimulation cannot produce discrimination.[24] It has been shown that activity in a single cutaneous afferent fiber could produce localized somatic sensations,[13,19,34] and frequency microstimulation of QA afferents linked to Meissner's corpuscles produced the sensation of flutter.[34] These observations strongly support the notion that the activity initiated in specific mechanoreceptors is read out by S1. This reading is then widely distributed to those anatomical structures that are linked to S1.[28] The whole sequence of events associated with this sensory discrimination task must depend on this distributed neural signal.[3,10,22,23,25,29]

ACKNOWLEDGMENTS

The research of R. Romo was supported in part by an International Research Scholars Award from the Howard Hughes Medical Institute and from grants from the Millennium Science Initiative, CONACyT and DGAPA-UNAM.

REFERENCES

1. Adrian, E.D. The basis of sensation. The action of the sense organs. London: Christophers, 1928.
2. Britten, K.H. and Wezel, R.J. Electrical microstimulation of cortical MST biases heading perception in monkeys. *Nat. Neurosci.* 1: 59-63, 1998.

3. Brody, C.D., Hernandez, A., Zainos, A., and Romo, R. Timing and neural encoding of somatosensory parametric working memory in macaque prefrontal cortex. *Cereb. Cortex.* 13:1196–1207, 2003.
4. Burton, H. and Fabri, M. Ipsilateral intracortical connections of physiologically defined cutaneous representations in areas 3b and 1 of macaque monkeys. *J. Comp. Neurol.* 355: 508-538, 1995.
5. Burton, H., Fabri, M., and Alloway, K. Cortical areas within the lateral sulcus connected to cutaneous representations in areas 3b and 1: a revisited interpretation of the second somatosensory area in macaque monkeys. *J. Comp. Neurol.* 355: 539-562, 1995.
6. Graciano, M.S., Taylor, C.S., and Moore, T. Complex movements evoked by microstimulation of precentral cortex. *Neuron.* 34:841–851, 2002.
7. Green, D.M. and Swets, J.A. Signal Detection Theory and Psychophysics. New York: Wiley, 1966.
8. Hernández, H., Salinas, E., Garcia, R., and Romo, R. Discrimination in the sense of flutter: new psychophysical measurements in monkeys. *J. Neurosci.* 17: 6391-6400, 1997.
9. Hernández, A., Zainos, A., and Romo, R. Neuronal correlates of sensory discrimination in the somatosensory cortex. *Proc. Nat. Acad. Sci. USA* 97: 6091-6096, 2000.
10. Hernández, A., Zainos, A., and Romo, R. Temporal evolution of a decision-making process in medial premotor cortex. Neuron 33: 959-972, 2002.
11. Johnson, K.O. Sensory discrimination: decision process. *J. Neurophysiol.* 43:1771–1792, 1980.
12. LaMotte, R.H. and Mountcastle, V.B. Capacities of humans and monkeys to discriminate between vibratory stimuli of different frequency and amplitude: correlation between neural events and psychophysical measurements. *J. Neurophysiol.* 38: 539-559, 1975.
13. Macefield, G., Gandevia, S.C., and Burke, D. Perceptual responses to microstimulation of single afferents innervating joints, muscles and skin of the human hand. *J. Physiol.* 429:113-129, 1990.
14. Mountcastle, V.B. Modality and topographic properties of single neurons of cat's somatic sensory cortex. *J. Neurophysiol.* 20: 408-434, 1957.
15. Mountcastle, V.B., LaMotte, R.H., and Carli, G. Detection thresholds for stimuli in humans and monkeys: comparison with thresholds events in mechanoreceptive afferent nerve fibers innervating the monkey hand. *J. Neurophysiol.* 25: 122-136, 1972
16. Mountcastle, V.B., Steinmetz, M.A., and Romo, R. Frequency discrimination in the sense of flutter: psychophysical measurements correlated with postcentral events in behaving monkeys. *J. Neurosci.* 10: 3032-3044, 1990.
17. Mountcastle, V.B., Talbot, W.H., Darian-Smith, I., and Kornhuber, H.H. Neural basis of the sense of flutter-vibration. Science 155: 597-600. 1967.
18. Mountcastle, V.B., Talbot, W.H., Sakata, H., and Hyvärinen, J. Cortical neuronal mechanisms in flutter-vibration studied in unanesthetized monkeys. Neuronal periodicity and frequency discrimination. *J. Neurophysiol.* 32: 452-484, 1969.
19. Ochoa, J. and Torebjörk, E. Sensations evoked by intraneural microstimulation of single mechanoreceptor units innervating the human hand. *J. Physiol.* 342: 633-654, 1983.
20. Powell, T.P.S. and Mountcastle, V.B. Some aspects of the functional organization of the cortex of the postcentral gyrus of the monkey: a correlation of findings obtained in a single unit analysis with cytoarchitecture. Bull. Johns Hopkins Hosp. 105:133-162, 1959.

21. Recanzone, G.H., Merzenich, M.M., and Schreiner, C.E. Changes in the distributed temporal response properties of SI cortical neurons reflect improvements in performance on a temporally based tactile discrimination task. *J. Neurophysiol.* 67: 1071-1091, 1992.
22. Romo, R., Brody, C.D., Hernández, A., and Lemus, L. Neuronal correlates of parametric working memory in the prefrontal cortex. Nature 399: 470-473, 1999.
23. Romo, R., Hernández, A., and Zainos, A. Neuronal correlates of a perceptual decision in ventral premotor cortex. Neuron, 41: 165-173, 2004.
24. Romo, R., Hernández, A., Zainos, A., Brody, C.D., and Lemus, L. Sensing without touching: psychophysical performance based on cortical microstimulation. Neuron 26: 273-278, 2000.
25. Romo, R., Hernández, A., Zainos, A., Lemus, L., and Brody, C.D. Neuronal correlates of decision-making in secondary somatosensory cortex. *Nat. Neurosci.* 5: 1217-1225, 2002.
26. Romo, R., Hernández, A., Zainos, A., and Salinas, E. Somatosensory discrimination based on cortical microstimulation. Nature 392: 387-390, 1998.
27. Romo, R., Hernández, A., Zainos, A., and Salinas, E. Correlated neuronal discharges that increase coding efficiency during perceptual discrimination. Neuron 38: 649-657, 2003.
28. Romo, R. and Salinas, E. Touch and go: Decision-making mechanisms in somatosensation. Annu. Rev. Neurosci. 24: 107-137, 2001.
29. Romo, R. and Salinas, E. Flutter discrimination: neural codes, perception, memory and decision making. *Nat. Rev. Neurosci.* 4: 203-218, 2003.
30. Salinas, E., Hernández, A., Zainos, A., and Romo, R. Periodicity and firing rate as candidate neural codes for the frequency of vibrotactile stimuli. *J. Neurosci.* 20: 5503-5515, 2000.
31. Salzman, D., Britten, K. H., and Newsome, W.T. Cortical microstimulation influences perceptual judgements of motion direction. Nature 346: 174-77, 1990.
32. Sur, M., Wall, J.T., and Kaas, J.H. Modular distribution of neurons with slowly adapting and rapidly adapting responses in area 3b of somatosensory cortex in monkeys. *J. Neurophysiol.* 51: 724-744, 1984.
33. Talbot, W.H., Darian-Smith, I., Kornhuber, H.H., and Mountcastle, V.B. The sense of flutter-vibration: comparison of the human capacity response patterns of mechanoreceptive afferents from the monkey hand. *J. Neurophysiol.* 31: 301-334, 1968.
34. Vallbo, A.B. Single-afferent neurons and somatic sensation in humans. In *The Cognitive Neurosciences* (ed. M.S. Gazzaniga), pp. 237-252.Cambridge: MIT Press, 1995.
35. Vallbo, A.B. and Johansson, R.S. Properties of cutaneous mechanoreceptors in the human hand related to touch sensations. Hum. Neurobiol. 3: 3-14, 1984.
36. Vazquez, P., Cano, M., and Acuna, C. Discrimination of line orientation in humans and monkeys. *J. Neurophysiol.* 83:2639–2648, 2000.
37. Vogels, R. and Orban, G.A. How well do response changes of striate neurons signal differences in orientation: a study in the discriminating monkey. *J. Neurosci.* 10:3543–3558, 1990.
38. Werner, G. and Mountcastle, V.B. Neural activity in mechanoreceptive cutaneous afferents: stimulus-response relations, Weber functions, and information transmission. *J. Neurophysiol.* 28: 359-397, 1965.

5 Perceptual Learning and Referral in the Tactile System

K. Sathian

CONTENTS

I. Overview....................97
II. Transfer of Tactile Learning....................98
A. Early Studies....................98
B. Task Specificity....................98
C. Transfer between Locations....................98
D. Neural Basis of Tactile Learning Effects....................101
III. Tactile Superiority in the Blind: A Manifestation of Perceptual Learning?....................101
IV. Inter-Manual Referral of Tactile Sensation....................103
Acknowledgments....................104
References....................104

I. OVERVIEW

Eleanor Gibson, in her classic monograph,[23] defined perceptual learning as "an increase in the ability to extract information from the environment, as a result of experience and practice with stimulation coming from it". An aspect of perceptual learning, that is frequently studied in the laboratory, is the improvement in perceptual performance that occurs with practice, for example, in the ability to distinguish by taste different wines[23] or teas. Although perceptual learning has not been studied as intensively in touch as it has been in vision, there has been a fair degree of interest in whether, and to what extent, perceptual learning effects transfer between different parts of the cutaneous surface and whether the rules governing such transfer differ from those that have been described in the visual system.

II. TRANSFER OF TACTILE LEARNING

A. Early Studies

The issue of specificity has plagued studies of perceptual learning in the tactile system for a long time. Volkmann's pioneering studies in the nineteenth century established that the threshold separation for discriminating two points from one declined with practice (illustrating perceptual learning) and also found that the effect of practice transferred between hands.[23] Further studies of two-point discrimination on the forearm also suggested that practice effects transfer outside the trained zone to homologous areas contralaterally[15,43] but rather surprisingly, not to surrounding areas.[15] However, the small numbers of subjects in these early studies (two each in the last two reports cited) and the generally low reliability of the two point discrimination task[85,86] are reasons to question these findings.

B. Task Specificity

In the modern era, studies of vibrotactile pattern identification on the finger pad, using the Optacon, a reading aid for the blind, have emphasized the generalizability of tactile learning. Learning effects transfer from one subset of the alphabet to another.[17] Similarly, Craig (1988), finding that training on a letter identification task improved performance on this as well as an untrained task (gap detection), suggested that tactile experience results in a general enhancement of tactile sensitivity. In contrast to these results, we found in our laboratory that perceptual improvement in the tactile discrimination of gratings (consisting of alternating ridges and grooves) was relatively specific for the trained grating set. Learning effects showed only limited transfer between sets distinguished by changes in their groove width and those distinguished by changes in their ridge width.[78] This is not surprising since neurophysiological[26,74] and psychophysical[8,74] studies suggest that the mechanisms underlying grating discrimination differ for the two sets. Spatial variables are primary for gratings varying in groove width, whereas temporal variables make an important contribution for gratings varying in ridge width. Such task specificity is in line with observations in the visual system,[20,35,50,55] which led to the suggestion that perceptual learning reflects neural plasticity in early sensory cortex.

C. Transfer between Locations

In our laboratory, we used a variety of stimuli and tasks to study the transfer of perceptual learning between fingers. Our studies converge on the finding that between-session learning effects transfer readily, substantially, and in some cases almost completely, between fingers of either hand. I will review each study in turn.

In our first study,[78] we employed the same periodic gratings referred to above in studies of task specificity. The task was to discriminate between gratings that varied either in their groove width or in their ridge width. Discrimination was performed with the fingerpad which was actively moved by the subject across the grating. Initial training used one index finger and progressed to the index or middle finger of the other hand. Learning was reflected in the progressive decline of the difference limen. We found for both grating sets, that initial difference limens were much higher, learning effects were larger, and learning was slower when the subjects were completely task-naïve compared to when they had already been trained on one finger. The transfer of learning between fingers was generally substantial and complete in some instances. Similar results were obtained for another task; discrimination of the orientation of hand-held gratings applied to the passive fingerpad[78] in which the performance measure was the groove width permitting threshold discrimination of a 90° difference in grating orientation. This finding suggested that inter-digit transfer in the first grating experiment was not simply because active motion was used. As an aside, the grating orientation test provides a highly reliable index of tactile spatial acuity.[85,86] Because performance on this test improves with practice whereas peripheral visual acuity does not,[91] the grating orientation test man depend more on complex factors than just spatial resolution or receptor density.

That a high propensity for perceptual learning effects to transfer between fingers is characteristic of the tactile system and not restricted to tasks using gratings was shown in a subsequent study from our laboratory.[76] In this study, the stimulus was a pattern consisting of three dots arranged in a row. The central dot in the pattern could be offset laterally. The subjects had to distinguish between a pattern with the offset and one without. This is a hyperacuity task because the offset was 1 mm or less, and the limit of tactile spatial acuity (resolution) at the fingerpad is about 1 mm,[34,86] which corresponds to the spacing of the relevant mechanoreceptors,. This task was originally used in studies of vision[90] and later adapted to studies of touch.[39] In our version of this hyperacuity task, subjects actively brought their fingerpad down onto the dot patterns without lateral motion. Learning was measured in terms of decrease in the magnitude of spatial offset required for reliable offset detection. Our subjects demonstrated virtually complete transfer of perceptual learning effects between fingers of either hand.[76]

We obtained similar results with a variant of this dot pattern task in which the central dot was always offset and the subjects had to distinguish whether the offset was to the left or the right. In this variant of the task, the offset was held constant (at 2 mm, well outside the hyperacuity range) and learning was measured as a decline in the stimulus duration required for reliable spatial discrimination. This study also precisely controlled stimulus amplitude and duration using an electromechanical stimulator.[92] Despite these changes of experimental design the results were identical: once again there was complete transfer of learning effects to fingers of either hand.[46]

Studies by others have reported similar findings to those from our laboratory. Thus, training on discrimination of the duration of the interval between two vibrotactile stimuli was found to generalize between hands.[44] Similarly, learning effects in discriminating a particular spatio-temporal sequence of tactile stimuli transferred between hands.[80] Transfer of tactile learning though, is not universally reported. One group, studying within session learning effects, found no transfer beyond the trained finger for vibratory frequency discrimination but a topographic gradient of transfer in two other tasks: discriminating the pressure of a punctate stimulus and discriminating surface roughness.[30] Transfer was maximal to the homologous finger on the contralateral hand and the adjacent finger (the ones we tested in all our studies), but lower for other fingers. This topographic nature of transfer was consistent with between-session learning effects reported by the same group on the rat whisker system.[29,31] However, we did not find evidence for such topographic transfer in one study where we tested for it.[46] Between-session learning effects in this study (in which directional discrimination of spatial offset was the dependent measure, see previous paragraph) transferred equally to the finger adjacent to the trained finger and to a more distant finger. Perhaps, in humans, a restricted transfer pattern emerges in a single session and then generalizes between sessions. It is well established that perceptual learning effects in the visual system require both time and sleep between sessions for proper consolidation.[36,82] Possibly the transfer of learning effects also develops its full glory over time and as a result of sleep.

Our results in the tactile system are clearly at odds with a large number of studies in vision which have found, in a variety of tasks, that perceptual learning effects are quite specific for the retinal location of the visual stimulus used for training.[3,13,21,35,45,51,53,79] The propensity for inter-digit and inter-manual transfer of tactile learning might be related to the use of multiple fingers in concert during tactile sensing and manipulation. However, the degree of location and orientation-specificity of visual learning in orientation pop out detection increases with increasing task difficulty; moreover, a single, prolonged exposure to the stimulus abolishes specificity.[2] There is the possibility that the lack of location specificity in previous tactile studies could have arisen because the tasks were too easy.

We investigated this by conducting an experiment with a difficult version of the grating discrimination task,[9] using gratings that varied in ridge width because they are more difficult to discriminate than those varying in groove width.[73,78] Tactile access to each grating was restricted to a single rapid sweep, whereas in our previous study,[78] subjects were allowed repeated scanning. We maximized task difficulty by using only a subset of gratings with small differences in ridge width rather than following the usual progression from large to small differences. These methodological changes did make learning more difficult, as evidenced by our having to reject many more subjects than in the previous study, because they did not improve their performance even after a number of sessions. Yet perceptual learning effects still transferred substantially from the right (initially trained) hand to the left hand, as evidenced by lower initial thresholds and faster improvement on the left hand. Inter-manual transfer was complete in some subjects, where initial thresholds on the left hand equaled final thresholds on the right hand. Note that in this study[9] and in previous studies, intrinsic asymmetry cannot account for the results because there were no significant inter-manual differences in final threshold. This fits with previous reports that manual asymmetries in normal tactile performance are absent or minor.[77,83] Thus, it appears that the readiness with which perceptual learning effects transfer between fingers and hands cannot be attributed simply to the tasks used being too easy. Hence, there might be a genuine difference between the tactile and visual systems in the transfer of perceptual learning effects between locations. Studies are currently under way in our laboratory to examine this more directly using stimuli and hyperacuity tasks that are closely matched between vision and touch.

D. Neural Basis of Tactile Learning Effects

The transfer patterns discussed above, suggest that the neural processes underlying tactile learning are probably not restricted to area 3b of the primary somatosensory cortex (SI) in which neurons normally have circumscribed receptive fields on a single digit[52] and callosal connections are sparse in the hand representation.[38] Candidate areas mediating tactile learning should contain neurons with multi-digit and bilateral receptive fields, and hand representations with abundant callosal connections. Areas 1 and 2 in posterior SI[33,38] and parietal opercular cortical areas, including second somatosensory cortex (SII),[40,64,65] meet these criteria. Of course, interactions between multiple areas may also be involved.

Very few studies have directly explored neural correlates of tactile learning. Merzenich's group, in a classic study, trained owl monkeys to discriminate vibratory frequency with one finger.[57] Subsequent neurophysiologic recordings under anesthesia revealed that the representation of the trained skin area had expanded markedly within the somatotopic map in area 3b[59] and extended into area 3a,[58] which was normally dominated by deep inputs. Further, receptive fields on the trained finger were both larger and more numerous in area 3b.[59] Moreover, temporal synchrony within the population discharge was higher,[60] and correlated well with psychophysical measures of between-session performance improvement.[57,60]

In musicians who play stringed instruments, magneto-encephalographic (MEG) recordings of somatosensory cortical activity evoked by tactile stimuli to the fingers indicate that the representation of the string-playing hand is enlarged.[16] This may reflect complex perceptual learning but could, alternatively, underlie intrinsic differences between musicians and others. Learning to discriminate one spatio-temporal sequence of tactile stimuli from other sequences was associated with a decrease in the amplitude of the MEG signal evoked by stimulation.[80] Certainly, more work is required to understand the neural basis of perceptual learning, but it is safe to say that changes in early sensory cortex are clearly involved. This does not exclude a role for processes in higher order sensory areas, which would fit with the possible contribution of attentional effects as reviewed elsewhere.[24,70]

III. TACTILE SUPERIORITY IN THE BLIND: A MANIFESTATION OF PERCEPTUAL LEARNING?

It is now quite well established that the blind, at least those who read Braille, are superior to the normally sighted on many tactile spatial capabilities.[72] Common to all the tactile tests on which blind individuals have been reported to outperform their sighted counterparts was the requirement for the processing of fine spatial detail on the fingerpad, around the limit of spatial resolution. The test that showed the largest difference between blind and sighted was the hyperacuity test discussed earlier in this chapter (distinguishing between a three-dot pattern with an offset of 1 mm or less and a pattern without an offset). Using this test, we found that early blind subjects averaged thresholds that were 50% lower than sighted controls.[27] Next in magnitude was a test in which blind subjects could distinguish a 90° difference in the orientation of lines that were nearly 40% shorter than in the sighted.[81] The grating orientation discrimination test has already been mentioned briefly earlier in this chapter. Its application in studies of the blind has revealed that blind subjects are about 30% better than the sighted,[25,84] although a study from our laboratory failed to find such a difference.[27] The blind are also able to detect gaps that are about 15% smaller than the sighted[81] and can identify some types of Braille-like patterns about 15% more accurately.[22] It is worth noting that the blind do not perform better than the sighted on some tests: discrimination of length[81] and of textures – both real world ones like sandpapers or abrasive stones[32] and laboratory constructs such as gratings.[27]

An interesting question is whether superior tactile spatial performance is a consequence of extended perceptual learning, associated with use-dependent neural plasticity. Evidence in favor of this is with only 3–4 d practice, sighted subjects can do just as well as the blind on the hyperacuity task.[27] This is consistent with reports that practiced sighted subjects could match blind Braille readers on the Optacon,[12] and the deaf-blind in decoding speech by feeling the speaker's face and neck.[62,63] Interestingly, blind subjects showed no improvement in hyperacuity over the 3-4 day period of practice that benefited the sighted[27] suggesting that the blind were already operating at maximal capability. Also favoring the idea of use dependent neural plasticity was the finding of a 30% lower threshold on the Braille reading finger than on other fingers tested,[84] although it must be noted that our study found no superiority of hyperacuity on the hand used for reading Braille[27] and another study found no significant difference on grating orientation discrimination between those who read Braille and those who did not.[25]

Another issue is the relative contribution of (intra-modal) plasticity in the somatosensory system versus cross-modal plasticity manifested as visual cortical involvement in tactile perception. The representation of the Braille reading finger in somatic sensorimotor cortex is expanded,[48] and varies dynamically as a function of short term use.[49] However, the relation of these aspects of sensorimotor cortical plasticity to changes in performance remains uncertain. Finger amputation in owl monkeys leads to expansion of the cortical representation of adjacent digits.[42] However, in humans, corresponding changes in spatial abilities are lacking.[87] Similarly, expansion of the cortical representation of the trained part of the receptor sheet does not correlate with perceptual learning in the tactile system,[51,59,60] although it does correlate in the auditory system.[61]

Many studies have found that the visual cortex is active during Braille reading in the blind.[5,6,41,67,68,69] Moreover, the visual cortex is functionally involved in Braille reading in the blind.[10] Such cross-modal functioning was reported to be more extensive in the early blind compared to the late blind,[10,11,67,68] but some workers find the opposite.[5,41] Whether cross modal and early developmental plasticity are relevant to measurable differences in tactile ability of the blind, is not clear. Two of the pyschophysical studies cited above[25,27] included both early-blind and late-blind subjects. Neither study revealed a significant performance difference between the early and late blind, although there were trends for the early blind to be better. Thus, at present there is no clear evidence that visual deprivation during a critical period in early life favors the acquisition of superior tactile capabilities.

Yet another possible trigger for improved acuity is visual deprivation. Remarkably, blindfolding sighted subjects for just a few days improves their ability to discriminate Braille characters, in the absence of training.[37] This was associated with increased involvement of visual cortex in tactile perceptual tasks.[47] Further, blindfolding sighted subjects for as little as 90 min lowers their thresholds on the grating orientation task by about 15%.[18] Light deprivation for a similar period enhances the excitability of the human visual cortex.[4] Whether the psychophysical and neural changes described in these studies are causally related remains unknown. However, these findings suggest that there are nonvisual sensory inputs into visual areas, consistent with recent anatomical[19,66] and earlier physiological[28] work in macaque monkeys as well as an accumulating body of evidence from human studies that many visual cortical areas are active during,[75] and functionally involved in,[92] tactile perception even in sighted subjects. Accordingly, exuberance of multi-sensory inputs into and recruitment of visual cortex during tactile perception in the blind could reflect quantitative rather than qualitative differences compared to the sighted.

IV. INTER-MANUAL REFERRAL OF TACTILE SENSATION

Following on observations, reviewed elsewhere,[1,56] that tactile stimuli can be referred ipsilaterally from intact body parts to phantoms of resected body parts, Ramachandran and colleagues reported that tactile stimuli applied to the intact hand of arm amputees could evoke contralateral percepts referred in a topographic manner to the phantom hand.[54] Analogous contralateral referral of sensations also occurs in patients affected by sensory loss resulting from neurological lesions. In these patients, touching the normal hand can elicit a referred percept in the anesthetic hand at a mirror symmetric location.[71] The referral is evoked for touch but typically not other kinds of somatosensory stimuli, such as pain or temperature.[54,71] Such referral is most probably transcallosally mediated. It may be related to performance improvement reported on grating orientation discrimination, associated with enhanced somatosensory evoked potentials, caused by acute somatosensory deprivation on the contralateral hand.[89]

The perceptual characteristics of the referred sensations were studied in some detail in patients with anesthetic hands,[71] with the findings that they were relatively high threshold, only crudely localized and lacked spatial organization. This led to the conclusion[71] that such referred percepts are unlikely to depend on neurons in area 3b, which exhibit fine grained spatial resolution and topographic organization.[14] Further, the hand representation in area 3b has sparse callosal connectivity[38] and under normal circumstances, its neurons do not possess bilateral receptive fields.[52] It is possible that transcallosal referral occurs more posteriorly in SI, in areas 1 and 2, or in parietal opercular cortex where bilateral somatosensory responsiveness[33,40,64,65] and callosal connectivity[38] are found in the hand representation. This would fit with the suggestion made earlier in this chapter, in Section IID on perceptual learning, that these areas which are higher than area 3b in the somatosensory cortical hierarchy, are more likely to be the locus of neural changes underlying perceptual learning and its transfer. However, a caveat is that under unusual circumstances, multi-digit receptive fields[88] as well as rapid inter-hemispheric transfer of receptive field plasticity[7] can emerge in area 3b. It is also interesting that, in both amputees[54] and patients with sensory loss,[71] visual input tended to strengthen referred percepts, suggesting a role for multisensory convergence. The exact mechanisms mediating inter-manual referral of tactile percepts and their relation to those involved in perceptual learning remain unknown but merit further study.

ACKNOWLEDGMENTS

I thank my present and former colleagues for their involvement in the studies reviewed here, which were supported by grants from the National Institute of Neurological Disorders and Stroke and the National Eye Institute at the NIH.

REFERENCES

1. Aglioti S (1999) "Anomalous" representations and perceptions: implications for human neuroplasticity. In: Neuronal Plasticity: Building a Bridge from the Laboratory to the Clinic. (Grafman J, Christen Y, Eds), pp 79-91. Berlin: Springer-Verlag.
2. Ahissar M, Hochstein S (1997) Task difficulty and the specificity of perceptual learning. Nature 387: 401-406.
3. Ball K, Sekuler R (1987) Direction specific improvement in motion discrimination. Vision Res 27: 953-965.
4. Boroojerdi B, Bushara KO, Corwell B, Immisch I, Battaglia F, Muellbacher W, Cohen LG (2000) Enhanced excitability of the human visual cortex induced by short term light deprivation. Cerebral Cortex 10: 529-534.
5. Büchel C, Price C, Frackowiak RSJ, Friston K (1998) Different activation patterns in the visual cortex of late and congenitally blind subjects. Brain 121: 409-419.
6. Burton H, Snyder AZ, Conturo TE, Akbudak E, Ollinger JM, Raichle ME (2002) Adaptive changes in early and late blind: A fMRI study of Braille reading. J Neurophysiol 87: 589-607.
7. Calford MB, Tweedale R (1990) Inter-hemispheric transfer of plasticity in the cerebral cortex. Science 249: 805-807.
8. Cascio C, Sathian K (2001) Temporal cues contribute to tactile perception of roughness. J Neurosci 21: 5289-5296.
9. Chevalier SJ, Sathian K (1998) Inter-manual transfer of tactile learning. Soc Neurosci Abstr 24: 68.9.
10. Cohen LG, Celnik P, Pascual-Leone A, Corwell B, Faiz L, Dambrosia J, Honda M, Sadato N, Gerloff C, Catala MD, Hallett M (1997) Functional relevance of cross modal plasticity in blind humans. Nature 389: 180-183.
11. Cohen LG, Weeks RA, Sadato N, Celnik P, Ishii K, Hallett M (1999) Period of susceptibility for cross modal plasticity in the blind. Ann Neurol 45: 451-460.
12. Craig JC (1988) The role of experience in tactual pattern perception: a preliminary report. Int J Rehabil Res 11: 167-183.
13. Crist RE, Kapadia MK, Westheimer G, Gilbert CD (1997) Perceptual learning of spatial localization: Specificity for orientation, position, and context. J Neurophysiol 78: 2889-2894.
14. DiCarlo JJ, Johnson KO, Hsiao SS (1998) Structure of receptive fields in area 3b of primary somatosensory cortex in the alert monkey. J Neurosci 18: 2626-2645.

15. Dresslar FB (1895) Studies in the psychology of touch. Am J Psychol 6: 313-368.
16. Elbert T, Pantev C, Wienbruch C, Rockstroh B, Taub E (1995) Increased cortical representation of the fingers of the left hand in string players. Science 270: 305-307.
17. Epstein W, Hughes B, Schneider SL, Bach-y-Rita P (1989) Perceptual learning of spatiotemporal events: evidence from an unfamiliar modality. J Exp Psychol: Hum Percept Perform 15: 28-44.
18. Facchini S, Aglioti SM (2003) Short term light deprivation increases tactile spatial acuity in humans. Neurology 60: 1998-1999.
19. Falchier A, Clavagnier S, Barone P, Kennedy H (2002) Anatomical evidence of multimodal integration in primate striate cortex. J Neurosci 22: 5749-5759.
20. Fiorentini A, Berardi N (1980) Perceptual learning specific for orientation and spatial frequency. Nature 287: 43-44.
21. Fiorentini A, Berardi N (1981) Learning in grating waveform discrimination: specificity for orientation and spatial frequency. Vision Res 21: 1149-1158.
22. Foulke E, Warm JS (1967) Effects of complexity and redundancy on the tactual recognition of metric figures. Percept Mot Skills 25: 177-187.
23. Gibson EJ (1969) Principles of perceptual learning and development. New York: Meredith.
24. Gilbert CD (1994) Early perceptual learning. Proc Natl Acad Sci USA 91: 1195-1197.
25. Goldreich D, Kanics IM (2003) Tactile acuity is enhanced in blindness. J Neurosci 23: 3439-3445.
26. Goodwin AW, John KT, Sathian K, Darian-Smith I (1989) Spatial and temporal factors determining afferent fiber responses to a grating moving sinusoidally over the monkey's finger pad. J Neurosci 9: 1280-1293.
27. Grant AC, Thiagarajah MC, Sathian K (2000) Tactile perception in blind Braille readers: A psychophysical study of acuity and hyperacuity using gratings and dot patterns. Percept Psychophys 62: 301-312.
28. Haenny PE, Maunsell JHR, Schiller PH (1988) State dependent activity in monkey visual cortex. II. Retinal and extraretinal factors in V4. Exp Brain Res 69: 245-259.
29. Harris JA, Diamond ME (2000) Ipsilateral and contralateral transfer of tactile learning. NeuroReport 11: 263-266.
30. Harris JA, Harris IM, Diamond ME (2001) The topography of tactile learning in humans. J Neurosci 21: 1056-1061.
31. Harris JA, Petersen RS, Diamond ME (1999) Distribution of tactile learning and its neural basis. Proc Natl Acad Sci USA 96: 7587-7591.
32. Heller MA (1989) Texture perception in sighted and blind observers. Percept Psychophys 45: 49-54.
33. Iwamura Y, Iriki A, Tanaka M (1994) Bilateral hand representation in the postcentral somatosensory cortex. Nature 369: 554-556.
34. Johnson KO, Phillips JR (1981) Tactile spatial resolution. I. Two point discrimination, gap detection, grating resolution and letter recognition. J Neurophysiol 46: 1177-1191.
35. Karni A, Sagi D (1991) Where practice makes perfect in texture discrimination: evidence for primary visual cortex plasticity. Proc Natl Acad Sci USA 88: 4966-4970.
36. Karni A, Tanne D, Rubenstein BS, Askenasy JJM, Sagi D (1994) Dependence on REM sleep of overnight improvement of a perceptual skill. Science 265: 679-682.
37. Kauffman T, Théoret H, Pascual-Leone A (2002) Braille character discrimination in blindfolded human subjects. NeuroReport 13: 571-574.
38. Killackey HP, Gould HJ, Cusick CG, Pons TP, Kaas JH (1983) The relation of corpus callosum connections to architectonic fields and body surface maps in sensorimotor cortex of New and Old World monkeys. J Comp Neurol 219: 384-419.

39. Loomis JM (1979) An investigation of tactile hyperacuity. Sens Processes 3: 289-302.
40. Manzoni T, Barbaresi P, Conti F (1984) Callosal mechanism for the inter-hemispheric transfer of hand somatosensory information in the monkey. Behav Brain Res 11: 155-170.
41. Melzer P, Morgan VL, Pickens DR, Price RR, Wall RS, Ebner FF (2001) Cortical activation during Braille reading is influenced by early visual experience in subjects with severe visual disability: A correlational fMRI study. Hum Brain Mapp 14: 186-195.
42. Merzenich MM, Nelson RJ, Stryker MP, Cynader MS, Schoppmann A, Zook JM (1984) Somatosensory cortical map changes following digit amputation in adult monkeys. J Comp Neurol 224: 591-605.
43. Mukherjee KC (1933) The duration of cutaneous sensation (I) and the improvement of its sensible discrimination by practice (II). J Exp Psychol 16: 339-342.
44. Nagarajan SS, Blake DT, Wright BA, Byl N, Merzenich MM (1998) Practice related improvements in somatosensory interval discrimination are temporally specific but generalize across skin location, hemisphere and modality. J Neurosci 18: 1559-1570.
45. Nazir TA, O'Regan JK (1990) Some results on translation invariance in the human visual system. Spat Vis 5: 81-100.
46. Offe K, Chevalier SJ, Sathian K (2000) Inter-manual and inter-digit transfer of tactile learning. Soc Neurosci Abstr 26: 264.9.
47. Pascual-Leone A, Hamilton R (2001) The metamodal organization of the brain. Prog Brain Res 134: 427-445.
48. Pascual-Leone A, Torres F (1993) Plasticity of the sensorimotor cortex representation of the reading finger in Braille readers. Brain 116: 39-52.
49. Pascual-Leone A, Wasserman EM, Sadato N, Hallett M (1995) The role of reading activity on the modulation of motor cortical outputs to the reading hand in Braille readers. Ann Neurol 38: 910-915.
50. Poggio T, Fahle M, Edelman S (1992) Fast perceptual learning in visual hyperacuity. Science 256: 1018-1021.
51. Polat U, Sagi D (1994) Spatial interactions in human vision: from near to far via experience dependent cascades of connections. Proc Natl Acad Sci USA 91: 1206-1209.
52. Pons TP, Wall JT, Garraghty PE, Cusick CG, Kaas JH (1987) Consistent features of the representation of the hand in area 3b of macaque monkeys. Somatosens Res 4: 309-331.
53. Ramachandran VS (1976) Learning-like phenomena in stereopsis. Nature 262: 382-384.
54. Ramachandran VS (1995) Touching the phantom limb. Nature 377: 489-490.
55. Ramachandran VS, Braddick O (1973) Orientation specific learning in stereopsis. Perception 2: 371-376.
56. Ramachandran VS, Hirstein W (1998) The perception of phantom limbs. The D.O.Hebb lecture. Brain 121: 1603-1630.
57. Recanzone GH, Jenkins WM, Hradek GT, Merzenich MM (1992a) Progressive improvement in discriminative abilities in adult owl monkeys performing a tactile frequency discrimination task. J Neurophysiol 67: 1015-1030.
58. Recanzone GH, Merzenich MM, Jenkins WM (1992) Frequency discrimination training engaging a restricted skin surface results in an emergence of a cutaneous response zone in cortical area 3a. J Neurophysiol 67: 1057-1070.

59. Recanzone GH, Merzenich MM, Jenkins WM, Grajski KA, Dinse HR (1992b) Topographic reorganization of the hand representation in cortical area 3b of owl monkeys trained in a frequency discrimination task. J Neurophysiol 67: 1031-1056.
60. Recanzone GH, Merzenich MM, Schreiner CE (1992) Changes in the distributed temporal response properties of SI cortical neurons reflect improvements in performance on a temporally based tactile discrimination task. J Neurophysiol 67: 1071-1091.
61. Recanzone GH, Schreiner CE, Merzenich MM (1993) Plasticity in the frequency representation of primary auditory cortex following discrimination training in adult owl monkeys. J Neurosci 13: 87-103.
62. Reed CM, Doherty MJ, Braida LD, Durlach NI (1982) Analytic study of the Tadoma method: Further experiments with inexperienced observers. J Speech Hearing Res 25: 216-223.
63. Reed CM, Rubin SI, Braida LD, Durlach NI (1978) Analytic study of the Tadoma method: Discrimination ability of untrained observers. J Speech Hearing Res 21: 625-637.
64. Robinson CJ, Burton H (1980a) Organization of somatosensory receptive fields in cortical areas 7b, retroinsula, postauditory and granular insula of M. fascicularis. J Comp Neurol 192: 69-92.
65. Robinson CJ, Burton H (1980b) Somatotopographic organization in the second somatosensory area of M. fascicularis. J Comp Neurol 192: 43-67.
66. Rockland KS, Ojima H (2003) Multisensory convergence in calcarine visual areas in macaque monkey. Int J Psychophysiol 50: 19-26.
67. Sadato N, Okada T, Honda M, Yonekura Y (2002) Critical period for cross modal plasticity in blind humans: a functional MRI study. NeuroImage 16: 389-400.
68. Sadato N, Pascual-Leone A, Grafman J, Deiber M-P, Ibanez V, Hallett M (1998) Neural networks for Braille reading by the blind. Brain 121: 1213-1229.
69. Sadato N, Pascual-Leone A, Grafman J, Ibanez V, Deiber M-P, Dold G, Hallett M (1996) Activation of the primary visual cortex by Braille reading in blind subjects. Nature 380: 526-528.
70. Sathian K (1998) Perceptual learning. Current Science 75: 451-457.
71. Sathian K (2000a) Intermanual referral of sensation to anesthetic hands. Neurology 54: 1866-1868.
72. Sathian K (2000b) Practice makes perfect: Sharper tactile perception in the blind. Neurology 54: 2203-2204.
73. Sathian K, Burton H (1991) The role of spatially selective attention in the tactile perception of texture. Percept Psychophys 50: 237-248.
74. Sathian K, Goodwin AW, John KT, Darian-Smith I (1989) Perceived roughness of a grating: correlation with responses of mechanoreceptive afferents innervating the monkey's finger pad. J Neurosci 9: 1273-1279.
75. Sathian K, Prather SC, Zhang M Visual cortical involvement in normal tactile perception. In: The Handbook of Multisensory Processes. (Calvert G, Spence C, Stein B, Eds), Cambridge, MA: MIT Press. pp. 703–709.
76. Sathian K, Zangaladze A (1998) Perceptual learning in tactile hyperacuity: complete inter-manual transfer but limited retention. Exp Brain Res 118: 131-134.
77. Sathian K, Zangaladze A (1996) Tactile spatial acuity at the human fingertip and lip: bilateral symmetry and inter-digit variability. Neurology 46: 1464-1466.
78. Sathian K, Zangaladze A (1997) Tactile learning is task specific but transfers between fingers. Percept Psychophys 59: 119-128.

79. Shiu L-P, Pashler H (1992) Improvement in line orientation discrimination is retinally local but dependent on cognitive set. Percept Psychophys 52: 582-588.
80. Spengler F, Roberts TPL, Poeppel D, Byl N, Wang X, Rowley HA, Merzenich MM (1997) Learning transfer and neuronal plasticity in humans trained in tactile discrimination. Neurosci Lett 232: 151-154.
81. Stevens JC, Foulke E, Patterson MQ (1996) Tactile acuity, aging and Braille reading in long term blindness. J Exp Psychol: Applied 2: 91-106.
82. Stickgold R, James L, Hobson JA (2000) Visual discrimination learning requires sleep after training. Nature Neurosci 3: 1237-1238.
83. Summers DC, Lederman SJ (1990) Perceptual asymmetries in the somatosensory system: a dichhaptic experiment and critical review of the literature from 1929 to 1986. Cortex 26: 201-226.
84. van Boven RW, Hamilton RH, Kauffman T, Keenan JP, Pascual-Leone A (2000) Tactile spatial resolution in blind Braille readers. Neurology 54: 2230-2236.
85. van Boven RW, Johnson KO (1994a) A psychophysical study of the mechanisms of sensory recovery following nerve injury in humans. Brain 117: 149-167.
86. van Boven RW, Johnson KO (1994b) The limit of tactile spatial resolution in humans: grating orientation discrimination at the lip, tongue and finger. Neurology 44: 2361-2366.
87. Vega-Bermudez F, Johnson KO (2002) Spatial acuity after digit amputation. Brain 125: 1256-1264.
88. Wang X, Merzenich MM, Sameshima K, Jenkins WM (1995) Remodelling of hand representation in adult cortex determined by timing of tactile stimulation. Nature 378: 71-75.
89. Werhahn K, Mortensen J, van Boven RW, Zeuner KE, Cohen LG (2002) Enhanced tactile spatial acuity and cortical processing during acute hand deafferentation. Nature Neurosci 5: 936-938.
90. Westheimer G (1977) Spatial frequency and light spread descriptions of visual acuity and hyperacuity. J Opt Soc Am A 67: 207-212.
91. Westheimer G (2001) Is peripheral visual acuity susceptible to perceptual learning in the adult? Vision Res 41: 47-52.
92. Zangaladze A, Epstein CM, Grafton ST, Sathian K (1999) Involvement of visual cortex in tactile discrimination of orientation. Nature 401: 587-590.

6 The Effects of Sensory Deprivation on Sensory Function of SI Barrel Cortex

Ford F. Ebner and Michael Armstrong-James

CONTENTS

I. The Effects of Sensory Deprivation on Sensory Function of SI Barrel Cortex 110
A. Overview 110
B. Introduction 110
C. Methodology of Sensory Deprivation 111
1. Definition of Sensory Deprivation 111
2. Critical or Sensitive Periods 111
3. Plucking vs. Trimming of Whiskers To Produce Sensory Deprivation 112
4. Complete (Global) vs. Restricted (Partial) Deprivation 112
5. Timing: When is the Sensory Deprivation Imposed? 113
6. Recovery Periods: What is the Time between Trimming and Beginning of Analysis? 114
7. Controls: What is a Valid Control for Sensory Deprivation? 114
D. Overview of the Whisker to Barrel Cortex System 115
1. Development of Normal Cortical Response Properties 115
2. Ascending and Recurrent Circuits 117
3. Intracortical Circuits 118
E. Effects of Early Sensory Deprivation on Barrel Cortex 119
1. *In vivo* Analysis of Global Sensory Deprivation During Development 119
2. *In vitro* Analysis of Global Sensory Deprivation During Development 122
3. Anatomical Changes Produced by Global Sensory Deprivation 122
4. Partial Sensory Deprivation during Postnatal Development 123
5. Comparison of Partial with Global Sensory Deprivation 125

6. Effects of Sensory Deprivation on the Adult Brain 126
F. Molecular Mechanisms Affected by Sensory Deprivation 127
1. Effects of Sensory Deprivation on Excitatory Neurotransmission 128
2. Effects of Sensory Deprivation on Inhibitory Neurotransmission 129
References 130

I. THE EFFECTS OF SENSORY DEPRIVATION ON SENSORY FUNCTION OF SI BARREL CORTEX

A. Overview

The somatic sensory system in rats and mice is very immature at the time of birth, and the final maturation of sensory processing mechanisms requires a certain level of sensory experience in the first few weeks after birth. Research has led to significant insights into the effect of low levels of postnatal activity arising from rodent whiskers on the development of cortical function. The data show that both excitatory and inhibitory processes are affected by sensory deprivation (SD), with the severity of effects depending upon the time of onset, the duration of the deprivation, and the length of the recovery period after deprivation ends. However, even after prolonged recovery periods some SD deficits do not recover completely even after the whiskers regrow to normal lengths. A major impact of SD leads to degraded circuit dynamics in intracortical connections: excitatory inputs do not modify cortical cell responses appropriately and inhibition becomes fixed at some level that is not adjusted up or down appropriately by neural activity. Neural transmission from thalamic input layer IV to more superficial layers II/III is a major site of synaptic dysfunction. Global deprivation (trimming all whiskers) produces a more uniform down-regulation of sensory transmission when compared to trimming a subset of whiskers, presumably because restricted deprivation creates competition between active and relatively inactive interconnected cell groups. This activity-based competition leads to more complex changes depending on the pattern of whisker trimming. In rat barrel cortex activity-based, changes in function can be induced by altered tactile experience throughout life. But early postnatal SD degrades neuronal plasticity in the mature brain and interferes with the ability to learn subtle tactile discriminations, presumably throughout life. The data that support this picture of SD effects follow.

B. Introduction

In this chapter we focus on deficiencies that are induced by controlled manipulations of sensory activity in one sensory system; specifically by producing changes in the level of activity arising from the mystacial vibrissae (a.k.a. facial whiskers or simply whiskers) of a mouse or rat which project to the whisker-recipient area of SI cortex (formally identified in the initial description as the posteromedial barrel subfield,[1] and here referred to simply as barrel cortex). Since the literature on this system has grown very large, with rare exceptions we further restrict the topic to studies of somatic sensory deprivation (SD) that do not involve any neural damage: that is, to reports in which brain changes are induced by changing the levels of receptor-initiated activity arising from the whiskers without any intended damage to the central or peripheral neural structures. The literature on SD is difficult to compare without some interpretation, since few experiments have been carried out in the same way, or over equivalent periods of development, by different investigators. A brief definition of sensory deprivation leads to a discussion of some of the methodological variables that affect SD results. We will provide a short description of the sensory pathways that convey whisker information to the cortex for those unfamiliar with the model system. The main review will summarize known effects of SD on the whisker-to-cortex sensory system from birth to adulthood. The final section will summarize our current understanding of the molecular mechanisms that are affected by SD. We assume that understanding inadequate activity levels will shed light on the neural functions that require typical levels and patterns of neural activity for normal development. Finally, the sparse evidence will be mentioned that describes the exploration of changes in subcortical structures after SD.

C. Methodology of Sensory Deprivation

1. Definition of Sensory Deprivation

Levels of neural activity in sensory pathways change moment by moment throughout life, and can be disrupted by events as different as amputation of a limb or solitary confinement in a space that is nearly devoid of novel sensory stimuli. Here, we have restricted our definition of SD to "the effects of simple sensory disuse, without injury to the central or peripheral nervous system." SD, therefore, requires no neural and/or other cellular injury.

Modifications of sensory processing, representational mapping and synaptic efficacy in neocortex are now well established to depend on the balance and intensity of sensory activity patterns at all ages, no longer being restricted to a functional critical period in early postnatal life. The cortex in particular depends on sensory experience for normal development and if levels of activity in sensory systems stay very low, organization and functional plasticity can be impaired in that sensory system, perhaps throughout life. Prolonged periods of sensory deprivation in early life interfere with the maturation of circuits and timing of synaptic events that underlie experience-dependent changes such as long-term potentiation (LTP) or long-term depression (LTD) of synapses. Therefore, a common theme in the discussion that follows will be the effect of SD on cortical cell receptive field properties, synaptic plasticity, LTP and LTD, and their relevance to learning from sensory experience.

2. Critical or Sensitive Periods

There is no universal definition for the term critical or sensitive period of development. Until recently, morphological critical periods have been confused with functional critical periods. The term is commonly generalized to all aspects of functional cortical development. As we describe below, the "critical period for functional somatosensory cortical plasticity" ranges from a few days postnatal (neonatal) to at least 3 months postnatal (often described as adult), or perhaps longer, depending upon the experimental strategy used for deprivation, and the precise function studied.

3. Plucking vs. Trimming of Whiskers To Produce Sensory Deprivation

SD in the whisker system can be produced by either whisker trimming with scissors to cut the whiskers close to the face (similar to shaving in people) or by plucking out the whiskers by grasping them with a forceps and pulling them out of the follicle (similar to a paraffin cast hair removal). However, mystacial vibrissa are also called sinus hairs because of the blood sinus that surrounds the root of the whisker in the whisker follicle. The function of this blood sinus is not entirely clear in land mammals, but it is assumed to be related to controlling the sensitivity of the receptors by pressing them against the root of the whisker when the sinus is engorged. In marine mammals, the sinus has been shown to keep the receptors warm enough to function when the seawater and hence the skin surface temperature can be as low as 10° C.[2,3]

Because of the blood sinus, plucking the whiskers is often associated with bleeding from the follicle, and after plucking the whiskers sometimes regrow in a misshapen and abnormally curved shape. Attempts have been made to determine the extent of damage caused by plucking, and the evidence is mixed. Li, et al.[4] showed that the average number of axons in the deep vibrissal nerve is not reduced after whisker plucking, while axon numbers are abnormally low after follicle cauterization. The levels of galanin and neuropeptide Y expression are not elevated after whisker plucking. Whether plucking produces a loss or replacement of one or another type of receptor in the follicle such as that seen after trigeminal nerve regeneration[5] is unknown, but the physiology of the primary sensory neurons has been reported to be abnormal after whisker plucking, as evidenced by abnormally high magnitude responses to whisker deflections and abnormal periods of prolonged discharge after whisker deflection.[6] This study compared responses after plucking with those in normal control animals, but, unfortunately, did not compare the effect of trimming with plucking the whiskers. In any event, we will indicate the manner of whisker shortening in this review to alert the reader to results that employed either whisker trimming or whisker plucking protocols.

4. Complete (Global) vs. Restricted (Partial) Deprivation

An inevitable question at the outset of deprivation studies is whether one should trim all or only a subset of whiskers, and if not all, then which ones should be trimmed and which left intact? At first glance, this decision appears to consist of more or less of the same thing. However, global deprivation (trimming all whiskers) produces different outcomes compared to restricted deprivation (trimming a subset of whiskers), with restricted deprivation producing more complicated changes. Global deprivation results in abnormally low activity from whisker follicle receptors in an entire whisker pad, which by many criteria projects to a cortical area with a regional boundary whose total area is not altered by, for example, peripheral manipulations, even in the presence of extensive reassignment of inputs within the barrel cortex region.[7,8] The area of the face representation even maintains its overall area after permanent elimination of the whisker inputs on postnatal day one (PND1) that eliminates whisker responses completely and reorganizes them with responses to nose and small hair receptors.[9] The central point is that there is no competition between active and inactive inputs after global deprivation. Disparate activity levels leading to competitive interactions are a cardinal feature of restricted whisker deprivation where normally high activity (if intact) whisker input competes for cortical domain and circuitry with low (if trimmed) activity whisker input circuits. Restricted deprivation could consist of trimming all whiskers except one row or one arc, or cutting every other whisker in a checkerboard pattern, or any other pattern. Whisker-circuit interactions would be expected to influence cortical cell response properties in novel ways during and after the period of whisker trimming, and would lead to the prediction that restricted deprivation would produce generally similar, but in some ways different, deficits in cortical function compared to global (total) deprivation.

5. Timing: When is the Sensory Deprivation Imposed?

Neural activity levels before, during, and after SD affect the functional consequences of SD. Another important variable is when the whiskers are trimmed; effects of trimming are different when the onset occurs at different ages. The first month after birth is a very sensitive period in the development of the whisker-to-cortex system in rodents, so special note needs to be taken of exactly when trimming was started and the period of time the whiskers were trimmed when the effects of SD are evaluated. Finally, after the trimming is carried out for a defined time period, some of the effects may be reversed spontaneously or recovered over time after the trimming was discontinued. If sufficient time elapses before analysis, the whiskers regrow, and sensory activity could be upregulated and/or restored. Analyses immediately after and long after discontinuation may lead to different conclusions, especially if the system is capable of any spontaneous reorganization. Therefore, only reports in which the period between the end of deprivation and the onset of analysis is specified will be discussed.

Trimming whiskers is awkward before birth, but quite straightforward from the day of birth to old age. There is not much incentive to trim whiskers prenatally because both spontaneous and whisker-driven activity is nearly undetectable in the barrel cortex *in vivo* until near the end of the first postnatal week.[10] The period from postnatal day 7 (PND 7) to PND 21 is highly dynamic in barrel cortex. Initial intracortical circuits are forming, and toward the end of this period with activity-dependent circuit refinement. In this two week period, the rat somatosensory cortex transitions from almost nonfunctional to nearly mature. It is a period of easy inducement of plasticity as measured by LTP and LTD and whisker trimming induced changes in the cortex.[11] When rats are weaned at 3 weeks (PND 21), their cortex is still immature, and continues to develop at a slower pace past sexual maturity at 6-weeks old. Two-to-three months old is typically considered a mature rat cerebral cortex, with a full range of rat behavior definitely in place by 3 months. Cortical function remains relatively stable until some time between 1- and 2-years old when deficits in cortical function have been reported and behavioral skills begin to slow and decline in most strains of mice and rats. These time points identify the periods when sensory deprivation has been studied. Early sensory deprivation can start at birth or and continue through the rapid period of cortical circuit development to 1–2 postnatal months. Usually deprivation is carried out for a shorter period since very robust effects have been reported after only a few days of trimming. The investigator defined hypothesis provides the rationale for optimal duration of SD. It is noteworthy that no two laboratories have ever reported SD results using exactly the same experimental protocol.

6. Recovery Periods: What is the Time between Trimming and Beginning of Analysis?

If the animal is analyzed immediately after SD, then in principle, no time is available for any type of activity-induced recovery. If the period of deprivation is, for example, from PND 7 to 14 and the animal is analyzed at 180 d, then deficits could have been produced that are diminished or erased over time by normal use of the completely regrown whiskers, which takes roughly 1 month after the discontinuation of trimming. Whiskers in mice and rats regrow at a rate of around 1 mm/d: faster in the very young and slower in older animals. Therefore, if the whiskers are trimmed as in the example above, and a 2-week recovery period is used, then the regrown whiskers would only be ~14 mm long, less than half the length of the longest adult whiskers. This means that the trimmed whiskers would still be shorter and blunter than either the untrimmed whiskers on the other side of the face or the untrimmed whiskers around the trimmed whiskers, and, hence, may not yet produce equivalent levels of cortical activity. This scenario has two main implications for analysis of responses: one is that if the recovery period is zero or very short then it may be necessary to glue an extension onto the stump to apply test stimuli to the whiskers at the same distance from the face as the intact whiskers to deliver equivalent transduction of the follicle receptors. Second, if the angle of deflection is not identical, then comparing the physiological response characteristics is problematical. In most cases, the direction (angle) of deflection at a specified angular velocity gives the best description of iso-stimulus intensity.[12] Whether the whisker tip is cut off or intact becomes important near threshold intensities needed to produce cortical responses.[13] The other implication is that the period of survival after discontinuing whisker trimming is a time of flux, and a potentially important SD variable. An illustration of how survival time could affect results would be if a litter of 8 trimmed animals are to be analyzed 1 week after the end of trimming and each analysis were to take, say, 2–3 d, then the first animal would be analyzed at 7 d after the end of trimming and the last animal at best almost a month after discontinuing trimming. Most reports to date have not been clear about the duration of the period between discontinuing trimming and beginning analysis for each animal in a group.

7. Controls: What is a Valid Control for Sensory Deprivation?

What is the proper control for SD studies based on the shortening of mouse and rat whiskers? A commonly used control group in the literature is to reduce a litter of

11 or 12 pups to 8, and trim the whiskers on one side (partial or complete) in 4 experimental animals, then pick up the other 4, lightly anesthetize or simply hold them for roughly the same period of time, and then put them down without trimming the whiskers. The idea is to make early experiences the same except for whisker trimming. Another type of control for unilateral whisker trimming is to record from the hemisphere ipsilateral to the trimming (which is contralateral to the intact side) to provide a within-animal control. The problem with this control is that it requires the assumption that there are no interhemispheric influences that would have an impact on barrel cortex cells in both right and left hemispheres. This assumption is troubling as more reports emerge showing that the inputs from the whiskers on the two sides of the face interact, presumably continuously, in the awake, behaving animal, in normal[14,15] and sensation manipulated animals.[16,17,18,19]. The final control, if the query is to show age-related differences in the effect of sensory deprivation, then a control would be to compare the effects of 2 weeks of trimming, from, e.g., PND 7 to PND 21, with a comparable duration of trimming starting at a later age, e.g., from PND 30 to PND 44. If the same duration of deprivation produces serious deficits during the first month and no detectable SD deficits for the same deprivation period after one month, then the results would indicate that the need for activity to promote maturation of normal cortical responses is greater during the earlier post-natal period. Then, if the question became when is the effect maximal, the early period could be subdivided, and moved earlier and later to pinpoint the onset time and duration of the epoch of maximum sensitivity to SD for a certain feature of cortical function. Studies carried out in this way have not yet been reported.

D. Overview of the Whisker to Barrel Cortex System

1. Development of Normal Cortical Response Properties

Anatomically, and physiologically, rat cortex is very immature at birth. Sensory deprivation starting during the first week after birth, affects highly dynamic maturational processes that include, in rough order, (1) the end of cell division and migration into superficial cortical layers II and III, (2) neuronal dendrite and axon growth to establish the initial cortical circuits, (3) massive synaptogenesis accompanied by the coincident appearance of postsynaptic dendritic spines on subsets of cortical neurons, (4) the onset of cortical responses to peripheral stimulation and finally (5) myelination of axons. For this review, the onset of physiological neuronal responses to natural stimuli is centrally relevant and they are first elicited around postnatal day 6 to natural whisker stimuli under urethane anesthesia.[10] This report identified several striking contrasts between 1 week old and adult response properties in the barrel cortex. First, at PND7, only 3% of cells encountered showed stable spontaneous activity (SA) of 1/sec (only 12% of the cells showed any SA at all), while the majority of cells in normal adult rat cortex show SA under the same recording conditions. Most cells (88%) that were responsive to peripheral stimulation were not spontaneously active. Second, at P7, cortical cells could not respond faithfully to repetition rates faster than 1/10–15 sec. Consistent responses were rarely encountered at stimulus rates faster than 4 times per minute, and the largest number of responsive silent cells were found in layer IV (40%), followed by layers III + V

(~23% each), and layer VI (12%). Third, receptive fields on average were considerably larger at PND7 than in the adult. Fourth, the mean response latency for all units at PND7 was 88 ms compared to ~14 msec averaged over all layers of adult cortex under the same conditions. The long latency presumably reflects the immaturity of synapses and the absence of myelination. Fifth, many PND7 cortical cells already showed responses sensitive to the direction of whisker stimulation, with a few showing a distinct off response without a corresponding on response. Finally, some PND7 neuron responses are prone to repeated episodes of excitation and inhibition similar to oscillations after a single stimulus.

Subsequent to these *in vivo* results slice preparations have added considerable detail to early phases of cortical development. For example, direct electrical stimulation of slices on or before the day of birth can activate cortical cells, but the responses are easily fatigued and labile.[20] Even with direct stimulation, reliable responses require interstimulus intervals of 30 to 60 s in brain slices.[21] Inhibitory currents are initially detected in slice preparations of barrel cortex around PND7-8. When inhibitory posrsynapric potentials (IPSP's) appear during the second postnatal week, the N-methyl-d-Aspartate (NMDA) currents decrease in magnitude during the same period,[22] although these two events are not necessarily linked.

The duration of the NMDA receptor currents in cortex is very prolonged during the first postnatal week when cells in layer IV are activated by stimulation of thalamocortical inputs in slice preparations,[23] and the duration decreases to adult values during the second and third postnatal weeks. The same report showed that LTP of synapses can be produced by 1/s stimulation in slice preparations; however, the upregulation to 50-75% above baseline has an onset time of 10–15 min during the first postnatal week. Some of the changes in LTP are likely to be related to changes in the subunit composition of NMDA receptors that occur during the first postnatal month, notably the increase in the expression of NMDA2A receptor subunits and their addition to the receptor complex concomitant with a reduction in NMDAR2B (rat visual cortex,[24]). One recent report suggests that the low activity in visual cortex produced by dark rearing decreases NMDAR2A expression in all layers of rat visual cortex,[25] and this decrease may indicate a perturbation in the normal developmental change in NMDAR subtype composition as a result of low postnatal cortical activity.

In addition, the conversion of silent synapses that initially show only NMDA currents to those with fast α-amino-3-hydroxy-5-methylisoxazole proprionic acid (AMPA) receptor currents may play a role in the development of mature sensory responses in barrel cortex.[26] The low levels of cortical synaptic activity during the first few weeks of rat postnatal life could be produced by any combination of these normal, activity-influenced, developmental mechanisms.

The development of the peripheral receptor numbers in whisker follicles appears to follow a similar time course to the development of cortical function. The number of axons entering the follicles on PNDo is not significantly different from the number entering the follicles in the adult.[27] However, there are successive waves of differentiation of the receptors in the follicle-sinus complex that continue for roughly 3 weeks after birth.[28] For example, at birth the innervation of the inner conical body, a superficial region of the follicle near the pivot point of the whisker in the skin, does not appear until 3-4 d after birth and then matures rapidly. Whether these developmental events are influenced by levels of stimulation of the whisker (or by plucking or trimming the whisker) starting at PNDo is unknown, but many protocols for SD overlap with this period of peripheral follicle maturation.

2. Ascending and Recurrent Circuits

In the mature (>2 months) mouse or rat, each mystacial whisker follicle is richly innervated by ~200 primary sensory nerve fibers that project centrally to a first synapse in the Brainstem Trigeminal Complex (TC).[29,30] The TC cells project to the contralateral thalamus. The thalamic nuclei project to SI (barrel field) and SII cortical areas in an anatomically topographic manner. The cortex, in turn, gives rise to a powerful projection back to each relay nucleus that transmits whisker information to it. Sensory-motor transformations occur at all levels in the system to initiate and guide behavior from simple actions such as moving the whiskers, to complex bilateral learning discriminations using the whiskers.[15,31,32]

Several reviews have addressed the anatomy, physiology, and dynamics of somatosensory processing in the cortical barrel columns in detail.[33,34,35]

To elaborate on the outline above, it is well documented that the entire array of whiskers is topographically represented in cell groups within the principal (PrV) and in three spinal trigeminal subnuclei (SpV_{oralis}, $SpV_{interpositus}$, and $SpV_{caudalis}$). In all but one subnucleus (SpV_{oralis}), the whisker representation can be seen in a pattern of Cytochrome Oxidase (CO) patches, one for each whisker. In PrV, where there is a clear correspondence between the distribution of primary afferent axon collaterals and a CO patch.[36] Most PrV units are driven primarily by 1 or 2 vibrissae with low response thresholds (high sensitivity): they have small receptive fields (1–2 vibrissa) with minimal convergence of primary afferents, and almost all are projection neurons to other structures. The morphological data indicate that the distribution and pattern of primary afferent collaterals and terminals subserving different submodalities is similar within and between subnuclei.[36,37,38] The physiological data suggest that primary afferents convey what appears to be identical information to each subnucleus. Most SpV neurons have a significant overlap of collaterals from different vibrissae, more numerous interneurons, and roughly 6–8 whisker receptive fields.[36,39,40,41,42,43]

Projection neurons in the TC terminate in many structures, but the TC pathway terminals of greatest interest for this discussion are those to the thalamus where they form discrete patches in ventral posterior medial nucleus (VPM) barreloids, each consisting of cells that respond best to a single whisker,[44,45,46,47] even though each VPM neuron is responsive at a low level to many whiskers.[48,49,50,51,52,53] PrV gives rise to ascending lemniscal projections that terminate mainly in the VPM barreloids with a much less dense projection to POm the medial division of the posterior necleus. The $SpV_{interpolaris}$ and $SpV_{caudalis}$ sub-nuclei also provide a significant projection to thalamic nuclei,[54] with the projections to VPM being more concentrated in a restricted tail portion of the nucleus somewhat separate from the PrV inputs. Deschenes and colleagues have recently clarified this anatomical feature, in which VPM barreloids appear to have two subdivisions, a larger core with PrV inputs that project to SI layer IV barrels, and a smaller tail region lying at the border between ventral posterior lateral (VPL) and VPM with SpV inputs and projections to SII and the septa around barrels of SI.[55,56,57,58,59] The commissural inputs from the opposite hemisphere project to the septal zones and more widespread fields above and below the septa.[60]

Projections from SI back to subcortical structures arise from infragranular layers V and VI. Several thalamic nuclei receive inputs from layers V and VI of SI whisker barrel cortex; mainly VPM, POm, and intralaminar nuclei.[61,62,63,64,65] The origin of corticothalamic projections can be further subdivided within each lamina, in that upper layer VI cells project to VPM and lower layer VI cells project to POm.[66] Layer V corticothalamic neurons give rise to collaterals with large terminals in POm and the intralaminar nuclei.[64,65] Cortical axons course in bundles on their way to the thalamus and give off collaterals to the inhibitory reticular thalamic nucleus before they terminate in the thalamus.[64,65]

3. Intracortical Circuits

Barrel cells and septal cells respond to whisker stimulation at short latency, with barrel cells responding most strongly to a single whisker.[67–71] Cells in septal zones commonly respond with equal maximum magnitude to two or more whiskers.[71,72,73] Adult rat barrel cells project to the septa around the barrel and strongly to layers II and III in the column directly above the barrel.[73–75] Septal cells are often pyramidal cells with longer dendrites than stellate cells, and their dendrites are not restricted to the septa: they extend into one or more surrounding barrels.[73] The latter study[73] also showed that after normal rearing the septal cells respond equally well to two or more whiskers to produce about the same amplitude of postsynaptic potential. Septal cells give rise to much more widespread connections than do barrel cells: septa project more densely down the septa along the row, but also project down septa between arc whisker barrels, and provide dense connections with SII and motor cortex and other cortical areas. This distinction is blurred on the output side of barrel cortex, although the layer V-VI cells beneath the barrels do not appear to be evenly distributed. That is, a continuous layer of cells in layer VIa under both barrels and septa project to the VPM. Nearly every cell in barrel field layer Vb has at least a collateral branch that terminates in the pons.[76] However, while POm receives input from a rather continuous layer of cells in Vb and VIb, input from VIa arises predominantly from cells lying beneath the septal zones.[66]

Several studies have identified specific physiological response properties of septal cells that provide important information for interpreting SD studies. An extensive study that separates septal cell responses and quantitatively compares them with barrel, supragranular, and infragranular cells was reported by Armstrong-James and Fox.[71] They concluded that the single greatest difference between barrel and septal cells is the much greater percentage of barrel cells that respond at the highest level (>2 spikes/stimulus) to a single whisker. Both barrel and septal cells showed receptive field sizes of around seven whiskers if one includes both high and low magnitude responses. Intracellular studies of postsynaptic potentials below the threshold for action potentials (spikes) suggest that nearly every whisker on the contralateral face can influence nearly every cell in barrel cortex.[73,77] Finally, there are more septal cells (13%) that show short latency spikes to several whiskers than do barrel cells (2%).[71] These differences appear to reflect the anatomical fact that the VPM nucleus provides massive inputs to barrel domains, and that some VPM cell axons also give rise to collaterals that spill over into the surrounding septa and occasionally into adjacent barrels.[55,78] The 10% or so of projections of VPM cells outside of a barrel may explain the higher frequency of short latency responses by septal cells, when at the same time their responses are not dominated by a single whisker input. Layer IV responses are expressed through AMPA as well as NMDA receptor activation, while layer II/III responses rely almost exclusively on NMDA receptor activation for their expression.[79] This background is crucial for interpreting the reported effects of SD on intra-cortical circuitry.

The typical response sequence of most cells in mature barrel cortex is excitation followed by inhibition for both fast spiking and regular spiking neurons.[80–82] A unique feature reported for septal cells is that they can show a type of stimulus-induced inhibition (decrease in ongoing firing rate) that is not preceded by excitation (in rabbit SI[83]; in rat SI[84]). This feature is more easily encountered while recording from septal cells in awake animals in which cortical cells have a typically higher spontaneous discharge rate than when they are under anesthesia, and to date the response has not been reported in barrel cells and/or under anesthesia. The cells showing inhibition-only responses tend to be located on the outer edge of the barrel, but never in the barrel proper.

A third feature of septal cells that is rarely seen in barrel cells is that 28% of the regular spiking septal neurons show facilitation of responses to stimulation of two whiskers at very short interstimulus intervals.[85] The peak facilitation occurs when one whisker is stimulated 3 to 8 msec (mean 5.3±2.3 ms) prior to the other, and characteristically only occurs when one of the two whiskers is stimulated first (i.e., the two whisker responses are asymmetrical). Shimegi, et al.,[86] later showed after careful analysis of the histology that all of the cells showing facilitation (less than 1/3 of all cells in the study) were histologically located at the edge of a barrel or clearly in the septum.

An important but unresolved question about intracortical barrel-to-barrel transmission is whether barrels talk to other barrels directly through the septa or only indirectly by projecting to layer II/III and then to the adjacent barrel column. There has been a preference in the recent literature to consider barrels in mice and rats as insular cell clusters that respond to thalamic inputs and project to surrounding cells with very little direct cross-talk between cells in different barrels.[72,75,87] However, it is our view that both surround receptive fields (SRF's) and the plasticity of SRF's in rat barrels depend upon connections between barrels as well as through more indirect intracortical routes.[88] There is experimental agreement that barrel columns are dominated by input from one principal whisker and responsive at least at the level of post-synaptic potentials (by intracellular recording at levels subthreshold for spikes) to activity generated by nearly all of the whiskers.[73,77] Starting with any whisker/barrel combination, the closer the surround whisker tested, the greater the effect on the cells in the reference whisker column.[33,71,94,95] Whisker integration also occurs subcortically in the trigeminothalamic pathway[43] and in thalamic circuits.[50,89,96]

E. Effects of Early SD Sensory Deprivation on Barrel Cortex

1. *In Vivo* Analysis of Global Sensory Deprivation during Development

An important topic is how experience-generated neural activity in sensory pathways during development leads to the ability of cortical cells to modify their response to sensory inputs throughout life. Early work on rearing rat pups in isolated or relatively more enriched environments from weaning to adulthood clearly demonstrated that isolation produced gross changes, such as smaller brains, thinner cortex, lower synapse to neuron ratio, slower learning and many other negative effects compared to animals reared with stimulating environmental and social challenges.[97] From the point of view of sensory cortex, the main difference between impoverished and enriched rearing conditions is the level of outside stimulation between birth and adulthood. The effects of isolation, once in place, have been very difficult to reverse at an older age. Data showing that sensory experience dramatically affects cellular development after cortical input activity is abnormally biased during circuit formation were provided by Hubel and Wiesel's classic analysis of monocular deprivation in kittens and young monkeys.[98] However, monocular deprivation studies are not comparable models to global deprivation in the whisker system because eye vs. eye competition in contiguous cortical domains is always present throughout the binocular sector of the VI cortex that they studied. Binocular deprivation would provide the comparable protocol to global whisker trimming, and after 1 month of binocular lid suture in monkeys the cortical cell responses to either right or left eye appear rather normal in their responses to test stimuli, but cortical cells failed to develop the circuitry needed to display binocular responses in cortical neurons.[99] Considerable evidence indicates that this deficit is due to the abnormal arrangement of intracortical connections within layers II/III such that ocular dominance columns for the same eye are abnormally interconnected only with each other through horizontal connections in layers II/III.[100]

We have found that after global whisker trimming from PNDo to PND3o, followed by a 60-d recovery period, the cells in barrel cortex were abnormal in several ways when compared to control animals that were deprived for the same duration of time but at a later period from PNDo40–P70 (analysis at P130 using urethane anesthesia). First, the average spontaneous firing of cortical cells in the P0-P30 animals was significantly reduced in the deprived cortex ($p<0.01$, MWU test) compared to the later PND40–P70 trimmed, or untrimmed control animals (0.4 ± 0.1 spikes/s vs. 0.8 ± 0.1 and 0.8 ± 0.2, respectively). Second, the response of cortical cells to test stimuli was also reduced for the neonatally trimmed group, with a pronounced decrement in the principal D2 whisker responses.[101] Finally, the same cortical cells did not show the expected shift in response to the active whiskers following 3 d of trimming all but two whiskers (whisker pairing) at the end of the recovery period. This result suggests that the development of the mechanisms that support synaptic modification in the adult brain require a certain level of experience-induced activity in barrel cortex during early postnatal life, a finding that fits with the theoretical idea that synaptic modification requires critical threshold levels of activity for synaptic strengthening[102] and that the biological mechanisms that generate the threshold require early experience to mature, in barrel cortex as they do in cat visual cortex.[103] In principle, the early low activity leaves cortical function permanently impaired. The principal D2 whisker and the intact surround whisker do show significant increases after 7 d of whisker pairing (at the $p<0.05$ level when controls have shifted much more at the $p<0.0001$ level), so response modification is not absent, just seven times more sluggish. This low capacity for synaptic modification would presumably be reflected in a slow capacity for learning on difficult behavioral tasks.

Deficits in behavioral learning of a modified gap-crossing task[104] have been documented by Carvell and Simons[105] after global bilateral whisker trimming in rats from PND 1 to 45. The animals were trained to make tactile discriminations using the middle C row whiskers beginning 60–70 d after the end of whisker trimming and their performance was compared to adult animals whose whiskers were untrimmed or trimmed with the same protocol beginning at PND 45. The resulting performance of the untrimmed and late trimmed adult animals was not different. The early SD animals showed a normal performance on the rough vs. smooth discrimination, but were severely deficient on the rough vs. different roughness (rough/rough) discrimination task. Four out of five early SD animals never even reached criterion levels of performance on the rough/rough task. A very interesting, but unexplained, correlation with the failed performance was a sharp reduction in whisking rates normally seen in the 6–12 Hz range and an increase in power at 13–22 Hz. These results show that extracting and comparing high acuity sensory information from environmental stimuli is greatly reduced after early SD and this deficit is not recovered even by extensive periods of typical experience plus months of training after the whiskers regrow to normal lengths. Further, the results implicate motor involvement in the abnormal sensory processing after SD that is reminiscent of eye movement impairments seen in dyslexic children who see gross objects in the outside world with little trouble, but have difficulty moving the eyes with precision along rows of fine print.

The issue of sensor- vs. motor-induced deficits was directly addressed by Diamond and colleagues[106] in deprivation studies in which cutting the motor nerve to the whiskers bilaterally was compared to bilateral global SD produced by whisker trimming. In these studies manipulations began on P8, either whisker trimming or VII nerve cut, and continued for 13–15 weeks before analysis was carried out without a recovery period. When infragranular cell clusters (layers V-VI) in barrel cortex were recorded with a 10×10-electrode array, there was no effect on overall topography of barrel field activation in either group, but there was a significant reduction in cortical response magnitudes after motor deprivation compared to normal rearing. Further, responses in SD animals were depressed significantly lower than after motor deprivation. Motor deprivation had no effect on adjacent whisker responsiveness, while SD caused a substantial decrease in the receptive field contrast (measured as the ratio of principal to surround whisker responses) with clear receptive field expansion after SD, but not after motor deprivation. When multiple whiskers were stimulated, the surround whisker inhibition was enhanced in the SD animals. These results taken together, suggest that the change in whisking rates found by Carvell and Simons[105] may not have been direct effects of motor problems but more related to changing whisking strategies to better integrate inadequate sensory information from the SD whiskers within impaired sensory cortex to allow the SD animals to solve the rough vs. rougher behavioral task. Further experiments will need to sort out alternative explanations, but two recent reports support the idea that rats can modify their whisking rate by briefly ramping up to 15–25 Hz whisking at the moment they commit to a behavioral response.[107,108]

2. *In Vitro* Analysis of Global Sensory Deprivation During Development

Global SD has been shown to affect the development of specific circuits in very specific ways. Perhaps the most stunning change in circuit organization was identified by Shepherd, et al.[109] in Svoboda's laboratory. After trimming all whiskers from PND 9 to 15 on one side, they analyzed brain slices without a recovery period. Previous results from the same lab had identified that global SD during this period disrupted receptive field structure in layers II/III of the barrel cortex.[110] However, Shepherd, et al, using laser scanning photostimulation, moved stimuli horizontally along layer IV in brain slices to activate the Layer IV-to-II/III circuits. They developed a strategy to allow separate barrel and septal stimulation, and found that the layer IV barrel-to-II/III circuits were strong in normal animals, but that stimulating septal cells between the barrels barely activated layer II/III cells above. After global SD from PND 9–15 (no recovery period) the normal pattern essentially reversed, so that septal projections became the dominant inputs to layer II/III and the barrel projections were weak. The basis for this change is currently being investigated in several laboratories, and may lead to the reinterpretation of previous results, such as those of Fox, et al.[111] in which plucking all but one whisker led to more widespread short latency responses to the intact whisker, predominating in the septa around the barrel corresponding to the intact whisker. The results are consistent with earlier slice studies of near-global SD which concluded that there is a significant weakening of synaptic strength in horizontal connections through layer II in deprived cortex after only five days of whisker trimming at a median age of ~P25.[112] The reorganization of layer IV-to-III connections from barrels to septa by SD, and the constriction of horizontal spread of activity in the supragranular layers above the barrels would likely also incur changes in the supragranular to layer V projections in cortex, and indeed, this has been reported. Schierloh, et al.[113] trimmed all whiskers on one side between PND 9 and 21 for 6 to 13 d, and analyzed without a recovery period. Brain slices were prepared to estimate the extent of connections between layers II/III and V using photostimulation of caged glutamate near cells in layers II/III. The normal cortex showed a strong preference for layer II/III cells above two or three separate septal zones to converge upon single layer V neurons beneath a septum. However this feature of intracortical circuitry was degraded by global SD, so that each layer V neuron received diminishing convergent inputs from layer II/III cells in direct proportion to lateral distance, independent of whether they were under barrels or septa.

3. Anatomical Changes Produced by Global Sensory Deprivation

The physiological changes after SD predict that anatomical connections also will be altered by the global SD. An important background note to anatomical changes caused by SD is that many intrinsic connections in barrel cortex develop in the second and third postnatal weeks by increasing the length of horizontal connections through the progressive elongation of more extensive axon arbors (Miller, et al.,[114] in mouse). The maximum impact of low activity rearing has been produced by peripheral nerve damage through cutting the infraorbital nerve on PND 7 and analyzing the spread of cortical connections on PND 42. This nerve damage procedure produces a drastic reduction in the horizontal spread of axon terminals in all layers of the barrel cortex.[115] However, global SD even without peripheral nerve damage has a strong impact: trimming all whiskers for 60 d without any recovery period produced a significant 44% reduction in numerical density of thalamocortical synapses, without any apparent change in barrel size.[116] The same study confirmed the GABAergic effects reported by Knott, et al.[117] in the observation that the numerical density of symmetrical synapses was significantly (55%) below controls in the cortex contralateral to the trimmed whiskers. Micheva and Beaulieu[118,119] first reported a reduction in GABAergic cells and GABA-immunoreactive axon terminals (puncta) in the row B, C, and D barrel columns contralateral to plucking almost all whiskers (B, C, D rows). They also observed an increase in GABA-positive elements in layer IV of the ipsilateral barrel cortex, a finding that contributes further evidence that the nondeprived cortex is also affected by unilateral deprivation. The plucking started just after birth and continued for 60 d, with no recovery period before analysis. It is not clear whether the ipsilateral changes represent some form of indirect compensation related to altered function in the deprived cortex, or perhaps an effect related to the greater use of the intact whiskers giving rise to greater activity in the cortex contralateral to the intact whiskers during unilateral SD. Whatever the correct interpretation turns out to be, the quantitative demonstration that both hemispheres were affected by a unilateral manipulation is important. A later paper from the same laboratory showed that while the density of dendritic spines was not changed in cortex following the SD, the total numbers were increased by 67% in ipsilateral barrels due to the greater thickness of layer IV.[120] Ipsilateral spines also showed an increase in spine head volume, suface area, and length of post-synaptic density coupled with shorter spine necks, compared to spines in the barrels contralateral to the trimmed whiskers. The authors interpret these changes as a product of activity-dependent plasticity driven by thalamocortical inputs from the intact whiskers, with little if any contribution from the events occurring in the contralateral barrel cortex. However, given the recent results described below following restricted SD, the possibility remains open that interhemispheric changes could contribute to bilateral effects. The question of the effect of a recovery period between end of deprivation and beginning of analysis on these bilateral changes remains to be explored.

4. Partial Sensory Deprivation During Postnatal Development

Restricted SD produces active columns of cortical neurons (nondeprived columns) in a pattern dictated by the array of intact whiskers after trimming. At the same time trimmed or plucked whiskers, produce deprived barrel columns by the resulting low activity arising from the whisker stub on the face. A clear demonstration of long lasting physiological changes in layer IV barrels following restricted SD was reported by Simons and Land in 1987[121] after restricted neonatal whisker trimming of either the C-row of whiskers, or all rows but the C-row, on one side of the face for 45–60 d. The long period of deprivation was followed by a comparably long recovery period (3–15 weeks), until the animals were unquestionably adult and the whiskers were completely regrown. Any residual alterations in cortical cell properties could be assumed to be permanent deficits that persist for a very long time. Persistent deficits were identified by physiological recording from single cells in layer IV barrels with a history of normal activity during development (normal or intact whisker barrels), or barrels that received abnormally low activity during development (deprived barrels). The animals were anesthetized with pentobarbital and maintained with fentanyl coupled with paralysis produced by pancuronium bromide. When Simons and Land recorded cortical responses after restricted SD, the deprived barrel cells showed quite different response properties from non-deprived barrel neurons. The deprived barrel cells showed: 1) elevated spontaneous activity, 2) greater responses to both principal and surround whisker stimulation (spontaneous activity was not subtracted from the total spikes), 3) enlarged receptive fields (more whiskers produced a response), and 4) reduced directional tuning.

Population responses to principal whiskers showed that both onset and offset responses to the 200 msec duration deflections were greater in the deprived barrels than in adjacent nondeprived barrel columns (or in normally reared animals). They proposed that the SD led to an increased whisker drive on cortical cells from thalamic inputs (increased excitation) coupled with a decreased efficacy of intracortical inhibitory mechanisms in the deprived barrels (see below).

One of the questions that these results raise is whether there is a critical period, and if so, does the sensitive period for SD end at the same time for cells in all cortical layers. Fox[122] reported that SD changes could be produced by raising rat pups with all whiskers plucked except one intact whisker (the D1 whisker) on one side of the face. Whiskers were plucked for 30 to 90 d starting at PND 0, 2, 4 or 7. Then, with a minimal recovery period only long enough to let the whiskers elongate to apply test stimuli, cortical cell responses were recorded to whisker stimulation. In this paradigm, the one intact whisker is used during daily exploration by the animal throughout the period when all surrounding whiskers are producing levels of neural activity well below normal. This produced cells in an expanded area outside of the nondeprived barrel (i.e., cells in deprived cortex) that showed short (5–10 ms) latency responses to the preserved D1 whisker if the SD started at PND 0, 2, or 4 (30%, 18% and 13% of recorded cells, respectively). The percentage of barrel cells that respond better to the D1 (preserved) whisker than their appropriate but deprived principal whisker fell from 37% to 23% to 12% at the same onset times, and then didn't fall lower. Effects of SD were seen in layers II/III over a longer period than in layer IV. This result suggested that SD leads to different effects in different layers, with layer IV stabilizing during the first postnatal week in a way that layers II/III do not. A follow-up study by the same author[92] provided evidence that the expanded cortical domains dominated by the single intact whisker when SD was initiated at PND 0 were diminished by small lesions placed in the D1 barrel at the end of the deprivation period, leading to the suggestion that the expansion depended upon intracortical connections rather than synaptic reorganization at the thalamic or brainstem levels. Curiously, although latencies changed, response magnitudes showed no changes for the preserved D1 whisker. Plasticity in layer IV was greatly diminished at later onsets of deprivation when only one whisker was left intact, which is unexpected when compared to plasticity from whisker trimming in adult animals, as discussed below.

Another important question is whether a short period of neonatal deprivation degrades the plasticity of cortical function to sensory challenges throughout life. In a recent study, 3-week periods of trimming two whiskers starting just after birth impaired the ability of the deprived cortex to produce plastic changes at maturity.[123] In this study two whiskers (D2 and D3) were trimmed on one side from P0 to P21 and barrel-column domains in the deprived cortical columns were analyzed. After a 2-month recovery period, all whiskers were trimmed except for D2 and D3 which normally induces whisker pairing plasticity (WPP)[88] and cells were analyzed in the D2 (a deprived) barrel column. The two whiskers left intact (paired) could either be the previously deprived whiskers D2 and D3 or one deprived (D2) and one spared (D1) whisker. Cells in the layer IV D2 barrel showed elevated responses to principal D2 whisker stimulation, but reduced responses to the early-intact D1 and particularly low responses to the early-deprived D3 surround whisker. The interesting finding was that the plasticity associated with WPP was normal in layer IV of the deprived barrel columns, but greatly reduced and delayed in the supragranular layers II/III that receive driving inputs from layer IV. The whisker with normal early experience developed a stronger influence on the deprived D2 neurons than did the deprived whisker D3 on the other side of D2. Unfortunately, these results came out just prior to those of Shepherd, et al.,[109] so a separate analysis of the septa around the deprived barrels was not carried out *in vivo*.

Removing all but one whisker in animals just after weaning (roughly 1 month old) for 7, 20 or 60 d does not reduce neuronal responses to the intact whisker in layer IV neurons surrounding the intact barrel cells.[124] However, for cells in the supragranular layers, responses to their regrown deprived whiskers were reduced significantly after 7 and 20 d of deprivation, but remain near normal after 60 d. Here again, with SD starting at this rather mature age, the deprivation effect on response magnitude in the deprived barrel columns is most pronounced in layers II/III. The recovery of responses is consistent with earlier results in normal adult rats by Armstrong-James, et al.[88] in which they showed that trimmed surround whisker responses after 30 d of deprivation were strongly reduced in layer IV cells. Principal whisker inputs in barrel columns were preserved.

5. Comparison of Partial with Global Sensory Deprivation

The essential difference between restricted and global SD is that global trimming reduces activity uniformly throughout the whisker barrel field domain, while partial trimming creates epicenters of high activity competing with adjacent barrel columns that have low sensory input activity. The distance over which competitive interactions occur is not known completely, since most studies were recorded in the barrels adjacent to the deprived or intact barrel, but Glazewski and colleagues[125,126] have provided evidence that potentiation of a spared whisker response and depression of responses to deprived whiskers decrease with distance from an active barrel column in cases of late (PND 28) onset plucking. In fact, if every other whisker is plucked in a checkerboard pattern for 7 d beginning at PND 28, then responses to deprived whiskers are depressed in layers IV and II/III to a greater extent than after plucking all whiskers but one.[127] However, the competitive interactions follow rules, and the rules are different for each layer of cortex, especially during the experience-sensitive developmental period from birth to 1 month of age in rats, and the rules apply even to fully adult animals.[88] In theory, this activity-based competition principle operates in the visual system after monocular deprivation (the BCM theory of Bienenstock, et al.[102]) and the same theory has been applied to explain plastic changes in whisker to barrel cortex responses of adults following whisker pairing[103,128] with and without insults to cortical development. In certain cases, such as prenatal alcohol exposure, barrel cortex circuits never develop normal responsive or spontaneous activity in the early postnatal period and cannot reach the threshold to be up-regulated by experience after the animals mature.[91]

When comparing restricted and global SD, an interesting difference between the 1987 Simons and Land report,[121] and the behavioral analysis by Carvell and Simons 10 years later,[105] is that sensory deprivation was carried out quite differently in the two sets of experiments. In 1987 some, but not all whiskers were trimmed on one side (C-row or all but C-row trimmed), while the behavioral deficits were the result of trimming all whiskers on both sides of the face starting at birth. Physiological analysis of the behavioral animals has not been reported. But a reasonable prediction is that total long term bilateral SD cortex would show a marked reduction in spontaneous as well as evoked activity throughout the barrel cortex, with reduced response levels similar to the results above where all of the whiskers on one side of the face were trimmed from PNDo to PND3o. The behavioral deficit in roughness discrimination produced by the global bilateral SD could be interpreted as a degradation of frequency discrimination, in that the animals could still discriminate grossly different discriminanda, but not fine differences.[105] It would be interesting to compare the response threshold, tactual acuity, and responses to different frequencies generated by the intact and deprived whiskers after early deprivation.

6. Effects of Sensory Deprivation on the Adult Brain

As stated earlier, the concept of a critical period of receptive field plasticity in the barrel cortex has required much reassessment in the face of multiple studies showing activity-dependent response plasticity in adult rat barrel cortex (>2 months of age). Plasticity of neurons to acute partial SD is produced robustly in all layers of the barrel cortex in normal adult rats by trimming all but two whiskers at maturity on one side of the face (the whisker pairing (WP) paradigm; Diamond, et al.[129]). This model produces significant changes in the response properties of cortical neurons in just a few hours, thus satisfying Hebb's hyopothesis. Layer IV neurons in the barrel column receiving input from one of the intact whiskers show a significant increase in response to stimulation of their principal whisker and also to the weaker surround whisker that remains intact on one side of the principal whisker, but a decrease in response to the cut, adjacent, surround row whisker on the other side of the principal whisker.[88,129] This type of WP plasticity (WPP) in normal adult rats occurs faster (<1 d) in layers II/III[130] than in layer IV, and is accelerated and enhanced by a novel and challenging (enriched) environment.[91]

The WPP response modifications develop significance in 2–4 h in awake animals,[131] and are detected easily under urethane anesthesia using single-unit analysis. Under similar conditions of testing and anesthesia used in our studies, surround receptive fields (SRFs) in layers I–IV have been shown by a number of criteria to depend for their expression on intracortical column to column relay, initiated by principal whisker discharges.[89,90,92] Thus, use-dependent alterations in information processing by multiple whiskers are presumably integrated through barrel to septum to more distant sites by linking barrel activity as well as indirect column to column circuits through the supragranular layers II and III. Either route would be expected to depend upon excitatory glutamatergic synapses and NMDA receptor activation,[79] and indeed, blocking NMDA receptors at the time of adult whisker trimming prevents WPP.[132,133] Response reductions to the cut surround whisker are first detectable after 7 d of deprivation with progressive response decrements for up to 30 d of deprivation (maximum time tested). Hence, with adult cortex the WPP paradigm generates Hebbian potentiation and competitive inhibition in at least the first four layers of the barrel cortex.

Several studies have analyzed with different techniques, the activity-based synaptic plasticity that occurs readily without tissue damage, in adult cortex as readily as in immature brains. Using 2-deoxyglucose autoradiography, several studies have documented the expansion of the cortical activation pattern of a single whisker after several weeks of trimming all but one whisker.[134,135] When only one whisker was plucked for 3 weeks (10 d recovery) the cortical area of 2-DG activation was greatly reduced for the plucked whisker. These changes are also reflected in the levels of cytochrome oxidase histochemical staining in the barrels appropriate to the trimmed (higher CO) and untrimmed (lower CO) whiskers.[136] An interesting later finding from the Kossut lab showed that the up-regulation effect appeared in all layers, including layer IV, in one month old mice, but only in layers II/III in adult (one year old) mice, not in layer IV.[137,138] Changes in the cortical modification pattern depend upon behavior as well as advancing age. Using chronic intrinsic signal optical imaging Frostig and colleagues showed that adult rats left in their home cage after plucking all but one whisker exhibited a similar spatial expansion and increased magnitude of cortical activation, but as little as four minutes per week of free exploration led to a contraction of the cortical activation locus and a reduction in peak signal.[139] After 4–6 weeks of the original home cage housing experience, the intrinsic signal imaging pattern was constricted by roughly half, but physiological recording indicated that the changes were in layers II/III, but not in layer IV.[140] It is not yet clear whether this experience-based constriction of signal in adult rat cortex is the same or different from the constriction of the intrinsic signal pattern that occurs in rats with all whiskers but D1 plucked from PND 0 for 5–7 weeks: the whisker plucked animals failed to show the restriction in activation pattern clearly seen in the control animals.[141]

F. Molecular Mechanisms Affected by Sensory Deprivation

As a result of these clear demonstrations that early postnatal whisker removal modifies the normal development and function of the barrel cortex, it is important to understand how the reduced postnatal sensory activity becomes translated into permanent deficits that appear to persist throughout life. There are changes in layer IV barrel cells whose principal whiskers have been trimmed for a longer or a shorter time if deprivation starts on the day of birth, but as reviewed above, a number of studies converge on the conclusion that the greatest damage is done to the intrinsic intra-cortical circuits that transfer information from layer IV to layers II/III.[109,122,123] In this regard we think Feldman's hypothesis for the decrement in layer II/III response has merit: namely, that one negative effect on the superficial layer cells arises as a consequence of abnormal timing of inputs from layer IV to layer II/III, as discussed below.

1. Effects of Sensory Deprivation on Excitatory Neurotransmission

The generation of receptive fields in barrel cortex neurons has been intensively studied and is heavily dependent upon excitatory glutamatergic circuits. In summary, excitatory projection neurons in VPM thalamic barreloids are dominated by inputs from a single whisker, but respond weakly to a number of other whiskers, especially in the awake animal.[53] The excitatory VPM projections activate excitatory and inhibitory barrel neurons at short latency. The excitatory barrel neurons project almost exclusively to layers II/III above the barrel and into the septa surrounding the barrel. The septal and supragranular (layer II/III) neurons are modulated by direct projections from the POm thalamic nucleus and by inputs from the contralateral hemisphere. The septal and supragranular cells, in turn, generate extensive horizontal excitatory circuits that link the single whisker dominated barrel columns together and integrate the inputs from the entire ensemble of whiskers on the contralateral face with those from the ipsilateral whiskers.[15]

An interesting feature of excitatory circuitry is that robust surround receptive fields to single-whisker stimulation are excitatory; i.e., single cells are activated by more than one whisker, and while thalamic neurons do respond weakly to more than one whisker, the majority of cortical surround receptive fields depend upon excitatory intra-cortical circuitry and NMDA receptor activation.[90,92] Blocking NMDA receptors in the cortex from birth by slow release of the NMDA receptor blocker D-2-amino-5-phosphonopentonoic acid (D-AP5) (APV) reduces by roughly half, but does not eliminate, barrel reorganization induced by cauterizing a single row of whisker follicles on P0–P2.[142] The D-AP5 procedure as reported did not directly block subcortical receptors, and thalamic neurons presumably remain only indirectly affected to the cortical blockade, which may account in part for the changes that remain. P2 is within a few days of when thalamocortical synapses stop exhibiting robust LTP in layer IV of the barrel cortex,[23] and before cortical cells can be induced to show robust responses to sensory stimulation (usually PND6-7,[10]). However, partial sensory deprivation early in life prevents whisker pairing plasticity at maturity in layers II/III, but not in layer IV[123] and also prevents other forms of potentiation in adolescent (35 day) rats.[125] One explanation for the layer II/III failure in plasticity could be that SD conditions during the early postnatal period can prevent LTP due to improper timing of PSP's generated by layer IV inputs and action potentials generated by the layer II/III neurons: excitatory postsynaptic potentials (EPSP's) leading back-propagating action potentials

(AP's) by up to 15 msec produce LTP, while LTD is induced when AP's lead EPSP's within a 50 msec window.[143] SD could then create cortical conditions that leave LTD as the dominant synaptic change, which would reduce layer II/III cell responses to a low level and interfere with activity-based plasticity.[144] Plasticity is blocked by NMDA receptor blockade even in adult cortex if blockade occurs during the several days when only two whiskers are intact for WP.[132]

In this context it is interesting that Dolan and Cahusac[145] were able to show that 24 d of whisker trimming in adult rats produced cells in the deprived barrels that were greatly reduced in their sensitivity to applied glutamate agonists: glutamate, NMDA, AMPA or kainate. In the same preparations thc responses to test/conditioning whisker stimuli from 30 to 200 msec did not show changes in intracortical inhibition.

2. Effects of Sensory Deprivation on Inhibitory Neurotransmission

Activity-based cortical excitation leading to plasticity can only take place when inhibitory processes do not block the ability of excitation to enhance synaptic strength. SD plasticity in the barrel cortex is blocked when all activity is suppressed by the release of a GABA receptor agonist, muscimol, onto the surface of the cortex.,[93] Under these conditions thalamocortical transmission remains unaffected. One result of sensory manipulations is to produce a decrease in GABAa receptor beta2 and beta3 receptor subunit immunoreactivity after whisker follicle damage, but not after neonatal or adult whisker trimming.[146] This result again raises the question of whether pulling out or plucking the whiskers with the possibility of attendant follicle damage is equivalent to whisker trimming. Akhtar and Land[147] made an important (but as yet unexplained) observation that cells in the neonatal sensory deprived barrels failed, at maturity, to show activity-induced up- and down-regulation of glutamic acid decarboxylase (GAD), the synthetic enzyme for GABA. In these studies, SD was also induced by C-row or all but C-row trimming for 5–10 weeks starting at PNDo (neonatal) or PND60 (adult), but without a recovery period after either starting age. The effect of early, partial sensory deprivation in these cases was that GABA mechanisms appear to develop to a certain level of efficacy after deprivation, but at the same time are incapable of responding appropriately to changes in activity levels when the system was challenged later in life. A final set of experiments confirmed this deficit when neonatal deprived animals had their whiskers trimmed for 6 weeks, allowed to regrow and stabilize for two months, then row C was retrimmed for six weeks as adults before assaying the GAD levels. The row C whisker barrels showed GAD staining as expected for normal animals (no reduction), an effect that is most easily interpreted as abnormal GAD regulation because of the early deprivation producing a permanent deficiency in ability to adjust GAD levels based on activity.

Micheva and Beaulieu[118,119] found bilateral changes in layer IV GABAergic circuitry after neonatal whisker plucking, combined with an increase number of GABA-positive axon terminals on dendritic spines (GABA-positive spine synapses are rare except in layer IV barrels, and even there are only a small percentage of the total). Welker and his colleagues recently reported that 24 h of whisker stimulation in adult mice, by itself, can increase synaptic density in barrels by 36%, and both excitatory and inhibitory synapses are elevated.[117]

One of the persistent questions about SD is whether the effects seen in the cortex originate in subcortical structures, either the thalamus or the brainstem trigeminal complex relays to cortex or even in the primary afferents and associated follicle receptors which are still immature on the day of birth. Subtle, but potentially important effects of trimming whiskers have been described affecting the development of trigeminal nuclei after whisker trimming for 60 d with no recovery period.[148] Here the primary sensory fibers in the trigeminal nerve appeared normal, but local circuit neurons in the spinal trigeminal nucleus ($SpV_{interpolaris}$) showed a significant increase in multi-whisker receptive fields and many projection neurons showed discontinuous receptive fields with convergent activation by other facial structures, such as the nose. Cytochrome oxidase staining and levels of glutamatic acid decarboxylase reflect the history of low levels of activity in the VPM nucleus of the thalamus after long term sensory deprivation,[149] but when thalamic plasticity was tested after trimming whiskers in a checkerboard pattern, predictable changes were again produced in the cortex, but there were no changes detected in the receptive field or other response properties assayed in the thalamus.[150] Simons and Land[151] found little effect of deprivation on thalamic cell responses, and suggested that abnormalities in deprived barrels reflect changes that take place in the cortex itself.

SD interferes with circuit development such that the deprived cortex never undergoes the last stage of development leading to normal behavior using the deprived cortex. The cellular and molecular basis of SD deficits is becoming better understood, but the basis for prolonged impairments into adulthood remains unexplained. A goal for future research is to determine whether SD impairments can be reversed by activity-rich challenges later in life and/or by pharmacological manipulations that facilitate synaptic plasticity at the low activity levels found in the deprived cortex.

ACKNOWLEDGMENT: We want to thank Dr. Karl Svoboda for his good suggestions on an earlier version of this manscript.

REFERENCES

1. Woolsey, T.A. and Van der Loos, H., The structural organization of layer IV in the somatosensory region (SI) of mouse cerebral cortex. The description of a cortical field composed of discrete cytoarchitectonic units. Brain Res., 17, 205, 1970.
2. Dehnhardt, G., Mauck, B., and Bleckman, H., Seal whiskers detect water movements. Nature, 394, 235, 1988a.
3. Dehnhardt, G., Mauck, B., and Hyvarinen, H. Ambient temperature does not affect the sensitivity of mystacial vibrissae in harbor seals. J. Exp. Biol., 201, 3023, 1988b.
4. Li, X., Glazewski, S., Lin, X., Elde, R., and Fox, K. Effect of vibrissae deprivation on follicle innervation, neuropeptide synthesis in the trigeminal ganglion, and SI barrel cortex plasticity. *J. Comp. Neurol.*, 357, 465, 1995.
5. Renehan, W.E. and Munger, B.L., Degeneration and regeneration of peripheral nervein the rat trigeminal system. II. Reseponse to nerve lesions. *J. Comp. Neurol.*, 249, 429, 1986.

6. Shetty, P., Shoykhet, M., and Simons, D.j., Whisker plucking alters responses of rat trigeminal ganglion neurons. Somatosens. Mot Res., 20, 233, 2003.
7. Woolsey, T.A. and Wann, J.R., Areal changes in mouse cortical barrels following vibrissal damage at different postnatal ages. *J. Comp. Neurol.*, 170, 53, 1976.
8. Waite, P. M. E., Rearrangement of neuronal responses in the trigeminal system of the rat following peripheral nerve section. *J. Physiol.* (Lond), 352, 425, 1984.
9. Waite, P.M.E. and Taylor, P.K., Removal of whiskers in young rats causes functional changes in cerebral cortex. Nature, 274, 600, 1978.
10. Armstrong-James, M., The functional status and columnar organization of single cells responding to cutaneous stimulation in neonatal rat somatosensory cortex. SI. *J. Physiol.*, 246,501, 1975.
11. Feldman, D.E., Nicoll, R.A., and Malenka, R.C., Synaptic plasticity at thalamocortical synapses in developing rat somatosensory cortex: LTP, LTD, and silent synapses. J. Neurobiol., 41, 92, 1999.
12. Gibson, J.M. and Welker, W.I., quantitative studies of stimulus coding in first-order vibrissa afferents of rats. 2. Adaptation and coding of stimulus parameters. Somatosens. Res., 1, 95, 1983.
13. 13. Ebner, F.F. and Armstrong-James, M., Effect of locus and oscillation damping during whisker stimulation on the response of barrel field cortex neurons in urethane anesthetized adult rats. Soc. for Neurosci. Abstr., 29, 481, 2003.
14. Pidoux, B. and Verley, R., Projections on the cortical somatic I barrel subfield from ipsilateral vibrissae in adult rodents. EEG and Clin. Neurophysiol., 46, 715, 1979.
15. Shuler, M.G., Krupa, D.J., and Nicolelis, M.A.L., Bilateral integration of whisker information in the primary somatosensory cortex of rats. *J. Neurosci.*, 21, 5251, 2001.
16. Calford, M.B., and Tweedale, R., Immediate and chronic changes in responses of somatosensory cortex in adult flying-fox after digit amputation. Nature, 332, 446, 1988.
17. Shin, H.C., Won, C.K., Jung, S.C., Oh, S., Park, S. and Sohn, J.H., Interhemispheric modulation of sensory transmission in the primary somatosensory cortex of rats. Neurosci. Lett., 230, 137, 1997.
18. Li, L., Rema, V., and Ebner, F.F., Chronic suppression of activity in barrel field cortex down regulates sensory responses in contralateral barrel field., Soc. Neurosci. Abstr., 27, 393, 2001.
19. Li, L., Rema, V., and Ebner, F.F., Inter-hemispheric neural activity is required for normal responsiveness in barrel field cortex. Soc. Neurosci. Abstr., 28, 45.4, 2002.
20. LoTurco, J.J., Blanton, M.G. and Kriegstein, A.R., Initial expression and endogenous activation of NMDA channels in early neocortical development. *J. Neurosci.*, 11:792, 1991.
21. Kim, H.G., Fox, K. and Connors, B.W., Properties of excitatory synaptic events in neurons of primary somatosensory cortex of neonatal rats. Cerebral Cortex, 2, 148, 1995.
22. Agmon A. and O'Dowd D.K., Receptor-mediated currents are prominent in the thalamocortical synaptic response before maturation of inhibition. *J. Neurophysiol.*, 68, 345, 1992.
23. Crair, M.C. and Malenka, R.C., A critical period for long-term potentiation at thalamocortical synapses. Nature, 375, 325, 1995.
24. Sheng, M., Cummings, J., Roldan, L.A., Jan, Y.M., and Jan, L.Y., Changing subunit composition of heteromeric NMDA receptors during development of rat cortex. Nature, 368, 144, 1994.

25. Tongiorgi, E., Ferrero, F., Cattaneo, A., and Domenici, L., Dark-rearing decreases NR2A N-Methyl-D-Aspartate receptor subunit in all visual cortical layers. Neuroscience, 119, 1013, 2003.
26. Isaac J.T., Crair M.C., Nicoll R.A., Malenka R.C., Silent synapses during development of thalamocortical inputs. Neuron, 18, 269, 1997.
27. Crissman, R.S., Warden, R.J., Siciliano, D.A., Klein, B.G., Renehan, W.E., Jacquin, M.F.,and Rhoades, R.W., Numbers of axons innervating mystacial vibrissa follicles in newborn and adult rats. *Somatosens. Mot. Res.*, 8, 103, 1991.
28. 28. Munger, B.L. and Rice, F.L., Waves of differentiation of cutaneous afferents in rat mystacial skin. *J. Comp. Neurol.*, 252, 404, 1986.
29. Rice, F. L., Mance, A., and Munger, B., A comparative light microscopic analysis of the sensory innervation of the mystacial pad. I. Innervation of the vibrissal follicle-sinus complex. *J. Comp. Neurol.*, 252, 154,1986.
30. Arvidsson, J. and Rice, F.L., Central projections of primary sensory neurons innervating different parts of the vibrissae follicles and intervibrissal skin on the mystacial pad of the rat. *J. Comp. Neurol.*, 309, 1, 1991.
31. Carvell, G. E. and Simons, D. J., Task-and subject-related differences in sensorimotor behavior during active touch. Somatosens. Motor Res., 12, 1, 1995.
32. Krupa, D.J., Matell, M.S., Brisben, A.J., Oliveira, L.M., and Nicolelis, M.A.L., Behavioral properties of the trigeminal somatosensory system in rats performing whisker-dependent tactile discriminations. *J. Neurosci.*, 21, 5752, 2001.
33. Armstrong-James, M., The nature and plasticity of sensory processing within adult rat barrelcortex. The barrel cortex of rodents. Volume 11., Eds. Jones, E.G. and Diamond, I.T., Plenum Press, New York, 1995, Chapter 8, p. 333.
34. Simons, D. J., Neuronal intergration in the somatosensory whisker/barrel cortex. The barrel cortex of rodents. Volume 11. Eds. Jones, E.G. and Diamond, I.T., Plenum Press, New York, 1995, Chapter 5, p. 263.
35. Sachdev, R.N.S., Jenkinson, E., Zeigler, H.P., and Ebner, F.F., Sensorimotor plasticity in the rodent vibrissa system. The mutable brain: dynamic and plastic features of the developing and mature brain., ed., Jon Kaas. Harwood Academic Publishers, Amsterdam, 2001, Chapter 4, p. 123.
36. Jacquin, M. F., Renehan, W. E., Rhoades, R. W. and Panneton, W. M., Morphology and topography of identified primary afferents in rat trigeminal subnuclei principalis and oralis. *J. Neurophysiol.*, 55, 1911, 1993.
37. Hayashi, H., Morphology of central terminations of intra-axonally stained large myelinated primary afferent fibers from facial skin in the rat. *J. Comp. Neurol.*, 237, 195, 1985.
38. Jacquin, M. F., Renehan, W. E., Mooney, R. D., and Rhoades, R. W., Structure-function relationships in rat medullary and cervical dorsal horns. 1. Trigeminal primary afferents. *J. Neurophysiol.*, 55, 1153, 1986.
39. Renehan, W. E., Jacquin, M. F., Mooney, R. D., and Rhoades, W. R., Structure-function relationships in rat medullary and cervical dorsal horns. II. Medullary dorsal horn cells. *J. Neurophysiol.*, 55, 1187, 1986.
40. Gibson, J. M., A quantitative comparison of stimulus-response relationships of vibrissa-activated neurons in subnucleus oralis and interpolaris of the rat's trigeminal sensory complex. Receptive field properties and threshold determinations. Somatosens. Mot Res., 5, 135, 1987.
41. Jacquin, M. F., Golden, J. P., and Panneton, W. M., Structure and function of barrel "precursor" cells in trigeminal nucleus principalis. Dev. Brain Res., 43, 309, 1988.

42. Jacquin, M. F. and Rhoades, R. W., Cell structure and response properties in the trigeminal subnucleus oralis. Somatosens. Mot Res., 7, 265, 1990.
43. Jacquin, M. F., Barcia, M. and Rhoades, R. W., Structure-function relationships in rat brainstem subnucleus interpolaris. IV. Projection neurons. *J. Comp. Neurol.*, 282, 45, 1989.
44. Waite, P. M. E., The responses of cells in the rat thalamus to mechanical movement of the whiskers. *J. Physiol.* (Lond), 228, 541, 1973.
45. Van der Loos, H., Barreloids in mouse somatosensory thalamus. Neurosci. Lett., 2, 1, 1976.
46. Sugitani, M. Y., Sugai, T., and Ooyama, H., Somatotopic organization and columnar structure of vibrissae representation in the rat ventrobasal complex. Exp. Brain Res., 81, 346, 1990.
47. Land, P. W., Buffer, S. A., and Yaskosky, J. D., Barreloids in adult rat thalamus: three-dimensional architecture and relationship to somatosensory cortical barrels. *J. Comp. Neurol.*, 355, 573, 1995.
48. Waite, P. M. E., Somatotopic organization of vibrissal responses in the ventro-basal complex of the rat thalamus. *J. Physiol.* (Lond), 228, 527, 1973.
49. Ito, M., Response properties and topography of vibrissa-sensitive VPM neurons in the rat. *J. Neurophysiol.*, 60, 1181, 1988.
50. Simons, D. J. and Carvell, G. E., Thalamocortical response transformation in the rat vibrissa/barrel system. *J. Neurophysiol.*, 61, 311, 1989.
51. Diamond, M.E., Armstrong-James, M., and Ebner, F.F., Somatic Sensory Responses in the Rostral Sector of the Posterior Group (POM) and in the Ventral Posterior Medial Nucleus (VPM) of the Rat Thalamus. J. Comp. Neur., 318, 462, 1992.
52. Friedberg, M.H., Lee, S.M., and Ebner, F.F., Modulation of receptive field properties of thalamic somatosensory neurons by the depth of anesthesia. *J. Neurophysiol.*, 81, 2243,1999.
53. Nicolelis, M.A.L., Baccala, L. A., Lin, R. C. S., and Chapin, J. K., Sensorimotor encoding by synchronous neural ensemble activity at multiple levels of the somatosensory system. Science, 268, 1353, 1995.
54. Erzurumulu, R. S., Bates, C. A. and Killackey, H. P., Differential organization of thalamic projection cells in the brain stem trigeminal complex of the rat. Brain Res., 198, 427, 1980.
55. Pierret, T., Lavallee, P., and Deschenes, M., Parallel streams for the relay of vibrissal information through thalamic barreloids. *J. Neurosci.*, 20, 7455, 2000.
56. Lu, S. M. and Lin, R. C., Thalamic afferent of the rat barrel cortex: A light and electron-microscopic study using Phaseolus vulgaris leucoagglutinin as an anterograde tracer. Somatosens. Mot Res., 10, 1, 1993.
57. Jensen, K.F. and Killackey, H.P., Terminal arbors projecting to the somatosensory cortex of the adult rat. I. The normal morphology of specific thalamocortical afferents. *J. Neurosci.*, 7, 3529, 1987.
58. Koralek, K. J., Jensen, K.F., and Killackey, H. P., Evidence for two complementary patterns of thalamic input to the rat somatosensory cortex. Brain Res., 463, 446, 1988.
59. Koralek, K. J., Olavarria, J., and Killackey, H. P., Areal and laminar organization of corticocortical projections in the rat somatosensory cortex. *J. Comp. Neurol.*, 299, 133, 1990.
60. Hayama, T. and Ogawa, H., Regional differences of callosal connections in the granular zones of the primary somatosensory cortex in rats. Brain Res. Bull., 43, 341, 1997.

61. Price, T. R. and Webster, K. E., The cortico-thalamic projection from the primary somatosensory cortex of the rat. Brain Res., 44, 636, 1972.
62. Wise, S. P. and Jones, E. G., Cells of origin and terminal distribution of descending projections of the rat somatic sensory cortex. *J. Comp. Neurol.*, 175, 129, 1977.
63. Chmielowska, J., Carvell, G. E. and Simons, D. J., Spatial organization of thalamocortical and corticothalamic projection systems in the rat Sm1 barrel cortex. *J. Comp. Neurol.*, 285, 325, 1989.
64. Bourassa, J., Pinault, D., and Deschenes, M., Corticothalamic projections from cortical barrel-field to the somatosensory thalamus in rats: A single fibre study using biocytin as an anterograde tracer. Eur. *J. Neurosci.*, 7, 19, 1995.
65. Hoogland, P. V., Welker, E., Van der Loos, H. and Wouterlood, F. G., Organization of the projections from the barrel cortex to thalamus in mice studied with Phaseolus vulgaris-leucoagglutinin and HRP. Exp. Brain Res., 68, 73, 1987.
66. Killackey, H.P. and Sherman, S.M., Corticothalamic projections from the rat primary somatosensory cortex. *J. Neurosci.*, 23, 7381, 2003.
67. Simons, D. J., Response properties of vibrissae units in rat SI somatosensory neocortex. *J. Neurophysiol.*, 41, 798, 1978.
68. Ito, M., Some quantitative aspects of vibrissa driven neuronal responses in rat neocortex. *J. Neurophyiol.*, 46, 705, 1981.
69. Ito, M., Processing of vibrissa sensory information within the rat neocortex. J Neurophysiol., 54, 479, 1985.
70. Chapin, J.K., Laminar differences in sizes, shapes and response profiles of cutaneous receptive fields in the rat SI cortex. Experimental Brain Research, 63, 549, 1986.
71. Armstrong-James, M. and Fox, K., Spatiotemporal convergence and divergence in the rat SI "barrel" cortex. *J. Comp. Neurol.* 263, 265, 1987.
72. Brumberg, J.C., Pinto, D.J., and Simons, D.J., Cortical columnar processing in the rat whisker-to-barrel system. *J. Neurophysiol.*, 82, 1808, 1999.
73. Brecht, M. and Sakmann, B., Dynamic representation of whisker deflection by synaptic potentials in spiny stellate and pyramidal cells in the barrels and septa of layer 4 rat somatosensory cortex. *J. Physiol.*, 543, 49, 2002.
74. Kim, U. and Ebner, F.F., Barrels and Septa: Separate circuits in rat barrel field cortex. J. Comp. Neurol, 408, 489, 1999.
75. Petersen, C.C. and Sakmann, B., Functionally independent columns of rat somatosensory barrel cortex revealed with voltage-sensitive dye imaging. *J. Neurosci.*, 21, 8435, 2001.
76. Legg, C. R., Mercier, B., and Glickstein, M., Corticopontine projection in the rat: the distribution of labeled cortical cells after large injections of horseradish peroxidase in the pontine nuclei. *J. Comp. Neurol.*, 286, 427, 1989.
77. Moore, C.I., Nelson, S.B., and Sur, M., Dynamics of neuronal processing in rat somatosensory cortex. Trends Neurosci., 22, 513, 1999.
78. Arnold, P.B., Li, C.X., and Waters, R.S., Thalamocortical arbors extend beyond single cortical barrels: an *in vivo* intracellular tracing study in rat. Exp. Brain Res., 136,152, 2001.
79. Armstrong-James, M., Welker, E., and Callahan, C.A., The contribution of NMDA and non-NMDA receptors to fast and slow transmission of sensory information in the rat SI barrel cortex. *J. Neurosci.*, 13, 2149, 1993.
80. McCormick, D.A., Connors, B.W., Lighthall, J.W., Prince, D.A., Correlations between electrophysiology of pyramidal and sparsely spiny stellate neurons of the neocortex. *J. Neurophysiol.*, 54, 782, 1985.

81. Connors, B.W, Gutnick, M.J., and Prince, D.A., Electrophysiological properties of neocortical neurons in vitro. *J. Neurophysiol.* 48, 1302, 1982.
82. Zhu, J.J. and Connors, B.W., Intrinsic firing patterns and whisker-evoked synaptic responses of neurons in rat barrel cortex. *J. Neurophysiol.*, 81, 1171, 1999.
83. Swadlow, H.A., Efferent neurons and suspected interneurons in SI vibrissa cortex of the awake rabbit: receptive fields and axonal properties. *J. Neurophysiol.*, 62, 288, 1989.
84. Sachdev, R.N.S., Sellien, H., and Ebner, F.F., Inhibition evoked by whisker stimulation in somatic sensory (SI) barrel field cortex of the awake rat. *J. Neurophysiol.*, 84, 1497, 2000.
85. Shimegi, S., Ichikawa, T., Akasaki, T., and Sato, H., Temporal characteristics of response integration evoked by multiple whisker stimulations in the barrel cortex of rats. *J. Neurosci.*, 19, 10164, 1999.
86. Shimegi, S., Akasaki, T., Ichikawa, T., and Sato, H., Physiological and anatomical organization of multi-whisker response interactions in the barrel cortex of rats. *J. Neurosci.*, 20, 6241, 2000.
87. Laaris, N. and Keller, A., Functional Independence of layer IV barrels. *J. Neurophysiol.*, 87, 1028, 2002.
88. Armstrong-James, M., Diamond, M., and Ebner, F. F., An innocuous bias in whisker use in adult rats modifies receptive fields of barrel cortex neurons. *J. Neurosci.*, 14, 6978, 1994.
89. Armstrong-James, M. and Callahan, C., Thalamocortical mechanisms in the formation of receptive fields of rat barrel cortex neruones. II The contribution of ventroposterior medial thalamic (Vpm) neurones. *J. Comp. Neurol.*, 303, 211, 1991.
90. Armstrong-James, M., Callahan, C. and Friedman, M., Thalamocortical mechanisms in the formation of receptive fields of rat barrel cortex neruones. I. Intracortical mechanisms. *J. Comp. Neurol.*, 303, 193, 1991.
91. Rema, V. and Ebner, F.F., Effect of enriched environment rearing on impairments in cortical excitability and plasticity after prenatal alcohol exposure. *J. Neurosci.*,19, 10993, 1999.
92. Fox, K., The cortical component of experience-dependent synaptic plasticity in the rat barrel cortex. *J. Neurosci.*,14, 7665, 1994.
93. Wallace, H., Glazewski, S., Liming, K., and Fox, K., The role of cortical activity in experience-dependent potentiation and depression of sensory responses in rat barrel cortex. *J. Neurosci.*, 21, 3881, 2001.
94. Armstrong-James, M., Fox, K., and Das-Gupta, A., Flow of excitation within barrel cortex on striking a single vibrissa. *J. Neurophysiol.*, 68,1345, 1992.
95. 95. Harris, J.A., Petersen, R.S. and Diamond, M.E., Distribution of tactile learning and its neural basis. Proc. Natl. Acad., Sci. USA 96, 7587, 1999.
96. Nicolelis, M.A., Chapin, J.K., and Lin, R.C., Thalamic plasticity induced by early whisker removal in rats. Brain Res., 561, 344, 1991.
97. Diamond, M.C., Enriching heredity: the impact of the environment on the anatomy of the brain. Free Press, New York, 1986.
98. Hubel, D.H., Wiesel, T.N., and LeVay, S., Plasticity of ocular dominance columns in monkey striate cortex. Philos. Trans. R. Soc. Lond. B, 278, 377, 1977.
99. LeVay, S., Wiesel, T.N., and Hubel, D.H., The development of ocular dominance columns in normal and visually deprived monkeys. *J. Comp. Neurol.*, 191, 1, 1980.
100. Lowel, S. and Singer, W., Selection of intrinsic horizontal connections in the visual cortex by correlated neuronal activity. Science, 255, 209, 1992.

101. Huang, W. and Ebner, F.F., The influence of early whisker sensory deprivation on cortical plasticity in the adult rat. Soc. Neurosci. Abstr., 54, 23, 1995.
102. Bienenstock EL, Cooper LN, and Munro PW., Theory for the development of neuron selectivity: orientation specificity and binocular interaction in visual cortex. *J. Neurosci.*, 2, 32, 1982.
103. Benuskova, L., Velayudhan, R., Armstrong-James, M., and Ebner, F., Theory for normal and impaired experience-dependent plasticity in neocortex of adult rats. Proc. Nat. Acad. Sci., 98, 2797, 2001.
104. Hutson, K.A. and Masterton, R.B., The sensory contribution of a single vibrissa's cortical barrel. *J. Neurophysiol.*, 56, 1196, 1986.
105. Carvell, G.E. and Simons, D.J., Abnormal tactile experience early in life disrupts active touch. *J. Neurosci.*, 16, 2750, 1996.
106. Erchova, I.A., Petersen, R.S., and Diamond, M.E., Effect of developmental sensory and motor deprivation on the functional organization of adult rat somatosensory cortex. Brain Res. Bull., 60, 373, 2003.
107. Berg, R. and Kleinfeld, D., Rhythmic whisking by rat: retraction as well as protraction of the vibrissae is under active muscular control. *J. Neurophysiol.,* 89, 104, 2003.
108. Mineo, L., Melzer, P., and Ebner, F.F., Whisker rate of blind rats is accelerated during object recognition. Soc. Neurosci. Abstr., 257, 9, 2002.
109. Shepherd GM, Pologruto TA, and Svoboda K., Circuit analysis of experience-dependent plasticity in the developing rat barrel cortex. Neuron, 38, 277, 2003.
110. Stern, E.A., Maravall, M., and Svoboda, K., Rapid development and plasticity of layer 2/3 maps in rat barrel cortex *in vivo.* Neuron, 31, 305, 2001.
111. Fox, K., Glazewski, S., Chen, C.M., Silva, A., and Li, X., Mechanisms underlying experience-dependent potentiation and depression of vibrissae responses in barrel cortex. J. Physiology, Paris, 90, 263, 1996.
112. Finnerty, G.T. and Connors, B.W., Sensory deprivation without competition yields modest alterations of short-term synaptic dynamics. *Proc. Natl. Acad. Sci. USA*, 97, 12864, 2000.
113. Schierloh, A., Eder, M., Zieglgansberger, W., and Dodt, H.-U., Sensory deprivation changes the pattern of synaptic connectivity in rat barrel cortex. NeuroReport, 14, 1787, 2003.
114. Miller, B., Blake, N.M., Erinjeri, J.P., Reistad, C.E., Sexton, T., Admine, P., and Woolsey, T.A., Postnatal growth of intrinsic connections in mouse barrel cortex. *J. Comp. Neurol.*, 436, 17, 2001.
115. McCasland, J.S., Bernardo, K.L., Probst, K.L. and Woolsey, T.A., Cortical local circuit axons do not mature after early deafferentation. Proc. Natl. Acad. Sci. USA, 89, 1832, 1992.
116. Sadaka, Y., Weinfeld, E., Lev, D.L., and White, E.L., Changes in mouse barrel synapses consequent to sensory deprivation from birth. *J. Comp. Neurol.*, 457, 75, 2003.
117. Knott, G.W., Quairiaux, C., Genoud, C., and Welker, E., Formation of dendritic spines with GABAergic synapses induced by whisker stimulation in adult mice. Neuron, 34, 265, 2002.
118. Micheva, K.D. and Beaulieu, C., Neonatal sensory deprivation induces selective changes in the quantitative distribution of GABA-immunoreactive neurons in the rat barrel field cortex. *J. Comp. Neurol.*, 361, 574, 1995.
119. Micheva, K.D. and Beaulieu, C., Development and plasticity of the inhibitory neocortical circuitry with an emphasis on the rodent barrel field cortex. Canadian *J. Physiol.* Pharmacol., 75, 470, 1997.

120. Vees, A.M., Micheva, K.D., Beaulieu, C., and Descarries, L., Increased number and size of dendritic spines in ipsilateral barrel field cortex following unilateral whisker trimming in postnatal rat. *J. Comp. Neurol.*, 400, 110, 1998.
121. Simons, D. J. and Land, P. W., Early experience of tactile stimulation influences organization of somatic sensory cortex. Nature, 326, 694, 1987.
122. Fox, K., A critical period for experience-dependent synaptic plasticity in rat barrel cortex. *J. Neurosci.*, 12, 1826, 1992.
123. Rema, V., Armstrong-James, M. and Ebner, F.F., Experience-dependent plasticity is impaired in adult rat barrel cortex after whisker disuse in early postnatal life. *J. Neurosci.*, 23, 358, 2003.
124. Glazewski, S. and Fox, K., Time course of experience-dependent synaptic potentiation and depression in barrel cortex of adolescent rats. *J. Neurophysiol.*, 75, 1714, 1996.
125. Glazewski, S., McKenna, M., Jacquin, M., and Fox, K., Experience-dependent depression of vibrissae responses in adolescent rat barrel cortex. Eur. *J. Neurosci.*, 10, 2107, 1998.
126. Glazewski, S., Herman, C., McKenna, M., Chapman, P.F., and Fox, K., Long-term potentiation *in vivo* in layers II/III of rat barrel cortex. Neuropharmacol., 37, 581, 1998.
127. Wallace, H. and Fox, K., Local cortical interactions determine the form of cortical plasticity. J. Neurobiol., 41, 58, 1999.
128. Benuskova, L., Ebner, F.F., Diamond, M.E., and Armstrong-James, M., Computational study of experience-dependent plasticity in adult rat cortical barrel column. Network: Comput. Neural Syst., 10, 303, 1999.
129. Diamond, M. E., Armstrong-James, M., and Ebner, F. F., Experience-dependent plasticity in adult rat barrel cortex. Proc. Natl. Acad. Sci., 90, 2082, 1993.
130. Diamond, M.E., Huang, W., and Ebner, F.F., Laminar comparison of somatosensory cortical plasticity. Science, 265, 1885, 1994.
131. Sellien, H. and Ebner, F.F., Short term changes in barrel field cortex neurons after whisker pairing in awake adult rats. Soc. Neurosci. Abstr., 27, 442, 2003.
132. Rema, V., Armstrong-James, M., and Ebner, F.F., Experience-dependent plasticity of adult rat S1 cortex requires local NMDA receptor activation. *J. Neurosci.,* 18, 10196, 1998a.
133. Rema, V., Armstrong-James, M., Jenkinson, E. W. and Ebner, F. F., Short-term enriched experience potentiates whisker-pairing plasticity of barrel cortex neurons in the adult rat. Soc. Neurosci. Abstr., 22, 320, 1998b.
134. Hand, P.J., Plasticity of the rat cortical barrel system. Changing concepts of the nervous system. eds., Strick, P. and Morrison, A., Academic Press, New York, 1982, p. 49.
135. Kossut, M., Glazewski, S., Siucinska, E., and Skangiel-Kramska, J., Functional plasticity and neurotransmitter receptor binding in the vibrissal barrel cortex. Acta Neurobiol. Exp., 53, 161, 1993.
136. Land, P.W. and Simons, D.J., Metabolic activity in SmI cortical barrels of adult rats is dependent on patterned sensory stimulation of the mystacial vibrissae. Brain Res., 341, 189, 1985.
137. Kossut, M., Effects of sensory deprivation upon a single cortical vibrissal column: a 2-DG study. Exp. Brain Res., 90, 639, 1992.
138. Skibinska, A., Glazewski, S., Fox, K., and Kossut, M., Age-dependent response of the mouse barrel cortex to sensory deprivation: a 2-deoxyglucose study. Exp. Brain Res., 132, 134, 2000.
139. Polley, D.B., Chen-Bee, C.H., and Frostig, R.D., Two directions of plasticity in the sensory-deprived adult cortex. Neuron, 24, 623, 1999.

140. Polley, D.B., Kvasnak, E., and Frostig, R.D., Naturalistic experience transforms sensory maps in the adult cortex of caged animals. Nature, 429, 67, 2004.
141. Yazawa, I., Sasaki, S., Mochida, H., Kamino, K., Momose-Sato, Y., and Sato, K., Developmental changes in trial-to-trial variations in whisker barrel responses studied using intrinsic optical imaging: comparison between normal and de-whiskered rats. *J. Neurophysiol.*, 86, 92, 2001.
142. Schlaggar, B.L., Fox, K., and O'Leary, D.M.M., Postsynaptic control of plasticity in developing somatosensory cortex. Nature, 364, 623, 1993.
143. Feldman, D.E., Timing-based LTP and LTD at vertical inputs to layer II/III pyramidal cells in rat barrel cortex. Neuron, 27, 45, 2000.
144. Allen, C.B., Tansu, C., and Feldman, D.E., Long-term depression induced by sensory deprivation during cortical map plasticity *in vivo. Nat. Neurosci.*, 6, 291, 2003.
145. Dolan, S. and Cahusac, P.M., Differential effect of whisker trimming on excitatory and inhibitory transmission in primary somatosensory cortex of the adult rat *in vivo.* Neuroscience, 70, 1714, 1996.
146. Land, P.W., de Blas, A.L., and Reddy, N., Immunocytochemical localization of GABAA receptors in rat somatosensory cortex and effects of tactile deprivation. *Somatosens. Mot. Res.*, 12, 127, 1995.
147. Akhtar, N.D., and Land, P.W., Activity-dependent regulation of glutamic acid decarboxylase in the rat barrel cortex: effects of neonatal versus adult sensory deprivation. *J. Comp. Neurol.*, 307, 200, 1991.
148. Jacquin, F., Rhoades, R. W., and Klein, B. G., Structure-function relationships in rat brainstem subnucleus interpolaris XI. Effects of chronic whisker trimming from birth. *J. Comp. Neurol.*, 356, 200, 1995.
149. Land, P.W. and Akhtar, N.D., Chronic sensory deprivation affects cytochrome oxidase staining and glutamic acid decarboxylase immunoreactivity in adult rat ventrobasal thalamus. Brain Res., 425, 178, 1987.
150. Fox, K., Wallace, H., and Glazewski, S., Is there a thalamic component to experience-dependent cortical plasticity? Philos. Trans. R. Soc. Lond. B, 357, 1709, 2002.
151. Simons, D.J. and Land, P.W., Neonatal whisker trimming produces greater effects in nondeprived than deprived thalamic barreloids. *J. Neurophysiol.*, 72, 1434, 1994.

7 Role of Plasticity in Sensorimotor Transformations

Linda Hermer-Vazquez,
Raymond Hermer-Vazquez,
and John K. Chapin

CONTENTS

I. Introduction: Conceptualizing Models of Sensorimotor Transformations 140
 A. Task Analyses 140
 B. Devising Task Analyses Based on Current Understandings of Neural Processing 141
II. Principle I: Sensory and Motor Processing are Constantly Influencing One Another 142
 A. Non-Hierarchical Views of Sensorimotor Processing 142
III. Principle II: Information from Multiple Spatial Scales is Processed Simultaneously 144
IV. Principle III: Laboratory Animals are Constantly Evaluating Information Across Multiple Time Scales in Order to More Accurately Predict What Will Happen in Their World 145
 A. Parallel Processing At Multiple Time Scales 146
V. Principle IV: Action Potentials are not the Only Causal Neural Activity, with Ramifications for Behavior 148
VI. Principle V: Behaviors Appear to be Organized into Survival-Related Repertoires that can be Adapted to Novel Circumstances 149
 A. Motor Organization in Terms of Survival-Related Behaviors 149
 B. Sensorimotor Learning According to this View 150
VII. Final Comments 151
References 151

I. INTRODUCTION: CONCEPTUALIZING MODELS OF SENSORIMOTOR TRANSFORMATIONS

What are the mechanisms by which the brain transforms sensory inputs into motor outputs? The fact that this remains a fundamentally unanswered question is attributable not only to the shear complexity of the sensory and motor systems themselves, but even more to the multiplicity of bridges between these systems in the central nervous system (CNS). Moreover, though it is clear that the development and maintenance of the sensorimotor transformation machinery must involve some sort of plasticity, it is not clear how or where this plasticity occurs. How then should one approach the study of these systems in awake animals performing behaviors involving sensorimotor transformations?

A. Task Analyses

Researchers in cognitive and behavioral neuroscience have realized for decades that in order to carry out insightful behavioral experiments, one needs a working model at the psychological level of how subjects execute their task. This working model, which is often referred to as a task analysis,[21,32] contains such information as the nature of the sensory information used to make decisions, the time scale over which this sensory information is evaluated, the cognitive transformations of those data before the animal signals indicates its choice with a motor output, and the kinematics and dynamics of this choice-signaling movement. A researcher's task analysis provides the framework in which she interprets the outcome of experiments using the modeled task.

Older task analyses tended to be based on a sequential, hierarchical, engineering-related logic that first, the animal's mind receives sensory inputs; next, a transfer function combines that information with information about the animal's memories and goals and transforms it into a motor command; and finally, this motor command is executed. This logic reflected the view of the day that perception, cognition and behavior themselves were organized hierarchically, with information processing progressing from perceptual analysis, to cognition, to motor preparation and execution. Moreover, this view assumed that subjects were passive receivers and processors of information – with mental activity consisting of little more than the stimulus-response reflexes envisioned by the behaviorist psychologists of the early 20th century.

Although such a view might have accurately described the function of early electrical circuits, findings in modern neuroscience have turned nearly every aspect of that view on its head with regard to how the mind-brain functions. During the course of the evolution of life over the last 4 billion years, selective pressures have shaped living organisms at nearly all levels of analysis – from the genomic and protein levels, to the levels of biochemical and neural networks, to the level of the survival- and reproductive-related behavior of individuals and groups within a species. An infinite number of strictly psychological models could describe a given behavioral process, such as how a rodent selects a morsel of food, grasps it, and brings it to its mouth. However, a correct psychological model for how the rodent mind actually performs this task is more likely to be constructed if it is built using knowledge about the structure and function of task-related neural circuits as well as incorporating the relevant cognitive and behavioral data.

B. Devising Task Analyses Based on Current Understandings of Neural Processing

In this chapter, we describe five principles that sketch the beginnings of a more nuanced conception of how sensorimotor behaviors are learned and performed, based on new findings from neuroscience. Although most of the principles we will describe apply to all species with nervous systems, we will focus on the rat for several reasons. First, rats are a good mammalian exemplar, because the earliest mammals of approximately 175 million years ago resembled rats in their size and their heavy reliance on olfaction and somatosensory perception.[25] (Figure 7.1). Also, most of the hypothesized early-mammal and current-rodent neuroanatomy and neurochemistry have been conserved throughout the mammalian radiation. For instance, all mammals including rats have primary and secondary somatosensory, auditory, and visual areas[23] that are neocortical in their layer structure[1] (i.e., they are 6-layered). Furthermore, they all have three-layered olfactory cortices; a highly similar limbic system;

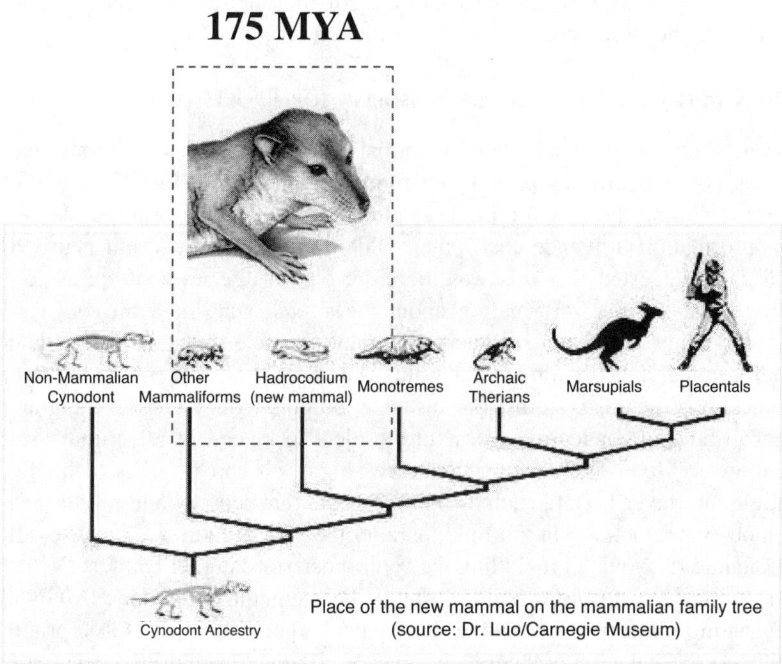

FIGURE 7.1 The evolution of mammals, from non-mammalian cynodonts (also known as protomammals), to the first known mammal (Hadrocodium wui) approximately 175 million years ago, to humans.

a basal ganglia and cerebellum; and four parallel motor pathways (the corticospinal, rubrospinal, vestibulospinal, and reticulospinal tracts). Additionally, there is extensive homology across rats and other mammals in terms of cytoarchitecture, cytochemistry, and the roles played by different neurotransmitters and neuromodulators. In this chapter, we will illustrate the five principles using data from all levels of the rat nervous system, from the sensory periphery and spinal level to higher cognitive levels. Most of our sensory examples will come from the somatosensory modality, although we will discuss other sensory modalities when they illustrate a point well.

II. PRINCIPLE I: SENSORY AND MOTOR PROCESSING ARE CONSTANTLY INFLUENCING ONE ANOTHER

A common view of information flow through the mammalian nervous system is first, sensory information ascends through sensory thalamic relay nuclei (for all modalities except olfaction, where information flows directly to the cortex); next, this information undergoes multiple transformations as it proceeds through various cortical areas processed via lateral layer V connections; and finally, the transformed information descends via multiple motor system tracts to control muscular output. At the behavioral level, this translates into task analyses of the data-in, transfer-function-performed, data out logic described in the introduction.

A. Non-Hierarchical Views of Sensorimotor Processing

However, a closer examination of mammalian brain anatomy suggests that this picture is not only oversimplified, but fundamentally wrong. At all levels of the neuraxis including the spinal cord, ascending sensory information and descending motor information influence each other. Older neuroanatomical and neurophysiological data suggested that this was the case; e.g., at the level of spinal reflexes, proprioceptive sensory information about the stretch state of a muscle has been known for many years to feed back to regulate muscle tension. However, newer neuroanatomical data demonstrate that this is an ever present theme in the structure and function of neural systems. For instance, outputs from the basal ganglia, once thought to be a motor learning structure, project to layer I of all primary sensory cortical areas,[27] likely influencing the processing of all cortical cells with dendrites extending into layer I. Data such as these indicate that sensory and motor areas are continuously interacting via multiple, parallel looping structures. Likewise, all sensory thalamic relay nuclei including the ventroposterior lateral thalamus (VPl) (processing somatosensory information), the medial geniculate thalamus (MGN) (processing auditory information) and the lateral geniculate thalamus (LGN) (processing visual data) receive axon collaterals of layer V efferents from primary motor cortex (M1) and the premotor cortex.[11] A fact that is not widely appreciated is that these axon collaterals constitute a larger portion of inputs to the thalamic relay nuclei than do the ascending sensory fibers! Thus, descending motor commands appear able to modulate or even drive thalamic sensory processing. These facts are among many more pieces of anatomical and physiological data suggesting that sensory and motor processing are constantly modulating each other.

The constant interaction of sensory and motor data streams is manifest in mammalian behavior in myriad instances. For example, the strength of an odor modulates sniffing intensity, even on the first sniff of the odor.[15] Furthermore, this modulation of sniffing intensity occurs so rapidly that it is thought that cortical processing cannot be involved; the modulation must take place at brainstem or spinal levels.[15] In our studies of rats performing an olfactory-driven, reach-to-grasp-food task,[39] the presence or absence of a food-related odor determines whether the rat will lift its paw, and guide the spatial accuracy of the reach.[13] Indeed, rats appear to sniff the target just prior to lifting the paw to gain the initial spatial coordinates for the impending reach, and then take several more update sniffs during the overt arm-movement phases of the task. These behavioral observations have a corresponding neurophysiology. For instance, while recording single units from the digit-wrist area of caudal M1 and the magnocellular red nucleus as rats perform this task, we have found that many neurons in both areas that code the overt arm-movement phases are strongly modulated by olfactory information taken in during the final sniff of the food target just prior to lifting the paw[12] (Figure 7.2).

Somatosensory processing also appears to constantly guide the rats' reaching maneuvers. In fact, the caudal M1 in rats is distinguished by having layer IV somatosensory inputs,[5] whereas most motor regions in mammals lack layer IV sensory inputs. Correspondingly, we have found that many rat M1 units are particularly active during phases of the reaching task in which somatosensory information is being evaluated, such as lifting the paw off the ground, brushing the paw against

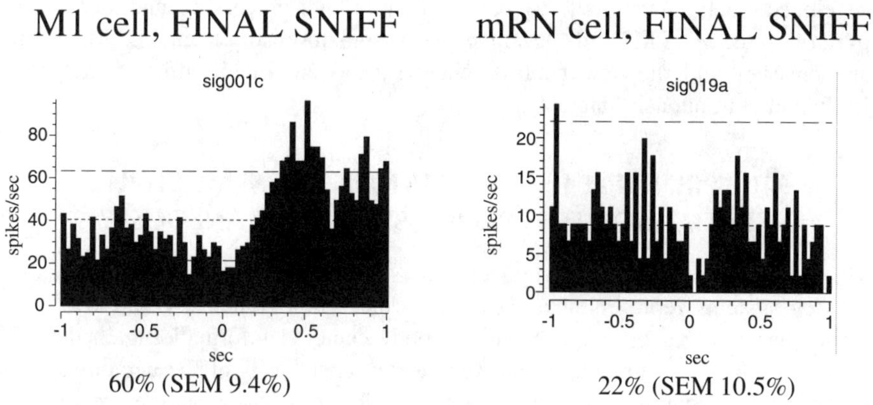

FIGURE 7.2 Perievent histograms for representative single cells in the rat caudal primary motor cortex (M1) and magnocellular red nucleus (mRN), centered around the final sniff of the food pellet before lifting the paw to initiate reaching. In each graph, the center horizontal line depicts the cell's firing rate, and the lines above and below it show two standard deviations from the mean (i.e., statistical significance of rate modulation). It can be seen that at the final sniff moment, each cell's firing rate is significantly depressed. The binwidth for these perievent histograms is 25 msec. The percentages below each graph show the proportion of single units recorded in each area that displayed such significant modulation on the final sniff.

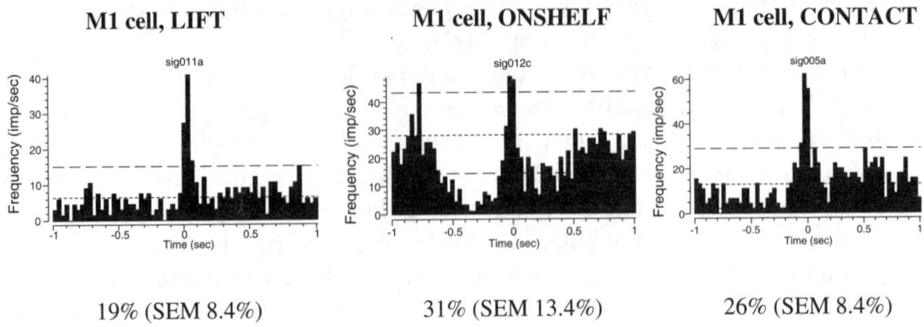

FIGURE 7.3 Perievent histograms for single cells recorded from the rat caudal M1 centered around each of three reaching-task events in which the cutaneous inputs to the paw change and are presumably evaluated by M1 cells: lifting of the paw off the ground (left), the paw making contact with the shelf (middle), and the paw making contact with the food pellet (right). Other elements of the graphs, such as the binwidth, are the same as in Figure 7.2.

the shelf on which the food target rests on the way to the target, and contacting the food pellet itself[12] (Figure 7.3). The role of somatosensory processing in motor cells' activity is especially clear once the task is well learned by the animal. When reaching tasks are first being learned, the latencies of spike rate increases in primary somatosensory cortex (S1) and M1 are relatively consistent with the data in, transfer function performed, data out task model, in that S1 units peak in their firing several tens of milliseconds prior to the M1 cells' peaking. However, in one study we conducted, the inadequacy of that model was strikingly illustrated. As animals became more proficient at the task, roughly one third of all recorded S1 units developed much longer latencies consistent with motor processing, and roughly one third of all recorded M1 cells developed early, somatosensory latencies.[5] These data are consistent with the view that task-related sensory and motor information dynamically and continuously interact.

III. PRINCIPLE II: INFORMATION FROM MULTIPLE SPATIAL SCALES IS PROCESSED SIMULTANEOUSLY

Three or four decades ago, many researchers advocated the view that mammalian sensory systems represent their sensory surfaces topographically and with high resolution, and that the high resolution, body centered information gained at the sensory surface was gradually transformed at higher levels of the nervous system by cells with progressively larger receptive fields into object-centered representations. However, it is now widely recognized that information is processed at multiple spatial scales simultaneously. Again, this is true at all levels of the mammalian nervous system. For instance, at the somatosensory surface, Type I cutaneous receptors process mechanical somatosensory data with high spatial resolution, while Type II receptors have larger, less well defined receptive fields.[38] The principle of simultaneous processing of sensory data at multiple spatial scales holds at higher levels of the nervous system, as it does for the parallel motor pathways, where for instance, the corticospinal pathway is hypothesized to be more critical for fine digit control, whereas the reticulospinal pathway is thought to be more critical for coordinating movements across the entire body.[29]

In many cases, different microcircuits in the CNS – defined by their cell types, neurochemistry, and connectivity – process information at their own, distinctive spatial scales. For instance, it has recently been discovered that in many thalamic nuclei, a matrix of calbindin-immunoreactive cells projects diffusely throughout the cortex, irrespective of sensory topography or even sensory domain.[17] In contrast, the precisely projecting, topographically ordered cells more classically associated with thalamic relay efferents stain positively for parvalbumin. For example, the relay nucleus for body somatosensory afferents, VPL, contains both parvalbumin-positive, precisely projecting, somatotopically organized afferents to layer IV of cortical area SI, with correspondingly small receptive fields, as well as much more widely projecting, calbindin-staining cells whose axons target the superficial cortical layers in multiple sensory modalities. It is now hypothesized that these diffusely projecting cells play a critical role in coordinating activity across brain regions, particularly in view of evidence that many layer IV corticothalamic feedback neurons target these transcortically projecting matrix cells.[17]

These new data on thalamo-cortico-thalamic processing reveal that cortical processing shares more space-related features with subcortical processing than previously realized. For example, much like the thalamic relay nuclei, the striatum of the basal ganglia contains topographically more ordered regions that process information with high spatial resolution (striosomes), as well more diffusely organized regions[26] (matrix cells). And the relative lack of topographical ordering in the cerebellum, where instead distant regions of the body surface are adjacently located in the cerebellar cortex,[19] is thought to promote sensorimotor learning across the whole body. Behavioral tasks therefore either need to be extremely well controlled if they are to be based primarily on the hypothesis of high resolution, topographic processing (e.g. by localizing perception and motion to a small and isolated portion of the body), or their corresponding task models need to account for processing occurring at multiple spatial scales.

IV. PRINCIPLE III: LABORATORY ANIMALS ARE CONSTANTLY EVALUATING INFORMATION ACROSS MULTIPLE TIME SCALES IN ORDER TO MORE ACCURATELY PREDICT WHAT WILL HAPPEN IN THEIR WORLD

It is now widely agreed that mammals – like nearly all species with nervous systems – are constantly re-evaluating the past at different time scales and combining the resulting knowledge with their current perceptions in order to predict the future at different time scales.[24] However, constructing an adequate model of how, for example, rats perform a given task is made more complicated by these insights, as they require modeling the co-occurrence of psychological processes at a variety of interacting time scales, sometimes producting nonlinearlities such as eureka moments.

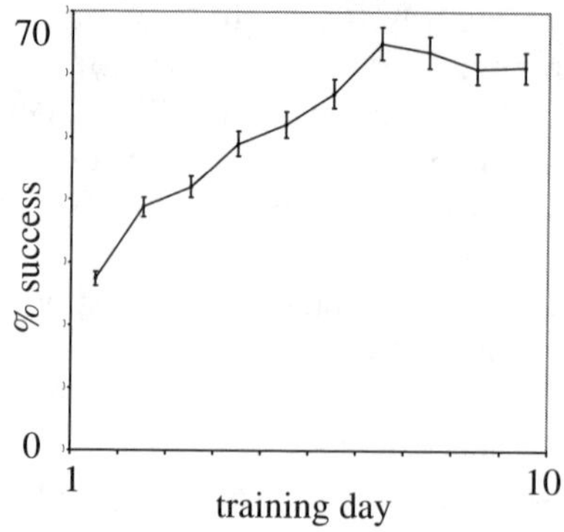

FIGURE 7.4 Percent correct (i.e., percent of trials on which the rat succeeded in grasping and retracting the pellet) over the course of the first 10 training days for a sample of 11 rats. The sample appears to gain skill in a linear fashion over the course of the first six training days, when it then begins to asymptote at its highest performance level.

Many older task models were based on the simplifying assumption that tasks were learned and maintained as a simple function of the number of training trials up through reaching an asymptotic performance level, and degraded as a simple function of lack of practice or interference by new memories.[2] This view suggested that time flows forward linearly, in a regular manner throughout task acquisition and performance, i.e., that the animal is using its experience on the current trial to shape its performance on the next trial, until some maximum level is reached and maintained. Clearly, there is substantial truth to this view. For example, rats in a recent study of ours[12] were trained daily on the skilled reach-to-grasp-food task[39] described earlier. On day one of training the sample of rats grasped the target correctly on 27% of trials, and improved daily and linearly in their performance until day six, when their success rate began to asymptote at approximately 68% correct (Figure 7.4). Consistent with this view of task learning, it was recently shown that the caudal digit-wrist motor cortex, which contains an overlapping somatosensory representation of the forepaw, expands and grows new synapses during the learning of this task, whereas no similar changes were detected in the (more strictly motor) rostral M1[20] (note that this finding also underscores Principal I). Moreover, the degree of long term potentiation in M1 synapses has been found to correlate with reaching skill acquisition.[34]

A. Parallel Processing At Multiple Time Scales

However, other neuroscientific data suggest that the information processing relevant to task learning and performance flows in a more complex manner, as rats or other animals attempt to better understand their past and better predict the future. These processes occur at multiple time scales throughout different levels of the nervous system. For example, at the somatosensory periphery, slowly adapting versus rapidly adapting cutaneous receptors in the mammalian glabrous skin[38] allow simultaneous perception of different aspects of the tactile world.[16] At the same time, even in the somatosensory motor periphery, multiple feedback processes are occurring, e.g., the

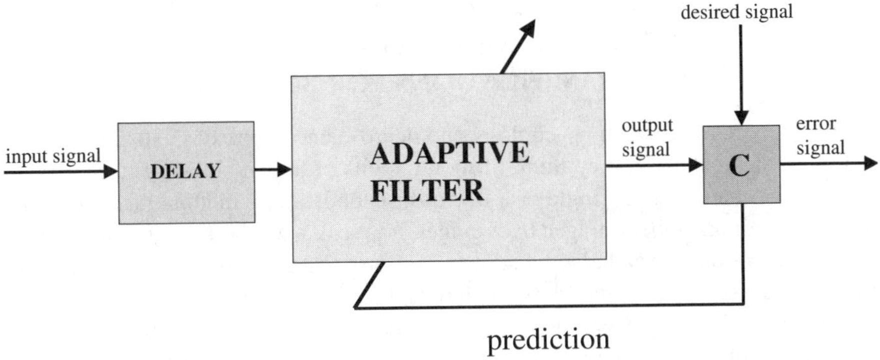

FIGURE 7.5 Schematic diagram of a system, natural or artificial, that improves its performance by evaluating the past outcome of behavior over multiple delay periods and modifying the behavior so that it increasingly resembles the desired output. In the figure, C stands for a comparator component, which compares the current behavioral output with the ideal output and then feeds that information back to the adaptive filter.

spinal stretch reflex illustrates how proprioceptive feedback is constantly influencing muscle tension. Modification of neural processing by evaluative feedback is seen at longer time scales as well, e.g., corticothalamic feedback projections modify subcortical sensory processing.[7,9] And at still longer time scales, such as hours, considerable data now suggest that rats re-evaluate their sensorimotor performance on task from the prior day during slow-wave sleep, developing better internal models for those tasks as evidenced by the fact that performance often improves after a complete sleep cycle – with no additional overt practice.[8,18] During this time, the animals appear to replay aspects of the day's experience[22] and consolidate their memories via interactions between the hippocampus and neocortex.[30] All of these processes occurring at distinctive time scales, help the rat use its past to predict its future more accurately. Figure 7.5 illustrates a model of an adaptive system that incorporates information about an animal's current prediction, the error in that prediction relative to some standard for good perception or performance, and the updating of the model, using information evaluated at different delay lengths (represented in the DELAY box).

Animals are also constantly using their current experience to develop better models of what might occur to them. For instance, it has been shown that rats trained in a fear conditioning task re-encode their memory of how to perform the task on each trial, indicated by the fact that maintenance of a high performance level after a given trial depends on protein synthesis.[36] Thus, the rat's internal model of how to perform a task does not appear to follow a simple forward, linear progression until asymptotic level is attained. We believe that for researchers interested in accurately modeling rat behavior, there is no short cut around knowing about and incorporating the psychology that corresponds to these various neural mechanisms for the animal's processes of learning, re-evaluating, and re-learning. These different processes simply must become a part of your task model.

V. PRINCIPLE IV: ACTION POTENTIALS ARE NOT THE ONLY CAUSAL NEURAL ACTIVITY, WITH RAMIFICATIONS FOR BEHAVIOR

When many researchers in psychology and neuroscience think of the neural activity that underlies behavior, they think of modulations of action potential rate. At very short time scales such as single or a few milliseconds, these modulations are called temporal coding, while averaged over longer time scales such as tens of milliseconds, these modulations are called rate coding.[33] Action potentials are clearly an important part of the neural code underlying behavior because they represent the output of neural networks. However, they are likely not the only causal aspect of neural activity. Subthreshold synaptic activity is constantly occurring, causing myriad current dipoles throughout the different compartments of a single neuron as well as between adjacent neurons. These subthreshold flows of current are important for understanding the neural basis of behavior for two reasons. First, they constitute the pre-processing that ultimately determines spike output. Second, in some cases subthreshold current fluctuations affect the spike firing rate of adjacent neurons.[3]

Spikes, because they occur in an all-or-none fashion, are a digital form of information processing. In the dendrites of the neurons, however, both action potentials and graded changes in voltage and in current flow can occur. At dendritic branches, nonlinear integration of incoming activity takes place whereas at the cell body, summation is linear.[31] One consequence is that subthreshold neural activity is constantly occurring at multiple frequencies simultaneously, and that these frequencies are continuously being modulated. Moreover, current sources and sinks occur at multiple places on the neuron and are used in neural processing at different times. This can be seen in recordings of local field potential activity using data analytical methods whose description is beyond the scope of this chapter. What these methods allow is the visualization of the constantly changing frequency of current fluxes (both subthreshold and suprathreshold) over time. Figure 7.5 presents suprathreshold and subthreshold activity in the rat's caudal M1 during performance of a single trial of the reaching task described earlier, between 0 and 100 Hz (cycles/second), with pixel color corresponding to amount of energy. Although we have found that significant modulations of action potential firing cluster in five kinematically defined task phases, it can clearly be seen in Figure 7.6 that the neural activity is continuously changing in terms of frequency and voltage.

A state of the art question in behavioral neuroscience concerns the relationships among continuous synaptic activity, phasic action potential firing, and behavior. For example, the continuous activity may underlie the gradually increasing probability, calculated by the animal that it is going to be put in the reaching box and given the opportunity to obtain chocolate pellets. First, the time of day that the rat is often run in the task approaches, next the experimenter may walk into the animal housing room and pull the animal's cage off the rack, etc. As these events are happening, we have observed that the animal increasingly salivates, oscillates its jaw, and increases its general activity level[13] – all processes associated with performing a skilled motor act to obtain a desired reward. These processes and behaviors do not appear to occur in an all-or-none, stepwise fashion, but rather, their probability gradually increases as it becomes more apparent to the animal what it is about to experience.

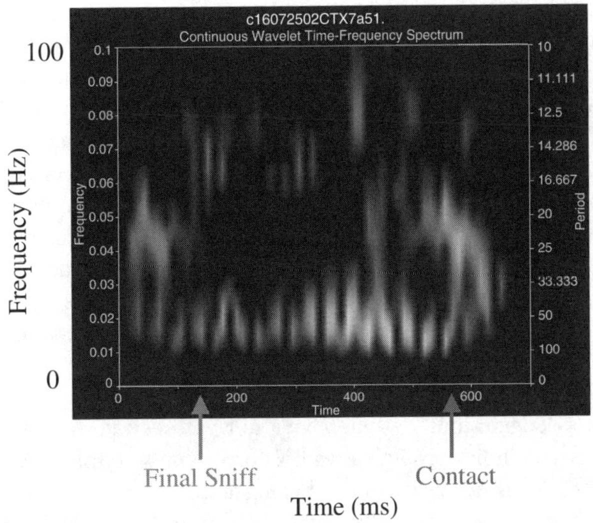

FIGURE 7.6 (See color figure following page 78) Sub- and suprathreshold neural activity in the vicinity of one M1 electrode during one trial of skilled reaching. The x-axis shows task-time, the y-axis shows frequency from 0 to 100 Hz, and pixel color represents amount of energy, with hotter colors indicating higher amounts of energy. See text for further description.

VI. PRINCIPLE V: BEHAVIORS APPEAR TO BE ORGANIZED INTO SURVIVAL-RELATED REPERTOIRES THAT CAN BE ADAPTED TO NOVEL CIRCUMSTANCES

Although not all researchers agree with this principle, an increasing wealth of data is suggesting that distinct regions of M1, the premotor cortex, and other motor areas, are organized according to the whole movements they produce, rather than being organized somatotopically.[10] Consistent with this newer view, while it used to be believed that individual motor cortical cells controlled individual muscles, it is becoming increasingly clear that individual cells code for synergies of muscles as a minimum.[4,6] For example, we have found that during the skilled reaching task described earlier, rat red nucleus cells appear to code for combined limb movements and postural shifts.[13]

A. Motor Organization in Terms of Survival-Related Behaviors

Even more surprisingly, given the older view, recent studies have strongly indicated that it is not only muscle synergies that are represented in motor areas, but whole, survival-related movements. For instance, studies with primates in which motor and premotor cortical areas are stimulated with stimulus trains in the natural range, approximately 500 msec long, are showing that stimulation in different subregions causes the animal to move its hand toward its mouth, or to defensively block objects from hitting its face, or to perform manual grasping movements in a frontal workspace, or to engage in other survival-related behaviors. Various subregions of M1 appear to code for fine manual movements in the manual workspace, whereas subregions of the premotor cortex appear to code for movements of the hand in the facial region or in the body's lateral space.[10] These studies are consistent with recent primate anatomical data showing that in M1 and the premotor cortex, there is not a precise body map as previously thought, but rather, the body map is fractured and intermingled although it contains a very rough somatotopy.[35] Subcortical as well as cortical areas in the primate may represent whole movements. Although most of these stimulation studies have been performed on motor cortical areas, there is evidence that stimulus trains of this length produce whole movements even when the stimulation occurs at the spinal level.[37]

To our knowledge, similar studies have not yet been performed with rats, but rat behavioral research has produced results consistent with this view of how movement control is organized in the brain. For instance, using the reach-to-grasp-food task described earlier, Metz and Whishaw (2000) found that rats did not adjust their grip size for food pellets of varying diameters. On the basis of these findings, they argued that manual movements in rats are organized in terms of stereotypic movements. Our observations of how rats learn the standard reach-to-grasp task are also consistent with this view. For instance, on the first few training trials rats often reach for the food pellet, and then, even if they succeed in grasping it, they do not retract the pellet toward their mouth, open their mouth, and place the pellet inside. Rather, after barely starting to retract their paw, they drop the pellet and let their arm go limp temporarily. Then, after several more trials they gradually pull the pellet closer and closer to their mouth. Similarly, on many initial trials rats fail to contact and grasp the target, and don't even extend their paw forward a sufficient distance, but still they open their mouths and retract their pelletless hand.[13] Thus, a crucial part of learning to perform the whole maneuver appears to be learning to conjoin two stereotypic movements – extending their paw outward and grasping the object, and then pulling it toward and inside their mouth.

B. Sensorimotor Learning According to this View

This may illustrate a more general point about how novel movements are learned by the rat and perhaps also the primate: the repertoire of stereotypic behaviors is gradually adapted to new circumstances. Both the Whishaw laboratory and our laboratory have evidence for this view. Iwaniuk and Whishaw (2000) performed a cladistic analysis of vertebrate skilled and unskilled forelimb movements, and concluded that skilled reaching emerged from more primitive early vertebrate wiping behaviors. We have evidence that in the rat M1, the neural control for skilled reaching and for the reach-like phase of locomotion, in which the forepaw is lifted off the ground, projected forward, and placed back down, is similar. In both cases, M1 cells preferentially encode the lift and paw-down phases of the movements. These findings suggest that the inter-joint timing required for one type of movement can be gradually shifted over the course of learning so that it produces a different, though related, movement.

The view that rats learn sophisticated motor behaviors by shaping a pre-established repertoire to new circumstances has dramatic implications for one's task analysis. When the rat is learning a new and difficult motor task, for instance, instead of learning a completely novel and lengthy series of joint torques and angles, the animal may start with a subset of its hardwired movements and then combine and subtly adapt them to the current circumstances.

VII. FINAL COMMENTS

Using recent findings from neuroscience, we have begun to sketch a new view of the psychological information processing that occurs as animals learn and execute a new task. This new view includes the facts that the animal's mind is continuously processing aspects of the task at multiple temporal and spatial scales, that sensory and motor processing are highly fused, and that whole, stereotypic movements are represented in the rat's brain and mind. We believe that if researchers in the areas of rat perception, cognition and behavior base their task designs and task analyses on the above principles, it will facilitate their development of a good psychological understanding of their subjects' performance. For example, if a researcher hypothesizes that rats will learn a new task by adapting a set of partly known pre-existing behaviors, it will greatly simplify his understanding of how to kinematically code the evolving motor sequence. However, researchers still must validate their task analyses by testing their hypotheses about how the animals perform their tasks. For example, although we argued earlier that rats are a good mammalian exemplar, they have strengths and limitations that are not always known *a priori* to the investigator, who is often biased by his own conscious experience, even when he has the above principles in mind.

REFERENCES

1. Aboitiz, F., Montiel, J. and Morales, D. (2003). The evolutionary origin of the mammalian isocortex: towards an integrated developmental and functional approach. *Behav. Brain Sci.* 26(5):535-52.
2. Baddeley, A. D. (1998). *Human Memory: Theory and Practice*. Boston, Allyn and Bacon.
3. Bullock, T.H. (1997). Signals and signs in the nervous system: the dynamic anatomy of electrical activity is probably information-rich. *Proc. Natl. Acad. Sci.* U S A. 94(1):1-6.
4. Capaday, C. (2004). The integrated nature of motor cortical function. *Neuroscientist* 10(3):207-20.
5. Chapin, J.K. and Woodward, D.J. (1986). Distribution of somatic sensory and active-movement neuronal discharge properties in the MI-SI cortical border area in the rat. *Exp. Neurol.* 91(3):502-23.
6. Cheney, P.D., Belhaj-Saif, A. and Park, M.C. (2004). Properties of Primary Motor Cortex Output to Forelimb Muscles in Rhesus Macaques. In press in *J. Neurophys.*
7. Engel, A.K., Fries, F. and Singer, W. (2001). Dynamic predictions: oscillations and synchrony in top-down processing. *Nat. Rev. Neurosci.* 2(10):704-16.

8. Gais, S., Plihal, W., Wagner, U., Born, J. (2000). Early sleep triggers memory for early visual discrimination skills. *Natl. Neurosci.* 3(12):1335-9.
9. Ghazanfar, A.A., Krupa, D.J. and Nicolelis, M.A.L. (2001). Role of cortical feedback in the receptive field structure and nonlinear response properties of somatosensory thalamic neurons. *Exp. Brain Res.* 141(1):88-100.
10. Graziano, M.S., Taylor, C.S., Moore, T. and Cooke, D.F. (2002). The cortical control of movement revisited. *Neuron* 36(3):349-62.
11. Guillery, R.W. and Sherman, S.M. (2002). The thalamus as a monitor of motor outputs. *Philos. Trans. R. Soc. Lond. B Biol. Sci.* 357(1428):1809-21.
12. Hermer-Vazquez, L., Hermer-Vazquez, R., Moxon, K.A., Kuo, K.-H., Viau, V., Zhan Y. and Chapin, J.K. (2004). Distinct temporal activity patterns in the rat M1 and red nucleus during skilled versus unskilled limb movement. *Behav. Brain Res.* 150(1-2):93-107.
13. Hermer-Vazquez, R., Hermer-Vazquez, L. and Chapin, J.K. (2004). Inhibitory mechanisms coupling olfactory and motor areas prior to voluntary movement. Manuscript submitted for review.
14. Iwaniuk, A.N. and Whishaw, I.Q. (2000). On the origin of skilled forelimb movements. *Trends Neurosci.* 23(8):372-6.
15. Johnson, B.N., Mainland, J.D. and Sobel, N. (2003). Rapid olfactory processing implicates subcortical control of an olfactomotor system. *J. Neurophysiol.* 90(2):1084-94.
16. Johnson, K. O. (2001). The roles and functions of cutaneous mechanoreceptors. *Curr. Opin. Neurobiol.* 11(4):455-61.
17. Jones, E.G. (2001). The thalamic matrix and thalamocortical synchrony. *Trends Neurosci.* 24(10):595-601.
18. Karni, A., Tanne, D., Rubenstein, B.S., Askenasy, J.J. and Sagi, D. (1994). Dependence on REM sleep of overnight improvement of a perceptual skill. *Science* 265(5172):679-82.
19. Kassel, J., Shambes, G.M. and Welker, W. (1984). Fractured cutaneous projections to the granule cell layer of the posterior cerebellar hemisphere of the domestic cat. *J. Comp. Neurol.* 225(3):458-68.
20. Kleim, J.A., Barbay, S., Cooper, N.R., Hogg, T.M., Reidel, C.N., Remple, M.S. and Nudo, R.J. (2002). Motor learning-dependent synaptogenesis is localized to functionally reorganized motor cortex. *Neurobiol. Learn. Mem.* 77(1):63-77.
21. Kosslyn, S. M. (1999). If neuroimaging is the answer, what is the question? *Philos. Trans. R. Soc. Lond. B Biol. Sci.* 29;354(1387):1283-94.
22. Lee, A.K. and Wilson, M.A. (2002). Memory of sequential activity during slow-wave sleep in the hippocampus. *Neuron* 36(6):1183-94.
23. Krubitzer, L. (1995). The organization of neocortex in mammals: are species differences really so different? *Trends Neurosci.* 18(9):408-17.
24. Llinas, R.R. (2001). *The I of the Vortex.* Cambridge, MA, MIT Press.
25. Luo, Z.-X., Crompton, A.W. and Sun, A.-L. (2001). A new mammaliaform from the early Jurassic and evolution of mammalian characteristics. *Science* 292:1535-1540.
26. Malach, R. and Graybiel, A.M. (1986). Mosaic architecture of the somatic sensory-recipient sector of the cat's striatum. *J. Neurosci.* 6(12):3436-58.
27. McFarland, N.R. and Haber, S.N. (2002). Thalamic relay nuclei of the basal ganglia form both reciprocal and nonreciprocal cortical connections, linking multiple frontal cortical areas. *J Neurosci.* 22(18):8117-32.

28. Metz, G.A. and Whishaw, I.Q. (2000). Skilled reaching an action pattern: stability in rat (Rattus norvegicus) grasping movements as a function of changing food pellet size. *Behav. Brain Res.* 116(2):111-22.
29. Mori, S., Iwakiri, H., Homma, Y., Yokoyama, T. and Matsuyama, K. (1995). Neuroanatomical and neurophysiological bases of postural control. *Adv. Neurol.* 67:289-303.
30. Poe, G.R., Nitz, D.A., McNaughton, B.L., Barnes, C.A. (2000). Experience-dependent phase-reversal of hippocampal neuron firing during REM sleep. *Brain Res.* 855(1):176-80.
31. Poirazi, P., Brannon, T. and Mel, B.W. (2003). Arithmetic of subthreshold synaptic summation in a model CA1 pyramidal cell. *Neuron* 37(6):977-87.
32. Posner, M. I. (1986): *Chronometric explorations of mind: the third Paul M. Fitts lectures, delivered at the University of Michigan, September 1976.* New York, Oxford University Press.
33. Rieke, F. et al. (1997). *Spikes: Exploring the Neural Code.* Cambridge, MA, MIT Press.
34. Rioult-Pedotti, M.S., Friedman, D., Hess, G. and Donoghue, J.P. (1998). Strengthening of horizontal cortical connections following skill learning. *Nat. Neurosci.* 1(3):230-4.
35. Sanes, J.N., Donoghue, J.P., Thangaraj, V., Edelman, R.R. and Warach, S. (1995). Shared neural substrates controlling hand movements in human motor cortex. *Science* 268(5218):1775-7.
36. Schafe, G.E. and LeDoux, J.E. (2000). Memory consolidation of auditory pavlovian fear conditioning requires protein synthesis and protein kinase A in the amygdala. *J. Neurosci.* 20(18):RC96.
37. Strick, P.L. (2002). Stimulating research on motor cortex. *Nat. Neurosci.* 5(8):714-5.
38. Vallbo, A.B. and Johanssen, R.S. (1984). Properties of cutaneous mechanoreceptors in the human hand related to touch sensation. *Hum. Neurobiol.* 3(1):3-14.
39. Whishaw, I.Q. and Pellis, S.M. (1990). The structure of skilled forelimb reaching in the rat: a proximally driven movement with a single distal rotatory component. *Behav. Brain Res.* 41(1):49-59.

8 Neural Plasticity in Adult Motor Cortex

Scott Barbay, Elena Zoubina, and Randolph J. Nudo

CONTENTS

I. Introduction 155
II. Learning-Induced Plasticity in the Motor Cortex 156
  A. Environmental Enrichment and Plasticity 156
  B. Motor Learning 158
III. Injury-Induced Plasticity in Motor Cortex 164
  A. The Concept of Spontaneous Recovery 164
  B. Reorganization of M1 Associated with Spontaneous Recovery After an Ischemic Infarct 166
  C. Directed Reorganization of Injury-Induced Plasticity in Motor Cortex 170
  D. Neuromodulation of Cortical Activity 174
References 177

I. INTRODUCTION

In its most general sense, functional mapping of the brain attempts to relate the activity of neurons or neuronal ensembles to observed functions of the organism. As sensation and movement are most easily observed, controlled, quantified and manipulated, maps of primary sensory and motor cortical regions are the ones most studied and better understood. Accordingly, studies of the mutability of sensory and motor maps of the cerebral cortex following various experimental manipulations have provided much of the information concerning neuroplasticity in the adult brain. As functional plasticity of cortical sensory regions is described in detail in other chapters of this book, here we will analyze neuroplasticity in the primary motor and premotor cortex of the adult mammal, and describe some of the putative mechanisms underlying these changes.

A number of methodological approaches have been used to assess functional organization in motor cortex ranging from functional brain imaging (e.g., functional magnetic resonance imaging or fMRI) to the recording of activity of single neurons and neuronal ensembles during the execution of motor tasks. One of the most common approaches, however, is the direct stimulation of selected brain regions by bursts of electrical current or magnetic fields and the relation of this stimulation to muscle or joint movement. In experimental animals, microelectrode stimulation in motor cortex has been utilized for decades. In this approach, a fine microelectrode is introduced into the motor cortex of an animal (typically anesthetized) so that the tip is located close to the cell bodies of the output neurons in layer V. Stimulation with a brief current train burst results in peripheral muscle contraction. By limiting the current amplitude and burst duration, the direct current spread can be limited to no more than 200 μm.[1] The major limitation of this technique, commonly known as intracortical microstimulation (ICMS) mapping, is that the polysynaptic spread of activity can extend for several millimeters,[2,3] somewhat obscuring the interpretation of plasticity mechanisms. However, its advantages lie in the fact that the functional organization of the motor cortex can be described in a single experimental session. The procedure can then be repeated both before and after experimental manipulations. The resulting comparisons can be used to relate neurophysiological to behavioral changes.

In this chapter, we will describe studies based primarily on the ICMS mapping technique applied to adult mammals, especially rats and nonhuman primates. These studies serve as a basis to describe the basic phenomenology of neuroplasticity in motor cortex under various conditions, and as a departure to speculate about the underlying mechanisms. The conditions that have been demonstrated to induce functional reorganization of the motor cortex include the acquisition of new motor skills, peripheral and central nervous system injury, and post-injury behavioral interventions (directed reorganization). Thus, neuroplasticity in the motor cortex has significance not only for understanding the neural bases of motor learning, but also for understanding the process of functional motor recovery following central nervous system disease and injury.

II. LEARNING-INDUCED PLASTICITY IN THE MOTOR CORTEX

A. Environmental Enrichment and Plasticity

It has been known for some time that differential experience with the environment can have varying effects on the cerebral cortex that results in changes in perceptual and behavioral competencies. Early experiments employed selective rearing techniques in which an animal's experience with the environment was restricted at various stages of development.[4,5] Environmental manipulations have also been used to demonstrate plasticity in the adult brain. Adult animals exposed to an enriched environment showed similar effects of plasticity in the cerebral cortex.[4] Subsequent research using the enriched environment paradigm has demonstrated activity-dependent plasticity at the synaptic level, focusing particularly on changes in dendritic morphology and the involvement of neurotrophins, and non-neuronal cells in the cerebral cortex and cerebellum as well as in subcortical areas such as the hippocampus and the amygdala.[6,7]

Animals exposed to an enriched environment (also referred to as complex environment by Greenough and associates) out-perform animals in the standard or impoverished housing conditions. Associated with enhanced behavioral competencies, is enhanced dendritic morphology as viewed by modified Golgi-Cox staining or electron microscopy. Due to the exploratory nature of the enriched environment, dendritic changes have been observed in the visual cortex[8,9] sensorimotor cortex[10,11] and hippocampus.[12] More recent experiments of Greenough and colleagues have used acrobatic environmental conditions that include a series of obstacles and ladders arranged for rats to traverse. This environment provides novel motor activity to enhance sensorimotor coordination. These manipulations have led to adaptive changes in dendritic morphology in the sensorimotor cortex and cerebellum.[13-16] These studies show that specific training and not generalized activity can direct dendritic plasticity to cortical areas relevant to a learning task. Additionally, Kolb and colleagues have shown that there may be qualitative differences between changes in dendritic morphology due to exposure to an enriched environment and those changes associated with specific training on a task.[17]

The importance of examining synaptic plasticity at the level of the dendrite is that changes in dendritic architecture are associated with the efficacy of neural transmission that can reflect enduring behavioral adaptations to an animal's environmental demands. Adaptive synaptic plasticity reflects the increase in new synapses, the strengthening of synapses and an expansion of potential influence from other groups of cells as dendritic arbors branch out within the cortex. These changes are driven by temporal activation of neurons that become associated during certain sensory or behavioral experiences. The neuronal connections characterizing specific experiences are formed within the constraints of pre-existing representational areas of the cortex established during ontogeny. These representations are constrained by the phylogenetic history of an organism and not influenced by typical environmental stimulation.[18] The ontogenetic history of the organism, the unique interaction of an individual with its environment, is represented as modification of existing connections through activity dependent mechanisms of synaptic plasticity.[19]

Also important for cortical plasticity are horizontal fibers arising from collateral projections from excitatory, pyramidal neurons. These horizontal fibers allow the activation of adjacent and nonadjacent cortical columns. This feature of cortical organization allows for the integration of cortical neurons within a network of sensory or motor representations. A cortical column refers to the smallest arrangement of neurons throughout the six cytoarchitectonic layers in the cerebral cortex that function together to define some feature of the environment. Vernon Mountcastle first described this in the somatosensory cortex.[20] However, subunits referred to as minicolumns have since been identified.[21] Evidence of columnar organization has also been observed in motor cortex during ICMS techniques pioneered by Asanuma and associates.[22-24] Horizontal fibers have been described in sensory and motor cortex.[25-27] These fibers have been shown to span over a millimeter and are thus capable of having a modulatory influence over many columns of neurons.[28] The influence of horizontal connections between distant areas within the motor cortex is mediated by both long-term potentiation (LTP)[28] and long-term depression (LTD).[29] These fibers have been shown to activate excitatory pyramidal cells as well as GABAergic interneurons.[30] Three characteristics of horizontal fibers: 1) dispersed connections within the cortex over several cortical columns, 2) capability of strengthening or weakening synapses through LTP and LTD and 3) inhibitory influences through GABAergic neurons, place these intrinsic cortical connections in an important position for modifying cortical representations associated with learning.[19,31]

The inhibitory effects of these corticocortical fibers are necessary for maintaining relatively stable, neural representations to protect the integrity of neural connections from relatively irrelevant environmental events.[19] The inherent plasticity that these neural representations undergo as an organism interacts with its environment has been well established in the sensory cortex and are more recently being recognized in motor cortex.[32,33]

B. Motor Learning

Our laboratory has been specifically interested in plasticity in the primary motor cortex (M1) during motor skill acquisition in both the normal and injured brain, and we have recently been investigating the relationship between M1 and the premotor areas (particularly the ventral premotor area, PMv ; see Figure 8.1). During the early stages of electrophysiology experiments such as those carried out by Sherrington and Penfield, M1 was thought to be involved with reflex activation of motor neurons and lacking plasticity necessary for volitional movements, especially since it was known that there were many corticospinal neurons that synapsed directly onto motor neurons. By the mid 20th century it was clear that the motor cortex was somatotopically organized.[34,35] With the advent of ICMS techniques developed by Asanuma and associates recording experiments in M1 began showing how M1 was organized with a finer resolution than the previous surface stimulation techniques allowed. Asanuma found that delivering a volley of short duration cathodal pulses at a high frequency would be effective when delivered through small tip electrodes using very weak currents ($\leq$ 60 μA) thus limiting the spread of direct excitation in the cortex.[1] Greater spatial resolution allowed investigators to uncover columnar organization in M1. The finer resolution allowed by ICMS has revealed that within the gross somatotopic organization of the motor cortex separating major body parts, specific areas such as the proximal and distal forelimb representations comprise a montage of movement representations due to extensive overlap of muscle representations.[36-44]

Motor representations have been shown to be dynamic as individual movement representations are idiosyncratic to each animal's personal experience with the environment while maintaining a common topography typical of a particular species.[41] Plasticity has been shown to be constrained, however, favoring adaptive changes in movement representations and remaining constant during the less significant adaptations to transient environmental requirements.[45-47] As mentioned above the mechanisms needed to maintain relatively stable neural connections in the cortex depend on a balance between excitatory and inhibitory interactions between pyramidal cells and local inhibitory interneurons. One way in which biologically important events can shift this balance towards strengthening new connections, and thus altering movement representations is through sensorimotor integration that occurs

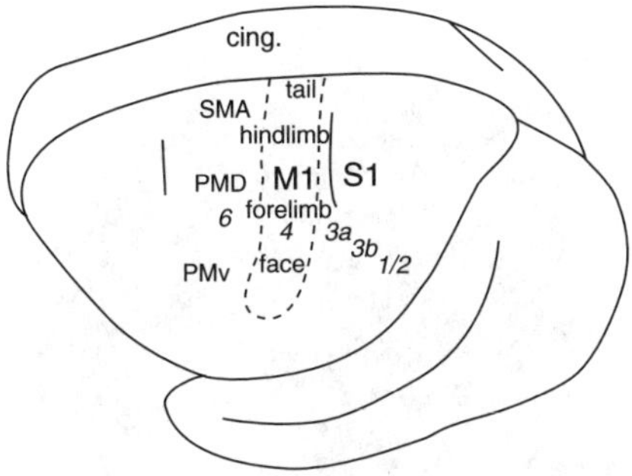

FIGURE 8.1 Somatotopic organization of motor areas (M1, PMv, Pmd and SMA) and somatosensory areas (S1: Areas 3a, 3b, 1 & 2) in primates.

from peripheral feedback loops while executing a new movement. Physiological evidence for this aspect of motor skill learning was demonstrated by Asanuma and colleagues using ICMS stimulation in cortical and subcortical sensory input areas while recording evoked potentials in M1.[48,49]

A theory of motor learning in M1 emerged from these early experiments based upon physiological and anatomical evidence of converging sensory input to M1 from area 2 of the primary somatosensory cortex (S1) and the ventrolateral nucleus of the thalamus (VL).[50-52] Mechanisms of synaptic plasticity associated with use-dependent motor learning was demonstrated by inducing LTP within motor cortex by stimulating both S1 and VL.[53,54] According to Asanuma's theory of motor learning in M1,[55] during a novel motor movement, specific sensory information associated with the movement is carried from somatosensory cortex (area 2 of S1) to the cortical motor columns engaged in the behavior while more diffuse sensory feedback associated with the movement innervates M1 from the thalamus (VL). This coactivation of sensory inputs induces LTP and strengthens synaptic connections with the motor columns engaged while a movement is performed. Once these connections are made, input from the primary somatosensory cortex is no longer necessary for the skilled movement (see Figure 8.2).

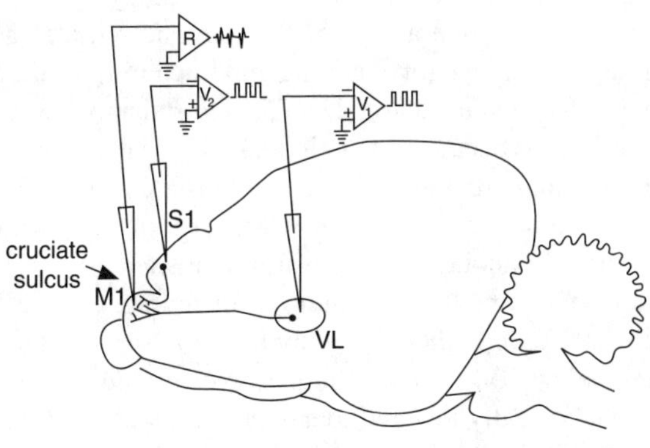

FIGURE 8.2 An Experimental preparation to test Asanuma's model of motor learning in M1. based on [55].

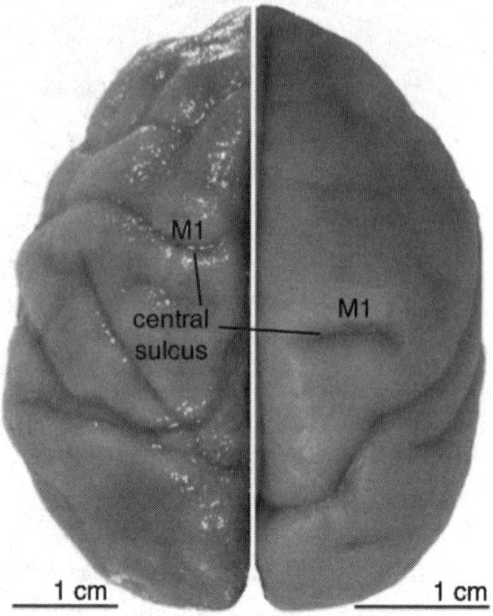

FIGURE 8.3 Comparison of sulcul patterns in non-human primate brains. Left: African green monkey (*Cercopithicus aethiops*), an Old World monkey. Right: Squirrel monkey (*Saimiri sciureus*), a New World monkey. The less convoluted brain of some New World monkeys (e.g., squirrel monkeys, marmoset monkeys) and prosimian primates (e.g., bushbabies) allows more accessibility to the primary motor cortex (M1). The central sulcus of the green monkey (and also macaque monkeys and humans) is significantly deeper rendering accurate topographic mapping more challenging.

Recent experiments in our laboratory have further characterized functional plasticity in the motor cortex using standard ICMS techniques as established by Asanuma and colleagues. This technique has been used by many investigators within rat, cat and monkey cerebral cortex to determine organization of movement representations with high spatial resolution. Because our interests have focused on skilled hand use and the consequences of motor learning within M1, we have concentrated our efforts on the distal hand and forearm representation in primates (see Figure 8.3). The dexterity and agility of the primate hand is ideal for studying motor skill learning, and because of the complex interaction of the intrinsic muscles and joints during motor skill learning, it is ideal for studying modifications in the central nervous system as optimal strategies are learned.[56] The squirrel monkey was chosen from among the other primates because of their relatively flat cortex that allows more precise placement of stimulating electrodes, and thus a nonambiguous identification of movement representations in M1. In our ICMS experiments, we use glass micropettes with tips averaging about 10 to 20 μm in diameter.

Current is delivered through a platinum wire immersed in a 3.5 molar NaCl solution: Thirteen, 200 μsec cathodal, monophasic pulses are delivered at 350 Hz and current is kept below 30 μA. The stimulating electrode is placed perpendicular to the cortex and mechanically lowered to a depth of about 1,750 μm targeting layer V of M1, the location of the corticospinal cell bodies. Each electrode penetration is recorded on a digital picture of the exposed cortical surface at 250 μm increments

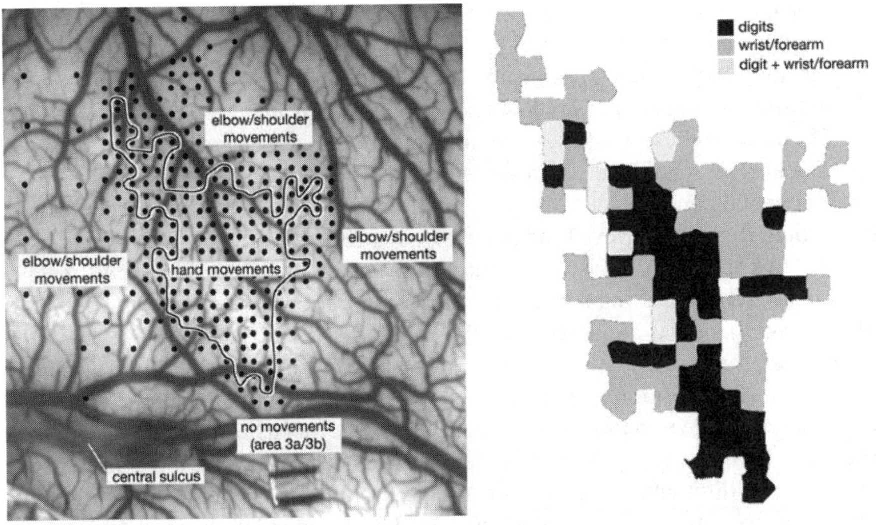

FIGURE 8.4 Microelectrode penetration sites in a typical intracortical microstimulation (ICMS) mapping experiment in a squirrel monkey. Left: Photograph of cortical vasculature within the motor cortex hand area. Thin black line circumscribes penetrations where ICMS evoked movements of the distal forelimb, or hand. Right: More detailed topography of distal forelimb movement representations in the same monkey. Note the mosaic appearance of interdigitated digit and wrist/forearm representations.

across the cortex (see Figure 8.4). This method allows us to create a detailed map, or topographic representation of movements in the tangential plane (see Figure 8.4). As this procedure induces negligible damage to the cortical tissue due to the small size of the microelectrodes, the mapping procedures can be repeated multiple times to detect changes in topography that may be associated with learning new motor-skills. Multiple maps in the same animal (within-subjects design) are necessary, since results from our motor mapping procedures demonstrate that baseline motor maps are highly individualized, presumably representing the personal history of each monkey before training.[57]

To examine motor learning dependent changes in topographic maps of the hand, we use a standard behavioral apparatus known as a Klüver board. This consists of a Plexiglas board attached to the monkey's home cage (monkeys are housed individually). The board contains five, evenly spaced food wells drilled into the surface. The wells differ in diameter ranging from wells large enough for the insertion of all digits, to wells that only allow one or two digits to be inserted. Flavored food pellets are delivered one at a time to one of the food wells. Monkeys are trained to make multiple retrievals of these food pellets for about one hour per day. The monkeys generally retrieve 500 to 600 food-pellets per day.

ICMS maps are derived prior to motor skill training to establish a normal baseline for each monkey and then again following extensive training on the Klüver board. The maps are composed of various digit and wrist/forearm movements (e.g., digit and wrist flexions and extensions and forearm supernations and pronations). These movement representations are bordered rostrally, medially, and laterally by proximal shoulder movements (see Figure 8.4). Caudally the hand representation is bordered by somatosensory cortex (specifically, area 3a) and no ICMS-evoked responses are elicited from stimulation at the low current levels used in our protocol.

The first of these studies to demonstrate the dynamic relationship between movement representations and how they reflect the behavioral demands of varying environmental contingencies was reported in 1996.[33] Three monkeys were trained to use their digits to retrieve food pellets from a Klüver board and a fourth monkey was trained to use his wrist and forearm to turn a rotatable eye-bolt to receive food pellets. The resulting movement representations in M1 reflected the demands of the task: there was an increase in digit representations for the monkeys trained on the Klüver board at the expense of wrist/forearm representations, and the opposite was true for the monkey trained to turn the bolt (see Figure 8.5). Another important finding from this study that addresses the mechanisms of use-dependant plasticity is the increase in multiple-joint movements involving the simultaneous execution of digit and wrist or digit and proximal movements form ICMS stimulation at low thresholds (less than 20 μA). These evoked multi-joint movements are referred to as dual responses and were only seen in abundance after digit training. The implication is that these simultaneous movements may become associated in the cortex through Hebbian-like synaptic mechanisms in which horizontal fibers connecting the two areas become strengthened through associated, repetitive activation.[19,28] This type of neural association has been well characterized within the context of plasticity within primary sensory maps and is referred to as the theory of temporal correlation of inputs. The results of these early experiments revealed common use-dependent plasticity shared by primary visual cortex, auditory cortex and somatosensory cortex.[58-63] Evidence of motor representations coupled through associated activity of inputs may indicate that the temporal correlation of inputs is a general guiding principle of use-dependent plasticity. The next question addressed in our laboratory was whether motor activity alone was sufficient to drive changes in motor cortex independent of learning new motor skills.

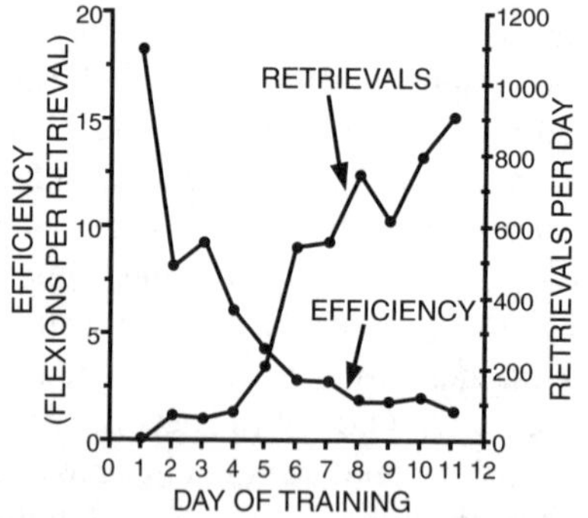

FIGURE 8.5 Typical learning curve for squirrel monkey retrieving food pellets from the smallest well of the Kluver Board. From [33].

Previous studies indicated that motor skill training led to learning induced changes in M1.[33] However, since these monkeys received repetitive training, it is possible that muscle activity alone (independent of learning a new skill) drove changes in M1. To address this issue, monkeys trained on the smallest well on the Klüver board were compared to monkeys trained on the largest well.[46] The largest well allows synergistic movements of all digits during retrieval of food pellets, but does not require the monkeys to learn a new skill, since the grasping of food items is already part of their normal repertoire of cage behaviors. The smallest well on the Klüver board requires the monkeys to use only one or two digits to remove the food pellets. Squirrel monkeys do not possess monosynaptic corticospinal projections to motor neurons as other primates (such as macaque monkeys, as well as humans). This anatomical trait may be related to a greater difficulty for squirrel monkeys to use their digits independently.[64] Nevertheless, over several days, squirrel monkeys can acquire the skill to retrieve pellets from the smallest well on the Kluver board. Motor maps acquired after monkeys retrieved pellets from the large well (for an equivalent number of finger flexions) did not differ from their baseline maps whereas the motor maps acquired from the monkeys trained on the small well showed an increase in digit area. The results of this experiment indicate that repetition of motor tasks is not sufficient to drive neurophysiological changes in motor cortex, and that skill-learning may be essential (see Figure 8.6).

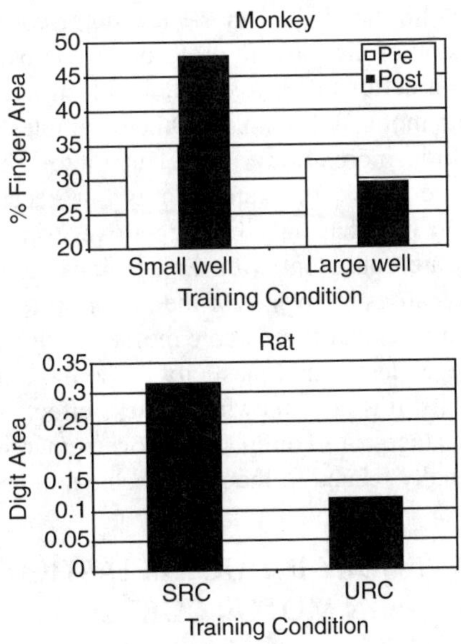

FIGURE 8.6 Top: Increase in finger representation area (expressed as % of total hand area) for small-well trained monkeys compared to large-well trained monkeys. From [46]. Bottom: Digit representation in caudal forelimb area of rats trained to retrieve pellets compared to caudal forelimb area of rats that merely pressed a bar to obtain food. From [65]. It would appear that motor representations are shaped by acquisition of new motor skills, but not by repetitive use alone.

In a related study, rats were trained to retrieved pellets from a rotating disk (400 reaches per day for 10 d).[47] The rats were positioned within a Plexiglas chamber designed to force the use of the rat's preferred limb. This task required rats to grasp the food pellet with their digits in order to retrieve the pellet over a gap placed between the rotating disk and the Plexiglas chamber. ICMS motor maps of the hindlimb, the caudal forelimb and the rostral forelimb areas of the sensorimotor cortex were derived for these rats. Motor maps were also derived from a paired control group in which rats were required to press a lever for food. This group served as the non-learning control group comparable to the large-well monkeys. There was an increase in hand representation within the caudal forelimb area in the skilled learning group but not in the nonlearning control group. No changes were observed in rostral forelimb or hindlimb areas. Further analysis of the brains of these animals demonstrated an increase in synapse number (per neuron) within layer V of the reorganized caudal forelimb area of skill-trained rats but not in the nonlearning control rats.[65] This was the first study to associate structural changes in dendritic morphology with physiological changes in movement representations in M1 resulting from motor skill learning.

Prior to the morphological/physiological correlation study, work from William Greenough's laboratory showed that motor skill training enhanced dendritic morphology in layer V and in layer II/III within the motor cortex of the rat after learning a reach and retrieval task.[66,67] These investigators demonstrated an increase in the number of synapses within layer II/III in the rat motor cortex following acrobatic training.[13] The expression of the c-fos (a proto-oncogene believed to play a role in cell growth) in the motor cortex that is associated with structural changes in neurons was also correlated with motor skill training. Other examples of motor skill training-enhancement of dendritic morphology in the rat motor cortex are found in the recovery of function literature. For example, rats that received acrobatic training have been compared to rats that only had repetitive exercise (non skill-learning control).[68] The results are consistent with the previous studies: acrobatic training increased number of synapses in layer V of the motor cortex and increased cortical volume compared to rats that had comparable motor activity, but not training. These studies describe use-dependent neural plasticity in the motor cortex associated with learning new motor skills. It is now known that these effects are quite specific since these morphological changes are confined to movement representations in motor cortex that are functionally related to the motor behaviors being learned.

III. INJURY-INDUCED PLASTICITY IN MOTOR CORTEX

A. The Concept of Spontaneous Recovery

The restoration of movements or cognitive abilities that have been impaired or lost as a result of an injury to the central nervous system is referred to as recovery of function. Usually the phrase recovery of function is used as a general description that may not refer to the exact reinstatement of lost functions, but refer to compensatory strategies to achieve goals that were no longer obtainable after injury. When recovery of function occurs unaided by therapeutic intervention it is referred to as spontaneous recovery. Theoretical interest in recovery of function evolved from research that supported the idea of localization of psychological and motor functions within the cerebral cortex.[69] Even though theories relating to localization of function eventually dominated, the theory had difficulty explaining how recovery of function could take place if the brain area mediating the function was destroyed. Thus, spontaneous recovery became a domain in which localizationist theory could be assessed.[69] Several theories have been able to explain spontaneous recovery. Two of these early theories that are still of relevance are diaschisis and vicariation of function.[70] The two major proponents of these theories, Constantine von Monakow and Hermann Munk (respectively), were not strict localizationists: they believed in a diffuse involvement of the brain for neurological functions that were more complex than primary motor and sensory functions.[71] However, their theories of spontaneous recovery differed in that diaschisis assumes that the temporary loss of functions is not entirely due to a permanent loss of neural tissue mediating that function. They felt a suppression of neural activity caused by loss of functional connections with the injured neural tissue, and vicariation assumes that the functional reorganization of surviving tissue is able to take on functions originally mediated by destroyed tissue.

Constantine von Monakow's theory of diaschisis stated that the loss of afferent input into a neural region would result in a transient suppression of activity (i.e., neural shock) in that region.[72] Accordingly, spontaneous recovery can be explained by a release from suppression of neural activity within remote areas of the brain that are functionally connected to that part of the brain that has sustained permanent injury. Behavioral or cognitive functions mediated by the suppressed system would regain function over time, whereas those functions mediated by the permanently damaged tissue would not regain function. Therefore, von Monakow could account for both the transient symptoms of brain damage that characterize spontaneous recovery as well as the chronic symptoms. Evidence supporting the concept of diaschisis has been observed in studies employing methods of quantifying metabolic activity in the brain following injury.[73,74] More recently, another form of diaschisis that results in the hyperexcitability of neural tissue associated with an area of infarct has been reported. *In vitro*, electrophysiological studies have demonstrated a phenomenon described as excitatory diaschisis resulting from ischemic damage induced by cortical photothrombosis in rats[75,76] or from middle cerebral artery occlusion in rats.[77,78] In these experiments it was found that cortical areas functionally associated with an ischemic infarct in the brain become transiently hyperexcitable for several days after the damage. This transient state of hyperexcitability was associated with suppression of GABAergic activity that is normally inhibitory. Unlike the suppression of neural activity assumed in von Monakow's theory of recovery, excitatory diaschisis may actually enhance recovery by allowing for an unrestrained or less restrictive state for functional reorganization in the intact, contralateral hemisphere as well as within intact areas of the damaged hemisphere.[75,79,80] However, there is also the possibility of exacerbating the original injury associated with hyperexcitability in the peri-infarct area immediately adjacent to the damaged tissue.[81] Therefore, although diaschisis does not imply the need for reorganizational plasticity in the cortex as a means for recovery, it is not incompatible with it.

Hermann Munk, using ablation techniques to damage certain areas of the cerebral cortex in experimental animals, contributed to the localization of sensory perception in the primary sensory corticies: visual perception in the occipital cortex, auditory perception in the temporal cortex and somatosensory perception in the postcentral gyrus.[71] In many of these experimental cases, it was noticed that the loss of behavioral or perceptual functions would recover relative to the extent of the damaged area. Munk hypothesized that the neural mechanisms mediating these recovered functions were relocalized in the cortex. Experiments that are more recent have supported Munk's theory of vicariation using methods to re-instate the functional deficits observed after the initial injury. This was done by destroying or suppressing cortical function in areas suspected of vicarious reorganization.[82] The therapeutic significance of reorganizational capacity in the cortex has been emphasized by the prominent Russian neurologist Alexander Luria's theory of retraining in which remaining brain tissue reorganizes to carry out a lost function in new ways.[83] This theory implies that functionally related areas of the cortex are capable of mediating similar if not the same functions. In this view of spontaneous recovery, corresponding neural areas of the cortex vicariously take on functional characteristics of damaged tissue as a consequence of neural activation associated with the lost function. Therapeutic intervention could be initiated to foster the process of retraining that may occur. Although Luria was more interested in higher cognitive functions affected by brain damage, this principle can apply to motor skill learning as well. In our laboratory, the possibility of directing mechanisms of cortical plasticity after brain damage is being addressed by implementing task specific behavioral rehabilitation, targeting skilled use of the hand in squirrel monkeys. This is accomplished by comparing functional reorganization in the M1 hand area during spontaneous recovery to differences in functional reorganization after behavioral training.

B. Reorganization of M1 Associated with Spontaneous Recovery After an Ischemic Infarct

The initial interest in spontaneous recovery after focal damage in M1 was to address the potential for reorganization of movement representations. Electrophysiological studies during the early part of the 20th century failed to find vicarious reorganization in undamaged cortex following ablation of forearm representations in non-human primates, even though significant use of the affected hand was observed. Years later, Glees and Cole reported the re-emergence of thumb representation in M1 associated with functional recovery after the original M1 thumb representation had been destroyed.[84] One explanation for this discrepancy is thought to be due to the coarse surface stimulation techniques used in these studies.[85] That is, perhaps injury-induced reorganization in M1 is too subtle to be detected by surface stimulation, but Glees and Cole were able to detect the re-emergence of the thumb representation in their experiment.[84] Therefore, it was thought that the more recently developed ICMS techniques which offer a much higher resolution of motor maps of M1 than the older surface stimulation techniques (250 μm inter-penetration distance between stimulation sites) would be able to detect compensatory reorganization in M1.

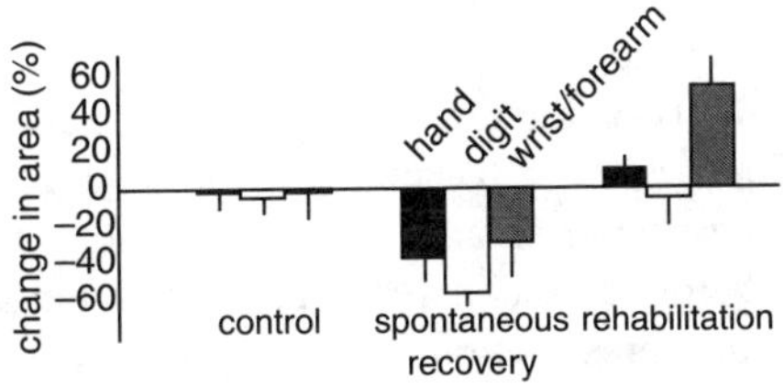

FIGURE 8.7 Reorganization of M1 hand area after subtotal infarct in squirrel monkeys. Control: No manipulation other than ICMS mapping procedure. Spontaneous recovery: No specific training after infarct. Rehabilitation: Daily repetitive training on Kluver board for two to three weeks after infarct Adapted from [86,120].

Nudo et al. (1996) used standard ICMS electrophysiological techniques to derive motor maps of M1 in squirrel monkeys as described in a previous section.[86] ICMS maps were derived before inducing an experimental ischemic infarct in M1, destroying approximately one third of the distal hand representation. The squirrel monkeys were allowed to recover unaided by any therapeutic interventions. A second ICMS map of the M1 hand area was then derived to determine the consequences of three to five months of spontaneous recovery. Even though the skilled use of the hand contralateral to the infarct spontaneously recovered by two months, the ICMS motor maps showed that over 50% of digit movement representations were replaced by proximal shoulder and elbow movement representations compared to baseline maps. The results of the Nudo et al. study did not confirm Glees and Cole's finding of a re-emergence of distal hand representation after a month of spontaneous recovery, but substantial reorganization — seemingly maladaptive reorganization — was observed in the intact cortex in M1 (see Figure 8.7).[86]

The functional reorganization observed in the study just discussed, reflected a further loss of hand representation than that originally destroyed.[86,87] These results are just the opposite of what was expected from the Glees and Cole study (1950), especially since the animals in this study also recovered use of the hand prior to the second motor map. Two questions arise from these results. 1. What is mediating recovery if the motor representations do not show the typical expansion observed with normal learning? 2. Why is there an expansion of proximal forearm representations within hand territory? A plausible explanation for these questions is that during spontaneous recovery the monkeys compensate for temporary impairment of the hand by relying on shoulder and forearm movements. Early lesion studies have documented behavioral compensation resulting from damage to CNS motor systems in primates and more recently in rats.[88,89]

In these studies more proximal arm and postural adjustments were made to compensate for impaired hand or forepaw movements. It is possible that similar compensatory strategies in the squirrel monkey could be responsible for use-dependent expression of more proximal movement representations in the squirrel monkey. Disuse of the digits and wrist combined with increased use of proximal movements would shift the balance between excitatory and inhibitory borders that maintain movement representations, and thus allow expression of formerly silenced proximal movement representations. To address this issue the preferred hand of monkeys was constrained for several weeks while the monkeys were left to behave freely within their home cage.[90] The elastic restraint used in this experiment prevented the monkeys from using their digits on the restrained hand, but allowed wrist and forelimb movements. Casual observation of the monkeys' home cage behavior indicated an increase of wrist and forearm movements compared to the nonrestrained hand. ICMS maps that were taken before and after restraint of the hand showed an increase in wrist/forearm movement representations and a related decrease in digit representations. The overall size of the motor maps remained the same. It is assumed that the increased activity of the wrist and forearm prevented the expression of proximal arm and elbow representations within the hand area as typically seen during spontaneous recovery. In a subsequent experiment in which the restraint was removed from the monkeys allowing use of the digits, digit movement representations were reinstated.[91] These studies indicate that functional borders between adjacent movement representations in M1 rely upon the relative use of the limbs..

The maintenance of functional boarders between representations has been demonstrated in rats, as movement representations in the cortex were shifted with the GABAergic antagonist bicuculline.[92] Imaging studies in humans using positron emission tomography or functional magnetic resonance imaging have also shown overlapping representation of distal and proximal arm movements.[42,93-95] These inhibitory/excitatory borders have also been well documented in sensory cortex.[96]

In addition to the reorganization of movement representations, somatosensory recordings in the M1 hand area also change after a small infarct in caudal M1. The motor hand representation of M1 in squirrel monkeys was first defined using the standard ICMS procedures.[97,98] Then areas within the M1 hand motor map that responded to cutaneous or proprioceptive stimulation of the hand were determined. A small ischemic infarct was created in the caudal M1 hand area and the monkeys were allowed to spontaneously recover. Each monkey was remapped at 1 and 4 months after the infarct. The pre-infarct sensory maps in M1 showed proprioceptive stimulation could be detected throughout the hand area and cutaneous stimulation could be detected in caudal M1 as expected from previous studies.[99,100] Virtually no sensory information could be detected in M1 after 1 month of spontaneous recovery and even 4 months after the infarct, very little sensory responses in M1 could be detected. By contrast, sensory responses in 3a and 3b of the primary somatosensory cortex, which is immediately caudal to M1, remained undisturbed by the infarct. It was surprising that the recovery of detectable sensory input into the M1 hand area did not completely occur by 4 months post-infarct, since typically the monkeys recover the ability to use their hand by one month after these small caudal lesions. It is possible that during spontaneous recovery cortical areas outside of M1 functionally compensate for loss of movement representations in the primary hand area.

There are several motor areas in the frontal cortex that contribute to different aspects of skilled use of the hand in primates and are reciprocally connected with M1 through direct cortico-cortical projections.[101] These areas such as the ventral premotor (PMv), the dorsal premotor (PMd) and the supplementary motor cortex (SMA) have a lot in common with M1. Each area has a complete hand representation and contributes a large number of corticospinal projections.[102,103] The degree of dexterity that squirrel monkeys can achieve is contingent upon the interaction of corticospinal projections with propriospinal and corticospinal neurons within the ventral horn of the spinal cord.[104] When a large number of corticospinal neurons are lost as assumed with the ischemic lesions we induce in M1, it is possible that the other motor areas in frontal cortex that send projections to the spinal cord, reorganize in compensation for loss of input from M1.

To address the possibility of spontaneous recovery being mediated by areas outside of M1 after ischemic damage, we began studying movement representations in PMv. It had already been suggested that the premotor cortex (PMd and PMv) was important for recovery of hand use in macaque monkeys after an M1 lesion.[105] Following the recovery of dexterity of the affected hand, the GABAergic agonist (muscimol) was used to suppress activity in the premotor cortex of both the intact and damaged hemispheres, the peri-infarct area surrounding the lesion and the intact hand representation in M1. The authors found that only suppression of the premotor area in the damaged hemisphere reinstated behavioral deficits. We, therefore, wanted to see if the hand representation in PMv reorganized in a way that would reflect this change.[106] The similarities between PMv in primates and humans concerning input and output connections, and task relevant functions of PMv made this an attractive area to search for compensatory reorganization following M1 damage.

Standard ICMS procedures were applied to map M1 and PMv in the hemisphere contralateral to the preferred hand of five squirrel monkeys before and after an experimental, ischemic infarct was induced in M1, directed at the total hand representation.[106] The results demonstrated that there was reorganization in PMv following 3 months of spontaneous recovery. All squirrel monkeys showed an increase in hand representation in PMv. Four of the five monkeys had an increase in digit representation and four of the five monkeys had an increase in wrist/forearm representation. Partial survival of tissue in M1 was observed for each monkey. This allowed for an interesting correlation showing that the amount of reorganization in PMv was negatively correlated with the amount of M1 hand representation that survived ischemic damage (see Figure 8.8). That is, the smaller the damage to M1 the less compensatory reorganization was seen in PMv. These lesions were larger than those used in the subtotal lesion studies, which were approximately 33% of the hand area. The smallest lesion in this study was over 50% of the M1 hand area.

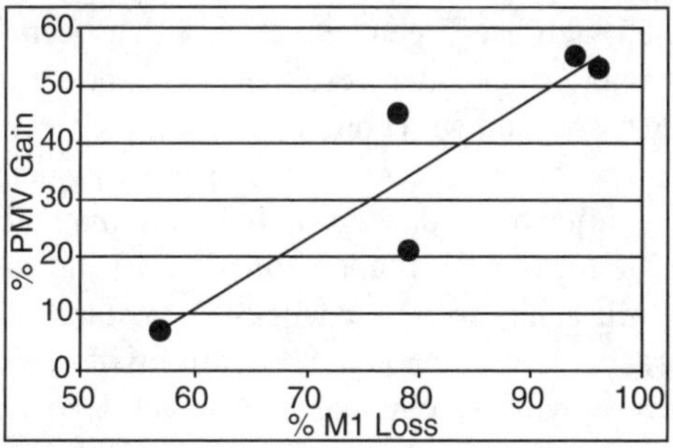

FIGURE 8.8 Relationship between M1 hand area lesion size and PMv hand area expansion. From [106].

Preliminary studies have demonstrated that destroying 33% of the M1 hand area does not promote observable changes in PMv as seen with the large lesions.[107] It has not been determined if there are subtle changes in the way spared M1 projections to the spinal cord or PMv projections to the spinal cord may have been altered or if other motor areas are involved. Anatomical evidence related to the large M1 lesions has shown interesting results, however. Novel axonal projections from PMv to the caudal somatosensory area (areas 1 and 2) have been reported in squirrel monkeys that survived large M1 lesions destroying a majority of M1 hand representation.[108] Survival time for these monkeys was at least several months. This finding is especially interesting because of the direct connections between somatosensory cortex and M1 in normal (uninjured) monkeys, and the corticospinal projections known to originate from somatosensory cortex.[49,103,109-111] It is believed that the novel axonal projections from PMv observed represent projections that are typically intended for rostral M1,[112] but are redirected towards somatosensory cortex: some of the labeled axons examined in the lesion cases bend around damaged M1 cortex. This implies an anatomical relationship between the physiologically characterized reorganization in PMv and the spontaneous improvement in dexterity demonstrated in previous studies.

One implication for these findings is that PMv may be taking on some of the functions or characteristics of the damaged M1 as connections are being formed with somatosensory cortex. A recent clinical case study was reported by Duffau et al. in which a 39-year-old man was diagnosed with a tumor described as an aterio-venous malformation or angioma in M1.[113] Pre- and post- surgical, direct stimulation of the cortex was used to examine tissue surrounding the tumor as a normal part of the procedure. Surgical removal of the tumor revealed redundant hand movement representations rostral to M1. Hand movements were also evoked prior to removal of the tumor in the somatosensory area of the cortex. These results demonstrate an immediate reorganization of cortical function following the removal of the tumor in M1 and a possible long term reorganization of cortical function in somatosensory cortex due to the development of the M1 tumor.

C. Directed Reorganization of Injury-Induced Plasticity in Motor Cortex

The previous studies cited above describe functional reorganization in motor areas of the cortex associated with use-dependent activity that occurs during motor skill learning in the normal brain and as a result of spontaneous recovery after brain injury, which has been associated with compensatory motor learning. The relearning that may occur during spontaneous recovery seems to be associated with behavioral compensation. That is, after brain damage individuals may adapt their behavioral strategies to compensate for impaired motor functions in such a way that pre-injury behavioral goals are still achievable by changes in postural adjustments or motor sequencing. Behavioral compensation is thus a learning of new motor activations or synergies to re-acquire previously obtainable behavioral goals. Instead of leaving this process to chance, therapeutic intervention or physical rehabilitation is thought to optimize the best strategies for recovery by directing the optimal use of the injured limb and thus directing optimal use-dependent plasticity.

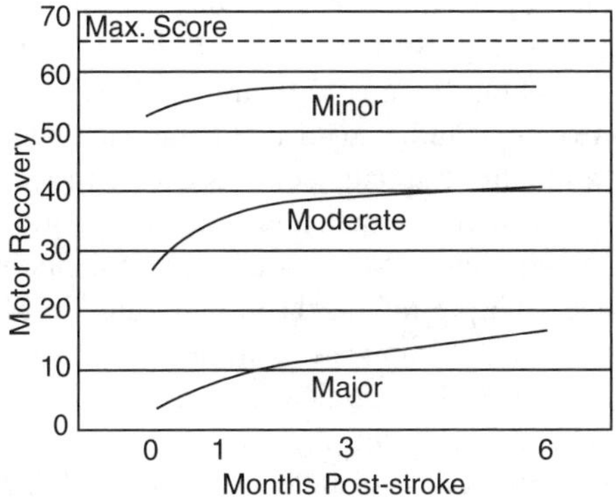

FIGURE 8.9 Functional recovery rates after stroke in humans as a function of initial severity. Adapted from [116].

The therapeutic efficacy of physical rehabilitation is currently being examined in the clinic given the mounting evidence of use-dependent plasticity in the brain. These studies have yielded promising results, especially concerning functional reinstatement of skilled use of the hand in patients that have entered treatment with chronic deficits. Motor deficits are considered to be chronic if no improvements are seen over several weeks or months. It has been reported that if no recovery is seen in 11 weeks, further improvements are doubtful.[114,115] Even when patients are improving, recovery seems to plateau about 6 months after the infarct (see Figure 8.9).[116] One rehabilitative therapy that is particularly promising is the constraint-induced movement therapy (CI therapy) introduced by Edward Taub and colleagues.[117,118] Briefly stated, CI therapy is based on the premise that unilateral damage to motor areas of the brain results in hyper-reliance on the nonparetic arm ipsilateral to the damaged hemisphere that is primarily controlled by the undamaged hemisphere. It is assumed that hyper-reliance on the nonparetic arm results in a motor neglect or disuse of the arm associated with the damaged hemisphere, and is referred to as learned nonuse. To encourage the use of the affected hand, patients are fitted with a sling or a mitten that restricts the nonparetic hand controlled by the intact hemisphere. Training consists of applying a behavioral training regimen in which the patients gradually relearn motor skills relevant to daily living through repetitive practice.[119] Several clinical studies have documented improvement in motor impairment and function scores after CI therapy.

In the first study to document neurophysiological plasticity associated with constraint-induced, repetitive, training of the impaired limb, squirrel monkeys were trained to retrieve food pellets from a Klüver board as described in the spontaneous recovery experiments presented in the previous section.[120] These monkeys also received a small infarct in caudal M1 as the spontaneous recovery monkeys from earlier studies (see previous section). However, unlike the spontaneous recovery monkeys, these monkeys were fitted with a soft mesh jacket that restricted the less affected hand (ipsilateral to the infarct) and were subsequently trained for 1 month on the Klüver board. Each monkey was required to retrieve at least 600 food pellets from the smallest well they could achieve per day. This training was conducted in two, 30 min sessions per day. By the end of 1 month of training, the monkeys returned to baseline levels on the most difficult food-well. The ICMS motor maps of M1 reflected improvements in behavior that contrasted with motor maps derived from monkeys after a month of spontaneous recovery. Instead of a further loss of distal hand representations and an expansion of proximal arm representations adjacent to the damaged hand area in M1, as seen after spontaneous recovery, motor skill training maintained the pre-infarct distal hand area in M1 initially spared by the infarct (see Figure 8.7). The implication of these results is that physical rehabilitation after stroke can drive physiological changes in the cortex associated with recovering skilled hand use. That is, the physiological changes observed after training more closely reflected the use of the distal hand in motor skill acquisition than those changes observed during spontaneous recovery.

Cortical reorganization following clinical rehabilitation has subsequently been studied in stroke patients. Three techniques for tracking cortical reorganization have been commonly used: positron emission tomography (PET), functional magnetic resonance imaging (fMRI), and transcortical magnetic stimulation (TMS). PET and fMRI can be used to produce functional maps of motor areas while subjects engage in relatively simple motor tasks. Typically the tasks consist of finger tapping (either simple or complex sequences of movements) or forced grip in which subjects may be required to apply varying amounts of force. The forced grip task allows examination of the moderately disabled subjects who cannot move their fingers independently. Movement related cortical activation is associated with increased metabolic activity as cellular demands for oxygen increase. TMS is a noninvasive way of stimulating cortical motor neurons through the scalp and skull. An electric current is passed through a stimulating coil that is positioned over the scalp, this generates a magnetic field that causes an electric current to flow through the cortex and depolarize neurons within a small area of the cortex. In several studies the TMS coil is moved in 1 cm increments across the scalp [121,122] however, it has been reported that mapping accuracy of 0.5 cm has been obtained.[123] When the TMS coil is held over motor areas of the cortex the stimulating current drives subtle muscle activation or motor evoked potentials that are recorded from the muscle using electromyography (EMG). Therefore, TMS can be used to derive muscle representations within motor cortex.[124]

Several clinical studies have examined stroke patients with damage restricted to one hemisphere and chronic motor deficits. The consistent findings of these studies using fMRI and PET is that after vascular damage to either cortical or sub-cortical motor structures, (e.g., capsular infarct destroying corticospinal fibers), task-related activation of cortex shows a recruitment of motor areas beyond that observed with normal subjects.[115,125-142] These areas include bilateral activation of primary and premotor cortex, and cerebellum. Several studies that evaluated the effects of CI therapy upon cortical reorganization in chronic stroke patients also found that after CI therapy the spread of activation was reduced compared to pretraining, baseline levels of activation. This reduction was towards typical activation patterns that have been observed in normal subjects.[115,135,143] Similar results were also reported with chronic stroke patients after an intensive form of rehabilitative training that did not require restriction of the non-paretic arm (task-oriented training).[131]

Studies using TMS have also shown reorganization in stroke patients with damage to the sensorimotor cortex after CI therapy. These studies have demonstrated CI therapy was successful in increasing dexterity and that this improvement was associated with an enlargement of distal hand muscle representation adjacent to the damaged cortex. An enlarged area of excitability reflects areas of the cortex that can be potentially involved in a motor task. These results are similar to those obtained from ICMS studies in animals. Several studies using fMRI or PET have demonstrated that the more difficult a task is, the more motor area is recruited within a localized functional area such as M1 or premotor cortex. There is also a nonlocalized spread of activation as multiple hand representations are being recruited.[144-147]

In general, these studies indicate that CI therapy facilitates recovery while reducing the number of spatially distinct hand representations that are recruited. Presumably, this reorganization of cortical activation occurs by strengthening motor areas that are directly affected by an ischemic infarct. That is, motor areas typically involved in the control of the paretic hand, located within the injured cortical hemisphere, regain functional significance in motor control. However, not all fMRI studies have reported reduced spread of bilateral activation associated with CI therapy. Some groups have reported the characteristic bilateral spread of activation seen with fMRI, but not the reduced activation others have reported after CI therapy.[148] Longitudinal studies have indicated that it may be necessary to track patients' recovery over time to detect a relationship between recovery and a reduced spread of cortical activation and increased laterality.[126,127,132,133,149]

Multiple motor representations of the hand, as described in the preceding section on spontaneous recovery, have been identified in primates both anatomically using tract tracing methods and physiologically using single cell recording and ICMS.[101,150] These multiple hand areas are spatially separated in the frontal cortex and include M1, PMv, PMd, SMA and the cingulate motor areas. All of these areas have direct corticospinal projections. About half of the corticospinal projection neurons are within M1, the remaining projection neurons are distributed among the nonprimary motor areas. In humans and most other primates, many of the corticospinal projections make direct contact to motor neurons. Furthermore, these areas are reciprocally connected through cortico-cortical projections within and between the hemispheres. Although each area has been shown to be more specialized for certain functions, in general, overlapping functions are well distributed throughout the various hand representations. The anatomical and functional organization between motor areas provides the substrate for a mutually supportive relation among the multiple motor representations in the frontal cortex. This is consistent with imaging results that indicate that the nonprimary motor areas are recruited at a time when M1 is maximally engaged during a complicated task or is compromised due to an infarction. Porter and Lemon (1995) consider the corticospinal tract, originating from the various motor cortices, as being involved in learning new motor skills, and not being restricted to just the execution of movements. It is this commonality of function between motor areas that supports the interpretation from imaging studies that when a major area contributing to the corticospinal tract is damaged, cortex associated with the remaining fibers are able to compensate for the lost output and promote relearning of motor function (i.e., functional recovery). Therefore, neural plasticity is likely to be a general characteristic of all motor areas of the cortex, including M1. As previously mentioned in this chapter, M1 was once considered to be relatively static, as Sherrington explained the final common pathway of cortical output, indicating that its role is limited to the execution of movements directed or learned by association motor areas. This view has changed in light of new anatomical and physiological evidence of shared involvement of premotor areas.[102]

D. Neuromodulation of Cortical Activity

The previous sections of this chapter presented evidence that cortical plasticity is initiated by use-dependent activation, resulting in functional and structural changes in connectivity; and, that these changes are associated with acquiring new motor skills or behavioral strategies, not merely motor activity. Also, as stated previously, the process for initiating cortical plasticity in areas functionally linked to the damaged cortical tissue is facilitated by a process described as excitatory diaschisis.[75] A consequence of excitatory diaschisis is the induction of LTP and LTD (synaptic mechanisms of plasticity). During this state of hyper-excitability, the cortex may become more receptive to novel, temporally associated inputs than under normal learning conditions in the intact brain.[151] In the normal brain, activity-driven cortical plasticity is modified by neurotransmitters that function as neuromodulators of synaptic activity, changing the way post-synaptic neurons respond to various afferent inputs. The monoamine neurotransmitters (norepinephrine, dopamine and serotonin) and acetylcholine, can have the effect of increasing the probability of synaptic excitation. These neuromodulators are released in response to changes in the environment. They allow for sensory-motor adaptations to biologically significant events and are thus related to motivational and attentional processes in learning. The monoamine neurotransmitters can directly influence cortical activity through ascending fiber tracts originating from the brainstem (norepinephrine from the locus coeruleus; serotonin from the raphe nucleus; and dopamine from the ventral tegmentum) or from the forebrain. Acetylcholine innervates the cortex from the nucleus basalis of Meynert in the basal forebrain. These cholinergic neurons in the forebrain can also be modified by norepinephrine and serotonin through the medial forebrain bundle.

Damage to the cholinergic system in the nucleus basalis of Meynert is associated with dementia as seen in Alzheimer's disease.[152] It has also been shown that damage to this system disrupts motor skill learning and cortical reorganization of motor maps associated with learning.[153] Artificial stimulation of the nucleus basalis of Meynert paired with a specific auditory stimulus has been shown to enhance the tonotopic, receptive fields in the primary auditory cortex associated with the paired tone.[154,155] Dopamine is associated with goal-directed behavior through cortical and subcortical systems referred to as the brain reward circuit; this includes the prefrontal cortex, the amygdala, the nucleus accumbens and the hippocampus.[156] Indirect activation of the motor cortex through mesolimbic dopaminergic stimulation of the nucleus accumbens and basal ganglia has been strongly implicated in motor learning.[157] Dopamine has also been shown to have specific modulatory effects on corticospinal tract neurons.[158] Noradrenergic and serotonergic inputs are seen throughout the cortex, and are associated with attention and arousal; both neurotransmitters have been shown to be necessary for synaptic plasticity in the visual cortex[159] and have been shown to modulate cortical excitability in the human motor cortex.[160-162]

Recent clinical studies have examined the effects of pharmacological agents that influence the activity of neuromodulators upon recovery of motor function after a stroke. The drug that has received the most attention has been dexamphetamine (d-AMPH); this drug increases monoaminergic activity by enhancing release of neurotransmitters and blocking their re-uptake, thus prolonging post-synaptic excitation. It is believed that enhanced activity of the monoamine neuromodulators (norepinephrine, dopamine and serotonin) will create a favorable environment in the cortex for adaptive, synaptic plasticity as seen early after an infarct when the cortex is in a state of hyper-excitability.[163-165] Adaptive changes in the motor systems associated with regained motor control or motor skill learning are thought to be enhanced when rehabilitative training is paired with d-AMPH administration. Many of the clinical studies have shown promising results,[166,167] but the mechanisms by which d-AMPH mediate functional recovery are still being examined.

Earlier animal studies have demonstrated that optimal recovery of motor function after cortical damage is enhanced by d-AMPH when it is paired with a related motor activity, and that improvement of motor function is mediated primarily through activation of the noradrenergic system.[168,169] Recovery of forelimb use in rats is enhanced by the combined treatment of coupling d-AMPH with a motor skill activity after a middle cerebral artery (MCA) occlusion, and this recovery is associated with synaptogenesis in the motor and parietal cortex of the injured hemisphere.[170] This finding is consistent with activity-dependent changes cited throughout this chapter and suggests that d-AMPH can facilitate neural communication that leads to adaptive sensory-motor integration necessary for motor skill acquisition. In addition to use-dependent strengthening of pre-existing connections, neural activity also leads to the release of neurotrophic factors that not only strengthens connections, but is also neuroprotective and induces neural sprouting and synaptogenesis. Amphetamine, as well as other drugs that stimulate noradrenergic activity, have been shown to increase neurotrophic factors in the brain.[171-175]

Several neurotrophic factors have been identified; the first to be identified was Nerve Growth Factor (NGF). Then, beginning in the 1980's Brain Derived Neurotrophic Factor (BDNF) and neurotrophin-3 (NT-3) were identified, later neurotrophin 4 and 5 (NT-4/5) were identified; each neurotrophic factor is a protein with approximately 120 amino acids.[176] NGF, BDNF, NT-3 and NT-4/5 are structurally related members of the same gene family, and are referred to as neurotrophins. The neurotrophins bind to a low affinity p75 receptor (a 75 kD glycoprotein receptor); each neurotrophin also binds to a high affinity tyrosine kinase receptor (trk). There are three tyrosine kinase receptors that bind neurotrophins: trkA binds NGF; trkB binds BDNF and NT-4/5; and trkC binds NT-3.[176,177] Originally, the neurotrophins were recognized for their role in promoting cell survival, but their role in activity-dependent, synaptic plasticity has been recently identified.[177,178]

BDNF is especially interesting regarding its possible role in synaptic plasticity within motor cortex. It has been found mostly in the cerebellum, hippocampus and neocortex and its high affinity trkB receptors have been found on neuronal dendrites providing a direct route for BDNF associated synaptic plasticity.[178,179] Accordingly, BDNF activity has been associated with dendritic growth and an increase in dendritic spine density.[177] Several studies have shown that BDNF is linked with LTP in the hippocampus, visual cortex and insular cortex.[177,178,180-183] In addition to activity-dependent plasticity, BDNF has been shown to be necessary for the maintenance of movement representations in the rat motor cortex.[184] Blocking BDNF synthesis with injections of BDNF antisense oligodexynucleotide within the forelimb motor area of the rat cortex resulted in a loss of forelimb movement representation. There was no loss of forelimb representation following a control injection of scrambled oligodexynucleotide that does not block BDNF synthesis. These results suggested that normal BDNF activity is necessary to maintain the functional integrity of cortical representations.

There is also evidence to suggest that BDNF may be partially responsible for D-AMPH mediated recovery from stroke. As stated earlier in this section, previous animal studies have indicated that D-AMPH facilitates behavioral recovery mainly through noradrenergic activation. Recent studies have shown a relationship between noradrenergic activity and BDNF/trkB activity. It has been shown that BDNF is synthesized in noradrenergic neurons and released anterogradely from noradrenergic terminals.[179] It is speculated that the corelease of nerurotransmitters with BDNF would have a more dramatic effect on post-synaptic neural activity than either stimuli alone.[173] Also, pharmacological activation of noradrenergic neurons with yohimbine (an α2-adrenergic antagonist) has led to rapid and transient trkB receptor activation in the rat cortex. When noradrenergic activity was suppressed by pretreatment with prazosin (an α1-adrenergic antogonist) the yohimbine effect was partially inhibited. These results are interpreted as evidence that pharmacological activation of noradrenergic neurons leads to a rapid release of BDNF upon noradrenergic target neurons in the cortex.[173] There is also evidence to suggest that the anti-depressants that increase noradrenergic and/or serotonergic activity (e.g., tricyclics and fluoxetine), stimulate the release of BDNF.[175,185-187]

These studies point toward the possibility that the molecular mechanisms that facilitate the success of post-stroke related rehabilitative therapy such as CI therapy may involve BDNF associated plasticity. Both motor activity and spatial learning have been shown to increase BDNF expression in the rat hippocampus[188] and in related cortical areas and is dependent upon noradrenergic activation.[189,190] This may be why motor activity alone seems to be neuroprotective: treadmill training prior to a focal ischemic cortical infarct has also been shown to be beneficial by reducing the loss of movement representations in the motor cortex.[184] It is also known that BDNF activity is increased following an MCA occlusion and is important for maintaining the survival of spared cortical neurons following a transient MCA occlusion.[191,192] Thus, physical activity especially associated with learning may have positive therapeutic benefits after brain damage due in part to enhancement of BDNF activity. However, two recent rat studies in which the beneficial influences of post-ishcemic exposure to an enriched environment did not find a positive relation between rehabilitative treatment and increased levels of BDNF as expected.[193,194]

In these two environmental enrichment studies BDNF levels found in peri-infarct cortical tissue, contralateral cortical tissue and hippocampus of rats exposed to the enriched environment after an MCA occlusion were lower than BDNF levels found in rats exposed to the standard housing conditions.[193] This same relationship between BDNF and housing conditions were found in a later study that examined cortical expression of BDNF in tissue affected by an MCA occlusion.[194] These results were unexpected since it has been shown that environmental enrichment improves behavioral recovery following focal ischemia.[11] It is not clear why environmental enrichment would reduce BDNF expression relative to rats in the standard housing conditions, but these studies indicate that the behavioral improvements associated with experience may not be entirely dependent on BDNF. Many other molecular events can mediate favorable recovery from cortical insults. What does seem to be clear on both a systems level and a molecular level is that induction of neural activation is necessary for establishing and maintaining adaptive cortical representations, and that dynamic changes in neural connections depend upon neuromodulating chemicals that are associated with learning and motivation. The relationship between learning, activity, and neural plasticity regarding the consequences of neural injury was summarized by Kesslak in the conclusion to their study demonstrating a relationship between motor activity, learning, and BNDF expression. In part the conclusion states, "These results present the interesting possibility that activity and learning may serve to store information and initiate mechanisms to protect and ensure the survival of the participating circuitry. … We suggest that a cascade of molecular events in learning, including protein synthesis, receptor binding, and other cellular activity, can induce growth and protect those cells from damage. … If activity and learning promote expression of neurotrophic factors, the results of the present study lend molecular evidence to the old adage of 'use it or lose it'."[188]

REFERENCES

1. Stoney, S. D., Jr., Thompson, W. D., and Asanuma, H., Excitation of pyramidal tract cells by intracortical microstimulation: effective extent of stimulating current, *J Neurophysiol* 31 (5), 659-69, 1968.
2. Asanuma, H. and Rosen, I., Spread of mono- and polysynaptic connections within cat's motor cortex, *Exp Brain Res* 16 (5), 507-20, 1973.
3. Matsumura, M., Chen, D., Sawaguchi, T., Kubota, K., and Fetz, E. E., Synaptic interactions between primate precentral cortex neurons revealed by spike-triggered averaging of intracellular membrane potentials *in vivo, J Neurosci* 16 (23), 7757-67, 1996.
4. Rosenzweig, M. R. and Bennett, E. L., Psychobiology of plasticity: effects of training and experience on brain and behavior, *Behav Brain Res* 78 (1), 57-65, 1996.
5. Yuste, R. and Sur, M., Development and plasticity of the cerebral cortex: from molecules to maps, *J Neurobiol* 41 (1), 1-6, 1999.
6. Grossman, A. W., Churchill, J. D., Bates, K. E., Kleim, J. A., and Greenough, W. T., A brain adaptation view of plasticity: is synaptic plasticity an overly limited concept?, *Prog Brain Res* 138, 91-108, 2002.

7. Mohammed, A. H., Zhu, S. W., Darmopil, S., JHjerling-Leffler, J., Ernfors, P., Winblad, B., Diamond, M. C., Eriksson, P. S., and Bogdanovic, N., Environmental enrichment and the brain., *Prog Brain Res* 138, 109-103, 2002.
8. Jones, T. A., Klintsova, A. Y., Kilman, V. L., Sirevaag, A. M., and Greenough, W. T., Induction of multiple synapses by experience in the visual cortex of adult rats, *Neurobiol Learn Mem* 68 (1), 13-20, 1997.
9. Black, J. E., Zelazny, A. M., and Greenough, W. T., Capillary and mitochondrial support of neural plasticity in adult rat visual cortex, *Exp Neurol* 111 (2), 204-9, 1991.
10. Kolb, B., Gorny, G., Soderpalm, A. H., and Robinson, T. E., Environmental complexity has different effects on the structure of neurons in the prefrontal cortex versus the parietal cortex or nucleus accumbens, *Synapse* 48 (3), 149-53, 2003.
11. Johansson, B. B. and Belichenko, P. V., Neuronal plasticity and dendritic spines: effect of environmental enrichment on intact and postischemic rat brain, *J Cereb Blood Flow Metab* 22 (1), 89-96, 2002.
12. Williams, B. M., Luo, Y., Ward, C., Redd, K., Gibson, R., Kuczaj, S. A., and McCoy, J. G., Environmental enrichment: effects on spatial memory and hippocampal CREB immunoreactivity, *Physiol Behav* 73 (4), 649-58, 2001.
13. Kleim, J. A., Lussnig, E., Schwarz, E. R., Comery, T. A., and Greenough, W. T., Synaptogenesis and Fos expression in the motor cortex of the adult rat after motor skill learning, *J Neurosci* 16 (14), 4529-35, 1996.
14. Kleim, J. A., Vij, K., Ballard, D. H., and Greenough, W. T., Learning-dependent synaptic modifications in the cerebellar cortex of the adult rat persist for at least four weeks, *J Neurosci* 17 (2), 717-21, 1997.
15. Kleim, J. A., Swain, R. A., Armstrong, K. A., Napper, R. M., Jones, T. A., and Greenough, W. T., Selective synaptic plasticity within the cerebellar cortex following complex motor skill learning, *Neurobiol Learn Mem* 69 (3), 274-89, 1998.
16. Kleim, J. A., Pipitone, M. A., Czerlanis, C., and Greenough, W. T., Structural stability within the lateral cerebellar nucleus of the rat following complex motor learning, *Neurobiol Learn Mem* 69 (3), 290-306, 1998.
17. Kolb, B. and Whishaw, I. Q., Brain plasticity and behavior, *Annu Rev Psychol* 49, 43-64, 1998.
18. Fuster, J. M., *Memory in the Cerebral Cortex: An Empirical Approach to Neural Networks in the Human and Non-Human Primate,* MIT Press, 1999.
19. Buonomano, D. V. and Merzenich, M. M., Cortical plasticity: from synapses to maps, *Annu Rev Neurosci* 21, 149-86, 1998.
20. Mountcastle, V. B., Modality and topographic properties of single neurons of cat's somatic sensory cortex, *J Neurophys* 20, 408-434, 1957.
21. Mountcastle, V. B., *Perceptual Neuroscience: The Cerebral Cortex*, Harvard Univ. Press, 1998.
22. Asanuma, H. and Sakata, H., Functional organization of a cortical efferent system examined with focl depth stimulation in cats, *J Neurophysiol* 30, 35-54, 1967.
23. Asanuma, H. and Ward, J. E., Patterns of contraction of distal forelimb muscles produced by intracortical stimulation in cats, *Brain Res* 27 (1), 97-109, 1971.
24. Rosén, I. and Asanuma, H., Peripheral afferent inputs to the forelimb area of the monkey motor cortex: input-output relations, *Exp Brain Res* 14, 257-273, 1972.
25. Lowel, S. and Singer, W., Selection of intrinsic horizontal connections in the visual cortex by correlated neuronal activity, *Science* 255 (5041), 209-12, 1992.
26. Huntley, G. W. and Jones, E. G., Relationship of intrinsic connections to forelimb movement representations in monkey motor cortex: a correlative anatomic and physiological study, *J Neurophysiol* 66 (2), 390-413, 1991.

27. Aroniadou, V. A. and Keller, A., The patterns and synaptic properties of horizontal intracortical connections in the rat motor cortex, *J Neurophysiol* 70 (4), 1553-69, 1993.
28. Hess, G. and Donoghue, J. P., Long-term potentiation of horizontal connections provides a mechanism to reorganize cortical motor maps, *J Neurophysiol* 71 (6), 2543-7, 1994.
29. Hess, G. and Donoghue, J. P., Long-term depression of horizontal connections in rat motor cortex, *Eur J Neurosci* 8 (4), 658-65, 1996.
30. Jones, E. G., GABAergic neurons and their role in cortical plasticity in primates, *Cereb Cortex* 3 (5), 361-72, 1993.
31. Rioult-Pedotti, M. S., Friedman, D., and Donoghue, J. P., Learning-induced LTP in neocortex, *Science* 290 (5491), 533-6, 2000.
32. Merzenich, M. M., Sources of intraspecies and interspecies cortical map variability in mammals, in *Comparative Neurobiology: Modes of Communication in the Nervous System*, Cohen, M. J. andStrumwaser, F. John Wiley, New York, 1985, pp. 105-116.
33. Nudo, R. J., Milliken, G. W., Jenkins, W. M., and Merzenich, M. M., Use-dependent alterations of movement representations in primary motor cortex of adult squirrel monkeys, *J Neurosci* 16 (2), 785-807, 1996.
34. Chang, H.-T., Ruch, C., and Ward, A. A., Topographical representation of muscles in motor cotex of monkeys, *J Neurophysiol* 10, 39-56, 1947.
35. Woolsey, C. N., Organization of somatic sensory and motor areas of the cerebral cortex., in *Biological and Biochemical Bases of Behavior*, Harlow, H. R. and Woolsey, C. N. University of Wisconsin Press, Madison, 1958, pp. 63-82.
36. Asanuma, H. and Rosén, I., Topographical organization of cortical efferent zones projecting to distal forelimb muscles in the monkey, *Exp Brain Res* 14, 243-256, 1972.
37. Kwan, H. C., MacKay, W. A., Murphy, J. T., and Wong, Y. C., Spatial organization of precentral cortex in awake primates. II. Motor outputs, *J Neurophys* 41 (5), 1120-1131, 1978.
38. Sessle, B. J. and Wiesendanger, M., Structural and functional definition of the motor cortex in the monkey (Macaca fascicularis), *J Physiol* 323, 245-65, 1982.
39. Strick, P. L. and Preston, J. B., Two representations of the hand in area 4 of a primate. II. Somatosensory input organization, *J Neurophys* 48 (1), 150-159, 1982.
40. Donoghue, J. P., Leibovic, S., and Sanes, J. N., Organization of the forelimb area in squirrel monkey motor cortex: representation of digit, wrist, and elbow muscles, *Exp Brain Res* 89 (1), 1-19, 1992.
41. Nudo, R. J., Jenkins, W. M., Merzenich, M. M., Prejean, T., and Gedela, R., Neurophysiological correlates of hand preference in primary motor cortex of squirrel monkeys, *J Neurosci* 12 (8), 2918-2947, 1992.
42. Sanes, J. N., Donoghue, J. P., Thangaraj, V., Edelman, R. R., and Warach, S., Shared neural substrates controlling hand movements in human motor cortex, *Science* 268, 1775-1777, 1995.
43. Poliakov, A. V. and Schieber, M. H., Limited functional grouping of neurons in the motor cortex hand area during individuated finger movements: A cluster analysis, *J Neurophysiol* 82 (6), 3488-505, 1999.
44. Schieber, M. H., Somatotopic gradients in the distributed organization of the human primary motor cortex hand area: evidence from small infarcts, *Exp Brain Res* 128 (1-2), 139-48, 1999.
45. Plautz, E. J., Nudo, R. J., Barbay, S., Friel, K. M., Kleim, J. A., and Frost, J. A., Persistence of neurophysiological changes associated with motor skill acquisition, in *Soc Neurosci*, New Orleans, 1999, pp. 1409.

46. Plautz, E. J., Milliken, G. W., and Nudo, R. J., Effects of repetitive motor training on movement representations in adult squirrel monkeys: role of use versus learning, *Neurobiol Learn Mem* 74 (1), 27-55, 2000.
47. Kleim, J. A., Barbay, S., and Nudo, R. J., Functional reorganization of the rat motor cortex following motor skill learning, *J Neurophysiol* 80 (6), 3321-5, 1998.
48. Kosar, E., Waters, R. S., Tsukahara, N., and Asanuma, H., Anatomical and physiological properties of the projection from the sensory cortex to the motor cortex in normal cats: the difference between corticocortical and thalamocortical projections, *Brain Res* 345 (1), 68-78, 1985.
49. Caria, M. A., Kaneko, T., Kimura, A., and Asanuma, H., Functional organization of the projection from area 2 to area 4gamma in the cat, *J Neurophysiol* 77 (6), 3107-14, 1997.
50. Yumiya, H. and Ghez, C., Specialized subregions in the cat motor cortex: anatomical demonstration of differential projections to rostral and caudal sectors, *Exp Brain Res* 53 (2), 259-76, 1984.
51. Strick, P. L., Light microscopic analysis of the cortical projection of the thalamic ventrolateral nucleus in the cat, *Brain Res* 55 (1), 1-24, 1973.
52. Strick, P. L., Anatomical analysis of ventrolateral thalamic input to primate motor cortex, *J Neurophysiol* 39 (5), 1020-31, 1976.
53. Iriki, A., Pavlides, C., Keller, A., and Asanuma, H., Long-term potentiation of thalamic input to the motor cortex induced by coactivation of thalamocortical and corticocortical afferents, *J Neurophysiol* 65 (6), 1435-41, 1991.
54. Aroniadou, V. A. and Keller, A., Mechanisms of LTP induction in rat motor cortex in vitro, *Cereb Cortex* 5 (4), 353-62, 1995.
55. Kimura, A., Melis, F., and Asanuma, H., Long-lasting changes of neuronal activity in the motor cortex of cats, *Neuroreport* 7 (4), 869-72, 1996.
56. Hepp-Raymond, M.-C., Huesler, E. J., and Maier, M. A., Precision grip in humans: temporal andspatial synergies, in *Hand and Brain: The Neurophysiology and Psychology of Hand Movements*, Wing, A. M., Haggard, P., and Flanagan, R. Academic Press, San Diego, 1996, pp. 37-62.
57. Nudo, R. J., Jenkins, W. M., and Merzenich, M. M., Repetitive microstimulation alters the cortical representation of movements in adult rats, *Somatosens Motor Res* 7 (4), 463-483, 1990.
58. Recanzone, G. H., Schreiner, C. E., and Merzenich, M. M., Plasticity in the frequency representation of primary auditory cortex following discrimination training in adult owl monkeys, *J Neurosci* 13 (1), 87-103, 1993.
59. Kilgard, M. P., Pandya, P. K., Vazquez, J., Gehi, A., Schreiner, C. E., and Merzenich, M. M., Sensory input directs spatial and temporal plasticity in primary auditory cortex, *J Neurophysiol* 86 (1), 326-38, 2001.
60. Singer, W., The formation of cooperative cell assemblies in the visual cortex, *J Exp Biol* 153, 177-97, 1990.
61. Gilbert, C. D., Plasticity in visual perception and physiology, *Curr Opin Neurobiol* 6 (2), 269-74, 1996.
62. Allard, T., Clark, S. A., Jenkins, W. M., and Merzenich, M. M., Reorganization of somatosensory area 3b representations in adult owl monkeys after digital syndactyly, *J Neurophysiol* 66 (3), 1048-58, 1991.
63. Recanzone, G. H., Merzenich, M. M., Jenkins, W. M., Grajski, K. A., and Dinse, H. R., Topographic reorganization of the hand representation in cortical area 3b owl monkeys trained in a frequency-discrimination task, *J Neurophysiol* 67 (5), 1031-56, 1992.

64. Bortoff, G. A. and Strick, P. L., Corticospinal terminations in two new-world primates: further evidence that corticomotoneuronal connections provide part of the neural substrate for manual dexterity, *J Neurosci* 13 (12), 5105-18, 1993.
65. Kleim, J. A., Barbay, S., Cooper, N. R., Hogg, T. M., Reidel, C. N., Remple, M. S., and Nudo, R. J., Motor learning-dependent synaptogenesis is localized to functionally reorganized motor cortex, *Neurobiol Learn Mem* 77 (1), 63-77, 2002.
66. Greenough, W. T., Larson, J. R., and Withers, G. S., Effects of unilateral and bilateral training in a reaching task on dendritic branching of neurons in the rat motor-sensory forelimb cortex, *Behav Neural Biol* 44 (2), 301-14, 1985.
67. Withers, G. S. and Greenough, W. T., Reach training selectively alters dendritic branching in subpopulations of layer II-III pyramids in rat motor-somatosensory forelimb cortex, *Neuropsychologia* 27 (1), 61-9, 1989.
68. Jones, T. A., Chu, C. J., Grande, L. A., and Gregory, A. D., Motor skills training enhances lesion-induced structural plasticity in the motor cortex of adult rats, *J Neurosci* 19 (22), 10153-63, 1999.
69. Rosner, B. S., Recovery of function and localization of function in historical perspective., in *Plasticity and Recovery of Function in the Central Nervous System*, Stein, D. G., J. J., R., and Butters, N. Academic Press, New York, 1974.
70. Xerri, C., Merzenich, M. M., Peterson, B. E., and Jenkins, W., Plasticity of primary somatosensory cortex paralleling sensorimotor skill recovery from stroke in adult monkeys, *J Neurophysiol* 79 (4), 2119-48, 1998.
71. Finger, S., *Origins of Neuroscience: A History of Explorations into Brain Functions,* Oxford University Press, New York, 1994.
72. Monakow, V., "Diaschisis", the localization in the cerebrum and functional impairment by cortical loci., in *Mood States and Mind*, Pribram, K. H. Penguin Press, Baltimore, 1969.
73. Stein, D. G., Brain injuries and theories of recovery., in *Restorative Neurology: Advances in Pharmacotherapy for Recovery After Stroke*, Goldstein, L. B. Futura Publishing Co. Inc., Armonk, New York, 1998, pp. 1-34.
74. Seitz, R. J., Azari, N. P., Knorr, U., Binkofski, F., Herzog, H., and Freund, H. J., The role of diaschisis in stroke recovery, *Stroke* 30 (9), 1844-50, 1999.
75. Buchkremer-Ratzmann, I., August, M., Hagemann, G., and Witte, O. W., Electrophysiological transcortical diaschisis after cortical photothrombosis in rat brain, *Stroke* 27 (6), 1105-9; discussion 1109-11, 1996.
76. Buchkremer-Ratzmann, I. and Witte, O. W., Pharmacological reduction of electrophysiological diaschisis after photothrombotic ischemia in rat neocortex, *Eur J Pharmacol* 320 (2-3), 103-9, 1997.
77. Reinecke, S., Lutzenburg, M., Hagemann, G., Bruehl, C., Neumann-Haefelin, T., and Witte, O. W., Electrophysiological transcortical diaschisis after middle cerebral artery occlusion (MCAO) in rats, *Neurosci Lett* 261 (1-2), 85-8, 1999.
78. Neumann-Haefelin, T. and Witte, O. W., Periinfarct and remote excitability changes after transient middle cerebral artery occlusion, *J Cereb Blood Flow Metab* 20 (1), 45-52, 2000.
79. Bury, S. D. and Jones, T. A., Unilateral sensorimotor cortex lesions in adult rats facilitate motor skill learning with the "unaffected" forelimb and training-induced dendritic structural plasticity in the motor cortex, *J Neurosci* 22 (19), 8597-606, 2002.
80. Witte, O. W. and Stoll, G., Delayed and remote effects of focal cortical infarctions: secondary damage and reactive plasticity, *Adv Neurol* 73, 207-27, 1997.

81. Kozlowski, D. A., James, D. C., and Schallert, T., Use-dependent exaggeration of neuronal injury after unilateral sensorimotor cortex lesions, *J Neurosci* 16 (15), 4776-86, 1996.
82. Slavin, M. D., Laurence, S., and Stein, D. G., Another look at vicariation, in *Brain Injury and Recovery*, Finger, S., Levere, T. E., Almli, C. R., and Stein, D. G. Plenum Press, New York, 1988, pp. 165-178.
83. Luria, A. R., Naydin, V. L., Tsvetkova, L. S., and Vinarskaya, E. N., Restoration of higher cortical function following local brain damage, in *Handbook of Clinical Neurology*, Vinken, P. J. and Bruyn, G. W. North-Holland, Amsterdam, 1969.
84. Glees, P. and Cole, J., Recovery of skilled motor functions after small repeated lesions of motor cortex in macaque, *J Neurophysiol* 13, 137-148, 1950.
85. Nudo, R. J., Plautz, E. J., and Frost, S. B., Role of adaptive plasticity in recovery of function after damage to motor cortex, *Muscle Nerve* 24 (8), 1000-19, 2001.
86. Nudo, R. J. and Milliken, G. W., Reorganization of movement representations in primary motor cortex following focal ischemic infarcts in adult squirrel monkeys, *J Neurophysiol* 75 (5), 2144-9, 1996.
87. Friel, K. M., Heddings, A. A., and Nudo, R. J., Effects of postlesion experience on behavioral recovery and neurophysiologic reorganization after cortical injury in primates, *Neurorehabil Neural Repair* 14 (3), 187-98, 2000.
88. Goldberger, M. E., Restitution of function in the CNS: the pathologic grasp in Macaca mulatta, *Exp Brain Res* 15 (1), 79-96, 1972.
89. Whishaw, I. Q., Pellis, S. M., Gorny, B. P., and Pellis, V. C., The impairments in reaching and the movements of compensation in rats with motor cortex lesions: an endpoint, videorecording, and movement notation analysis, *Behav Brain Res* 42 (1), 77-91, 1991.
90. Milliken, G. W., Plautz, E. J., Gardner, G. A., Raiszadeh, R., and Nudo, R. J., Reorganization of movement representations in primary motor cortex of adult sqpurrel monkeys following distal forelimb restriction., in *Soc Neurosci*, Washington D.C., 1994.
91. Milliken, G. W., Plautz, E. J., and Nudo, R. J., Recovery of finger movement representation afterr distal forelimb restriction in adult squirrel monkeys, in *Soc Neurosci*, Miami, Florida, 1995.
92. Jacobs, K. M. and Donoghue, J. P., Reshaping the cortical motor map by unmasking latent intracortical connections, *Science* 251 (4996), 944-7, 1991.
93. Colebatch, J. G., Deiber, M. P., Passingham, R. E., Friston, K. J., and Frackowiak, R. S., Regional cerebral blood flow during voluntary arm and hand movements in human subjects, *J Neurophysiol* 65 (6), 1392-401, 1991.
94. Grafton, S. T., Woods, R. P., Mazziotta, J. C., and Phelps, M. E., Somatotopic mapping of the primary motor cortex in humans: activation studies with cerebral blood flow and positron emission tomography, *J Neurophysiol* 66 (3), 735-43, 1991.
95. Rao, S. M., Binder, J. R., Hammeke, T. A., Bandettini, P. A., Bobholz, J. A., Frost, J. A., Myklebust, B. M., Jacobson, R. D., and Hyde, J. S., Somatotopic mapping of the human primary motor cortex with functional magnetic resonance imaging, *Neurology* 45 (5), 919-24, 1995.
96. Barbay, S., Peden, E. K., Falchook, G., and Nudo, R. J., Sensitivity of neurons in somatosensory cortex (S1) to cutaneous stimulation of the hindlimb immediately following a sciatic nerve crush, *Somatosens Mot Res* 16 (2), 103-14, 1999.
97. Humphrey, S. M., Gardner, G. A., Raiszadeh, R., and Nudo, R. J., Loss of sensory, but not motor responsiveness in intact cortex surrounding a focal ischemic infarct in area 4, in *Society for Neuroscience*, Washington D.C., 1994, pp. 179.

98. Nudo, R. J., Plautz, E. J., and Milliken, G. W., Adaptive plasticity in primate motor cortex as a consequence of behavioral experience and neuronal injury, *Semin Neurosci* 9, 13-23, 1997.
99. Strick, P. L. and Preston, J. B., Sorting of somatosensory afferent information in primate motor cortex, *Brain Res* 156 (2), 364-8, 1978.
100. Tanji, J. and Wise, S. P., Submodality distribution in sensorimotor cortex of the unanesthetized monkey, *J Neurophysiol* 45 (3), 467-81, 1981.
101. Passingham, R. E., *The Frontal Lobes and Voluntary Action*, 2nd ed. Oxford University Press, New York, 1997.
102. Dum, R. P. and Strick, P. L., The origin of corticospinal projections from the premotor areas in the frontal lobe, *J Neurosci* 11 (3), 667-89, 1991.
103. Nudo, R. J. and Masterton, R. B., Descending pathways to the spinal cord, III: Sites of origin of the corticospinal tract, *J Comp Neurol* 296 (4), 559-83, 1990.
104. Nakajima, K., Maier, M. A., Kirkwood, P. A., and Lemon, R. N., Striking differences in transmission of corticospinal excitation to upper limb motoneurons in two primate species, *J Neurophysiol* 84 (2), 698-709, 2000.
105. Liu, Y. and Rouiller, E. M., Mechanisms of recovery of dexterity following unilateral lesion of the sensorimotor cortex in adult monkeys, *Exp Brain Res* 128 (1-2), 149-59, 1999.
106. Frost, S. B., Barbay, S., Friel, K. M., Plautz, E. J., and Nudo, R. J., Reorganization of remote cortical regions after ischemic brain injury: a potential substrate for stroke recovery, *J Neurophysiol* 89 (6), 3205-14, 2003.
107. Dancause, N., Needham, K. D., Barbay, S., Zoubina, E. V., Frost, S. B., Plautz, E. J., Stowe, A. M., Eisner-Janowicz, I., and Nudo, R. J., Minimal lesion size in the primary motor cortex for the initiation of recovery associated physiological changes in the ventral premotor cortex., in *4th Forum of European Neuroscience*, Lisbon, 2004.
108. Dancause, N., Barbay, S., Plautz, E. J., Friel, K. M., Stowe, A. M., Zoubina, E. V., and Nudo, R. J., Redistribution of premotor cortical conections after an ischemic lesion in primary motor cortex, in *Society for Neuroscience* Viewer/Itenerary Planner, Washington D.C., Orlando, FL, 2002.
109. Widener, G. L. and Cheney, P. D., Effects on muscle activity from microstimuli applied to somatosensory and motor cortex during voluntary movement in the monkey, *J Neurophysiol* 77 (5), 2446-65, 1997.
110. Stepniewska, I., Preuss, T. M., and Kaas, J. H., Architectonics, somatotopic organization, and ipsilateral cortical connections of the primary motor area (M1) of owl monkeys, *J Comp Neurol* 330 (2), 238-71, 1993.
111. Ralston, D. D. and Ralston, H. J., 3rd, The terminations of corticospinal tract axons in the macaque monkey, *J Comp Neurol* 242 (3), 325-37, 1985.
112. Muakkassa, K. F. and Strick, P. L., Frontal lobe inputs to primate motor cortex: evidence for four somatotopically organized 'premotor' areas, *Brain Res* 177 (1), 176-82, 1979.
113. Duffau, H., Sichez, J. P., and Lehericy, S., Intraoperative unmasking of brain redundant motor sites during resection of a precentral angioma: evidence using direct cortical stimulation, *Ann Neurol* 47 (1), 132-5, 2000.
114. Nakayama, H., Jorgensen, H. S., Raaschou, H. O., and Olsen, T. S., Recovery of upper extremity function in stroke patients: the Copenhagen Stroke Study, *Arch Phys Med Rehabil* 75 (4), 394-8, 1994.

115. Levy, C. E., Nichols, D. S., Schmalbrock, P. M., Keller, P., and Chakeres, D. W., Functional MRI evidence of cortical reorganization in upper-limb stroke hemiplegia treated with constraint-induced movement therapy, *Am J Phys Med Rehabil* 80 (1), 4-12, 2001.
116. Duncan, P. W., Lai, S. M., and Keighley, J., Defining post-stroke recovery: implications for design and interpretation of drug trials, *Neuropharmacology* 39 (5), 835-41, 2000.
117. Taub, E., Miller, N. E., Novack, T. A., Cook, E. W., 3rd, Fleming, W. C., Nepomuceno, C. S., Connell, J. S., and Crago, J. E., Technique to improve chronic motor deficit after stroke, *Arch Phys Med Rehabil* 74 (4), 347-54, 1993.
118. Taub, E. and Wolf, S. L., Constraint induced movement techniques to facilitate upper extremity use in stroke patients., *Topics in Stroke Rehabilitation* 3, 38-61, 1997.
119. Taub, E., Crago, J. E., Burgio, L. D., Groomes, T. E., Cook, E. W., 3rd, DeLuca, S. C., and Miller, N. E., An operant approach to rehabilitation medicine: overcoming learned nonuse by shaping, *J Exp Anal Behav* 61 (2), 281-93, 1994.
120. Nudo, R. J., Wise, B. M., SiFuentes, F., and Milliken, G. W., Neural substrates for the effects of rehabilitative training on motor recovery after ischemic infarct, *Science* 272 (5269), 1791-4, 1996.
121. Liepert, J., Miltner, W. H., Bauder, H., Sommer, M., Dettmers, C., Taub, E., and Weiller, C., Motor cortex plasticity during constraint-induced movement therapy in stroke patients, *Neurosci Lett* 250 (1), 5-8, 1998.
122. Liepert, J., Graef, S., Uhde, I., Leidner, O., and Weiller, C., Training-induced changes of motor cortex representations in stroke patients, *Acta Neurol Scand* 101 (5), 321-6, 2000.
123. Krings, T., Buchbinder, B. R., Butler, W. E., Chiappa, K. H., Jiang, H. J., Cosgrove, G. R., and Rosen, B. R., Functional magnetic resonance imaging and transcranial magnetic stimulation: complementary approaches in the evaluation of cortical motor function, *Neurology* 48 (5), 1406-16, 1997.
124. Wassermann, E. M., Cohen, L. G., and Hallett, M., Transcranial magnetic stimulationas a tool for detecting changes in the organization of the human motor system after central and peripheral lesions., in *Cerbral Reorganization of Function After Brain Damage*, Levin, H. S. and Grafman, J. Oxford Press, New York, 2000, 304-317.
125. Cramer, S. C., Nelles, G., Benson, R. R., Kaplan, J. D., Parker, R. A., Kwong, K. K., Kennedy, D. N., Finklestein, S. P., and Rosen, B. R., A functional MRI study of subjects recovered from hemiparetic stroke, *Stroke* 28 (12), 2518-27, 1997.
126. Marshall, R. S., Perera, G. M., Lazar, R. M., Krakauer, J. W., Constantine, R. C., and DeLaPaz, R. L., Evolution of cortical activation during recovery from corticospinal tract infarction, *Stroke* 31 (3), 656-61, 2000.
127. Feydy, A., Carlier, R., Roby-Brami, A., Bussel, B., Cazalis, F., Pierot, L., Burnod, Y., and Maier, M. A., Longitudinal study of motor recovery after stroke: recruitment and focusing of brain activation, *Stroke* 33 (6), 1610-7, 2002.
128. Jang, S. H., Han, B. S., Chang, Y., Byun, W. M., Lee, J., and Ahn, S. H., Functional MRI evidence for motor cortex reorganization adjacent to a lesion in a primary motor cortex, *Am J Phys Med Rehabil* 81 (11), 844-7, 2002.
129. Johansen-Berg, H., Dawes, H., Guy, C., Smith, S. M., Wade, D. T., and Matthews, P. M., Correlation between motor improvements and altered fMRI activity after rehabilitative therapy, *Brain* 125 (Pt 12), 2731-42, 2002.

130. Foltys, H., Krings, T., Meister, I. G., Sparing, R., Boroojerdi, B., Thron, A., and Topper, R., Motor representation in patients rapidly recovering after stroke: a functional magnetic resonance imaging and transcranial magnetic stimulation study, *Clin Neurophysiol* 114 (12), 2404-15, 2003.
131. Jang, S. H., Kim, Y. H., Cho, S. H., Lee, J. H., Park, J. W., and Kwon, Y. H., Cortical reorganization induced by task-oriented training in chronic hemiplegic stroke patients, *Neuroreport* 14 (1), 137-41, 2003.
132. Jang, S. H., Kim, Y. H., Cho, S. H., Chang, Y., Lee, Z. I., and Ha, J. S., Cortical reorganization associated with motor recovery in hemiparetic stroke patients, *Neuroreport* 14 (10), 1305-10, 2003.
133. Ward, N. S., Brown, M. M., Thompson, A. J., and Frackowiak, R. S., Neural correlates of motor recovery after stroke: a longitudinal fMRI study, *Brain* 126 (Pt 11), 2476-96, 2003.
134. Ward, N. S., Brown, M. M., Thompson, A. J., and Frackowiak, R. S., Neural correlates of outcome after stroke: a cross-sectional fMRI study, *Brain* 126 (Pt 6), 1430-48, 2003.
135. Wittenberg, G. F., Chen, R., Ishii, K., Bushara, K. O., Eckloff, S., Croarkin, E., Taub, E., Gerber, L. H., Hallett, M., and Cohen, L. G., Constraint-induced therapy in stroke: magnetic-stimulation motor maps and cerebral activation, *Neurorehabil Neural Repair* 17 (1), 48-57, 2003.
136. Chollet, F., DiPiero, V., Wise, R. J., Brooks, D. J., Dolan, R. J., and Frackowiak, R. S., The functional anatomy of motor recovery after stroke in humans: a study with positron emission tomography, *Ann Neurol* 29 (1), 63-71, 1991.
137. Weiller, C., Chollet, F., Friston, K. J., Wise, R. J., and Frackowiak, R. S., Functional reorganization of the brain in recovery from striatocapsular infarction in man, *Ann Neurol* 31 (5), 463-72, 1992.
138. Weiller, C., Ramsay, S. C., Wise, R. J., Friston, K. J., and Frackowiak, R. S., Individual patterns of functional reorganization in the human cerebral cortex after capsular infarction, *Ann Neurol* 33 (2), 181-9, 1993.
139. Nelles, G., Spiekermann, G., Jueptner, M., Leonhardt, G., Muller, S., Gerhard, H., and Diener, H. C., Reorganization of sensory and motor systems in hemiplegic stroke patients. A positron emission tomography study, *Stroke* 30 (8), 1510-6, 1999.
140. Nelles, G., Spiekramann, G., Jueptner, M., Leonhardt, G., Muller, S., Gerhard, H., and Diener, H. C., Evolution of functional reorganization in hemiplegic stroke: a serial positron emission tomographic activation study, *Ann Neurol* 46 (6), 901-9, 1999.
141. Nelles, G., Jentzen, W., Jueptner, M., Muller, S., and Diener, H. C., Arm training induced brain plasticity in stroke studied with serial positron emission tomography, *Neuroimage* 13 (6 Pt 1), 1146-54, 2001.
142. Calautti, C., Leroy, F., Guincestre, J. Y., Marie, R. M., and Baron, J. C., Sequential activation brain mapping after subcortical stroke: changes in hemispheric balance and recovery, *Neuroreport* 12 (18), 3883-6, 2001.
143. Johansen-Berg, H. and Matthews, P. M., Attention to movement modulates activity in sensori-motor areas, including primary motor cortex, *Exp Brain Res* 142 (1), 13-24, 2002.
144. Catalan, M. J., Honda, M., Weeks, R. A., Cohen, L. G., and Hallett, M., The functional neuroanatomy of simple and complex sequential finger movements: a PET study, *Brain* 121 (Pt 2), 253-64, 1998.

145. Rao, S. M., Binder, J. R., Bandettini, P. A., Hammeke, T. A., Yetkin, F. Z., Jesmanowicz, A., Lisk, L. M., Morris, G. L., Mueller, W. M., Estkowski, L. D., and et al., Functional magnetic resonance imaging of complex human movements, *Neurology* 43 (11), 2311-8, 1993.
146. Remy, P., Zilbovicius, M., Leroy-Willig, A., Syrota, A., and Samson, Y., Movement- and task-related activations of motor cortical areas: a positron emission tomographic study, *Ann Neurol* 36 (1), 19-26, 1994.
147. Kawashima, R., Matsumura, M., Sadato, N., Naito, E., Waki, A., Nakamura, S., Matsunami, K., Fukuda, H., and Yonekura, Y., Regional cerebral blood flow changes in human brain related to ipsilateral and contralateral complex hand movements — a PET study, *Eur J Neurosci* 10 (7), 2254-60, 1998.
148. Schaechter, J. D., Kraft, E., Hilliard, T. S., Dijkhuizen, R. M., Benner, T., Finklestein, S. P., Rosen, B. R., and Cramer, S. C., Motor recovery and cortical reorganization after constraint-induced movement therapy in stroke patients: a preliminary study, *Neurorehabil Neural Repair* 16 (4), 326-38, 2002.
149. Calautti, C., Leroy, F., Guincestre, J. Y., and Baron, J. C., Dynamics of motor network overactivation after striatocapsular stroke: a longitudinal PET study using a fixed-performance paradigm, *Stroke* 32 (11), 2534-42, 2001.
150. Porter, R. and Lemon, R. N., *Corticospinal Function and Voluntary Movement* Clarendon Press, Oxford, 1995.
151. Castro-Alamancos, M. A. and Connors, B. W., Short-term synaptic enhancement and long-term potentiation in neocortex, *Proc Natl Acad Sci U S A* 93 (3), 1335-9, 1996.
152. Ezrin-Waters, C. and Resch, L., The nucleus basalis of Meynert, *Can J Neurol Sci* 13 (1), 8-14, 1986.
153. Conner, J. M., Culberson, A., Packowski, C., Chiba, A. A., and Tuszynski, M. H., Lesions of the Basal forebrain cholinergic system impair task acquisition and abolish cortical plasticity associated with motor skill learning, *Neuron* 38 (5), 819-29, 2003.
154. Bakin, J. S. and Weinberger, N. M., Induction of a physiological memory in the cerebral cortex by stimulation of the nucleus basalis, *Proc Natl Acad Sci USA* 93 (20), 11219-24, 1996.
155. Kilgard, M. P. and Merzenich, M. M., Cortical map reorganization enabled by nucleus basalis activity, *Science* 279 (5357), 1714-8, 1998.
156. Cooper, D. C., The significance of action potential bursting in the brain reward circuit, *Neurochem Int* 41 (5), 333-40, 2002.
157. Wickens, J. R., Reynolds, J. N., and Hyland, B. I., Neural mechanisms of reward-related motor learning, *Curr Opin Neurobiol* 13 (6), 685-90, 2003.
158. Awenowicz, P. W. and Porter, L. L., Local application of dopamine inhibits pyramidal tract neuron activity in the rodent motor cortex, *J Neurophysiol* 88 (6), 3439-51, 2002.
159. Brocher, S., Artola, A., and Singer, W., Agonists of cholinergic and noradrenergic receptors facilitate synergistically the induction of long-term potentiation in slices of rat visual cortex, *Brain Res* 573 (1), 27-36, 1992.
160. 160· Plewnia, C., Bartels, M., Cohen, L., and Gerloff, C., Noradrenergic modulation of human cortex excitability by the presynaptic alpha(2)-antagonist yohimbine, *Neurosci Lett* 307 (1), 41-4, 2001.
161. Plewnia, C., Hoppe, J., Hiemke, C., Bartels, M., Cohen, L. G., and Gerloff, C., Enhancement of human cortico-motoneuronal excitability by the selective norepinephrine reuptake inhibitor reboxetine, *Neurosci Lett* 330 (3), 231-4, 2002.
162. Ilic, T. V., Korchounov, A., and Ziemann, U., Complex modulation of human motor cortex excitability by the specific serotonin re-uptake inhibitor sertraline, *Neurosci Lett* 319 (2), 116-20, 2002.

163. Liepert, J., Storch, P., Fritsch, A., and Weiller, C., Motor cortex disinhibition in acute stroke, *Clin Neurophysiol* 111 (4), 671-6, 2000.
164. Manganotti, P., Patuzzo, S., Cortese, F., Palermo, A., Smania, N., and Fiaschi, A., Motor disinhibition in affected and unaffected hemisphere in the early period of recovery after stroke, *Clin Neurophysiol* 113 (6), 936-43, 2002.
165. Ilic, T. V., Korchounov, A., and Ziemann, U., Methylphenidate facilitates and disinhibits the motor cortex in intact humans, *Neuroreport* 14 (5), 773-6, 2003.
166. Goldstein, L. B., Effects of amphetamines and small related molecules on recovery after stroke in animals and man, *Neuropharmacology* 39 (5), 852-9, 2000.
167. Goldstein, L. B., Amphetamines and related drugs in motor recovery after stroke, *Phys Med Rehabil Clin N Am* 14 (1 Suppl), S125-34, x, 2003.
168. Goldstein, L. B. and Davis, J. N., Post-lesion practice and amphetamine-facilitated recovery of beam-walking in the rat., *Restor Neurol and Neurosci* 1, 311-314, 1990.
169. Feeney, D. M., Mechanisms of noradrenergic modulation of physical therapy: effects on functional recovery after cortical injury, in *Restorative Neurology: Advances in Pharmacotherapy for Recovery After Stroke*, Goldstein, L. B. Futura Publishing Co. Inc., Armonk, NY, 1998, pp. 35-78.
170. Stroemer, R. P., Kent, T. A., and Hulsebosch, C. E., Enhanced neocortical neural sprouting, synaptogenesis, and behavioral recovery with D-amphetamine therapy after neocortical infarction in rats, *Stroke* 29 (11), 2381-93; discussion 2393-5, 1998.
171. Semkova, I., Wolz, P., Schilling, M., and Krieglstein, J., Selegiline enhances NGF synthesis and protects central nervous system neurons from excitotoxic and ischemic damage, *Eur J Pharmacol* 315 (1), 19-30, 1996.
172. Flores, C., Rodaros, D., and Stewart, J., Long-lasting induction of astrocytic basic fibroblast growth factor by repeated injections of amphetamine: blockade by concurrent treatment with a glutamate antagonist, *J Neurosci* 18 (22), 9547-55, 1998.
173. Aloyz, R., Fawcett, J. P., Kaplan, D. R., Murphy, R. A., and Miller, F. D., Activity-dependent activation of TrkB neurotrophin receptors in the adult CNS, *Learn Mem* 6 (3), 216-31, 1999.
174. Meredith, G. E., Callen, S., and Scheuer, D. A., Brain-derived neurotrophic factor expression is increased in the rat amygdala, piriform cortex and hypothalamus following repeated amphetamine administration, *Brain Res* 949 (1-2), 218-27, 2002.
175. Saarelainen, T., Hendolin, P., Lucas, G., Koponen, E., Sairanen, M., MacDonald, E., Agerman, K., Haapasalo, A., Nawa, H., Aloyz, R., Ernfors, P., and Castren, E., Activation of the TrkB neurotrophin receptor is induced by antidepressant drugs and is required for antidepressant-induced behavioral effects, *J Neurosci* 23 (1), 349-57, 2003.
176. Kuno, M., *The Synapse: Function, Plasticity and Neurotrophism*, Oxford Univ Press, New York, 1995.
177. Lim, K. C., Lim, S. T., and Federoff, H. J., Neurotrophin secretory pathways and synaptic plasticity, *Neurobiol Aging* 24 (8), 1135-45, 2003.
178. McAllister, A. K., Katz, L. C., and Lo, D. C., Neurotrophins and synaptic plasticity, *Annu Rev Neurosci* 22, 295-318, 1999.
179. Fawcett, J. P., Bamji, S. X., Causing, C. G., Aloyz, R., Ase, A. R., Reader, T. A., McLean, J. H., and Miller, F. D., Functional evidence that BDNF is an anterograde neuronal trophic factor in the CNS, *J Neurosci* 18 (8), 2808-21, 1998.
180. Akaneya, Y., Tsumoto, T., Kinoshita, S., and Hatanaka, H., Brain-derived neurotrophic factor enhances long-term potentiation in rat visual cortex, *J Neurosci* 17 (17), 6707-16, 1997.

181. Alder, J., Thakker-Varia, S., Bangasser, D. A., Kuroiwa, M., Plummer, M. R., Shors, T. J., and Black, I. B., Brain-derived neurotrophic factor-induced gene expression reveals novel actions of VGF in hippocampal synaptic plasticity, *J Neurosci* 23 (34), 10800-8, 2003.
182. Ernfors, P. and Bramham, C. R., The coupling of a trkB tyrosine residue to LTP, *Trends Neurosci* 26 (4), 171-3, 2003.
183. Escobar, M. L., Figueroa-Guzman, Y., and Gomez-Palacio-Schjetnan, A., *In vivo* insular cortex LTP induced by brain-derived neurotrophic factor, *Brain Res* 991 (1-2), 274-9, 2003.
184. Kleim, J. A., Jones, T. A., and Schallert, T., Motor enrichment and the induction of plasticity before or after brain injury, *Neurochem Res* 28 (11), 1757-69, 2003.
185. Russo-Neustadt, A., Ha, T., Ramirez, R., and Kesslak, J. P., Physical activity-antidepressant treatment combination: impact on brain-derived neurotrophic factor and behavior in an animal model, *Behav Brain Res* 120 (1), 87-95, 2001.
186. Dias, B. G., Banerjee, S. B., Duman, R. S., and Vaidya, V. A., Differential regulation of brain derived neurotrophic factor transcripts by antidepressant treatments in the adult rat brain, *Neuropharmacology* 45 (4), 553-63, 2003.
187. Garza, A. A., Ha, T. G., Garcia, C., Chen, M. J., and Russo-Neustadt, A. A., Exercise, antidepressant treatment, and BDNF mRNA expression in the aging brain, *Pharmacol Biochem Behav* 77 (2), 209-20, 2004.
188. Kesslak, J. P., So, V., Choi, J., Cotman, C. W., and Gomez-Pinilla, F., Learning upregulates brain-derived neurotrophic factor messenger ribonucleic acid: a mechanism to facilitate encoding and circuit maintenance?, *Behav Neurosci* 112 (4), 1012-9, 1998.
189. Cotman, C. W. and Berchtold, N. C., Exercise: a behavioral intervention to enhance brain health and plasticity, *Trends Neurosci* 25 (6), 295-301, 2002.
190. Garcia, C., Chen, M. J., Garza, A. A., Cotman, C. W., and Russo-Neustadt, A., The influence of specific noradrenergic and serotonergic lesions on the expression of hippocampal brain-derived neurotrophic factor transcripts following voluntary physical activity, *Neuroscience* 119 (3), 721-32, 2003.
191. Kokaia, Z., Andsberg, G., Yan, Q., and Lindvall, O., Rapid alterations of BDNF protein levels in the rat brain after focal ischemia: evidence for increased synthesis and anterograde axonal transport, *Exp Neurol* 154 (2), 289-301, 1998.
192. Saarelainen, T., Lukkarinen, J. A., Koponen, S., Grohn, O. H., Jolkkonen, J., Koponen, E., Haapasalo, A., Alhonen, L., Wong, G., Koistinaho, J., Kauppinen, R. A., and Castren, E., Transgenic mice overexpressing truncated trkB neurotrophin receptors in neurons show increased susceptibility to cortical injury after focal cerebral ischemia, *Mol Cell Neurosci* 16 (2), 87-96, 2000.
193. Zhao, L. R., Mattsson, B., and Johansson, B. B., Environmental influence on brain-derived neurotrophic factor messenger RNA expression after middle cerebral artery occlusion in spontaneously hypertensive rats, *Neuroscience* 97 (1), 177-84, 2000.
194. Zhao, L. R., Risedal, A., Wojcik, A., Hejzlar, J., Johansson, B. B., and Kokaia, Z., Enriched environment influences brain-derived neurotrophic factor levels in rat forebrain after focal stroke, *Neurosci Lett* 305 (3), 169-72, 2001.

9 Reorganization of Motor Cortex after Damage to the Motor System

Jon H. Kaas

CONTENTS

I. Introduction 189
II. Motor Cortex Reorganization after Amputations, Nerve Injury, and Spinal Cord Damage in Mature and Developing Rats 190
III. Motor System Plasticity after Cortical Lesions in Rats 193
IV. Motor Cortex Reorganization after the Loss of a Limb in Mature and Developing Primates 194
V. Do New Connections Contribute To Motor Cortex Reorganization? 197
VI. Reorganization after Focal Lesions of Motor Cortex in Monkeys 198
VII. Motor Cortex Reorganization in Humans 199
VIII. Plasticity after Lesions of Motor Cortex in Humans 200
IX. Summary and Conclusions 200
References 202

I. INTRODUCTION

Sensory and motor systems of primates and other mammals reorganize after damage.[1] Much of the evidence comes from changes in the somatotopy of primary somatosensory or primary motor cortex after peripheral nerve injury that alters the pattern of afferent input and the motor outflow. Some of the adjustments are the immediate result of a rebalancing of patterns of activation and inhibition as sources of drive on excitatory and inhibitory neurons are altered. Such rebalancing of dynamic systems reconfigures sensory and motor maps in minor ways, which reveal latent neural circuit properties. Other changes emerge over seconds to hours due to a range of activity-dependent cellular mechanisms that affect synaptic strengths. Over somewhat longer periods of days to weeks, synaptic sites might be lost or gained as local circuits grow and rearrange. Over weeks to months, considerable new growth of axons and synaptic contacts can occur over distances that considerably alter the functional organization of sensory and motor systems, sometimes in ways that promote behavioral recovery, and sometimes in ways with undesirable outcomes.

One goal of research on sensorimotor plasticity is to understand the mechanisms of change, how to manipulate them in order to maximize recovery after sensory and motor loss, and minimize negative outcomes in human patients with neural damage.

While many aspects of sensorimotor plasticity have been studied, this review focuses on changes in the motor systems that follow a particularly severe type of motor system damage, the loss of a forelimb or hindlimb. In humans, badly damaged limbs might require amputation, and it is important to determine what happens to the somatosensory and motor systems as a result of the major loss of both the sensory afferents from the limb and the motor neuron outflow to the muscles of that limb. We now know from noninvasive studies in humans that long-standing amputations alter the representation of the body and body movements in somatosensory and motor cortex. Important aspects of the reorganization process have been determined in animal studies. Several basic discoveries on motor cortex reorganization have been derived from studies after motor nerve damage in rats,[2,3,4] and the results of such experiments are reviewed here. Other studies have been on prosimian and simian primates that have been accidentally injured, and have required the therapeutic amputation of a limb. Fortunately, such injuries have been rare in captive primates, but a few have occurred, and some of these primates have become available for study years after the amputations. Thus, the consequences of limb amputation on the motor system have been studied in these primates. The bulk of this review focuses on the results from these few, but informative, investigations. Consequences of motor cortex, nerve, and spinal cord injury are also noted. The final section of this review relates some of the findings from animal studies to results of cortical stimulation studies in humans.

II. MOTOR CORTEX REORGANIZATION AFTER AMPUTATIONS, NERVE INJURY, AND SPINAL CORD DAMAGE IN MATURE AND DEVELOPING RATS

The sensorimotor system of even small-brained mammals is complex, and involves a number of nuclei, cortical areas, and interconnecting pathways.[5] However, all placental mammals have a primary motor area of cortex (M1) that is located just anterior to somatosensory cortex. Neurons in M1 project to pools of neurons in the brainstem and spinal cord where they directly or indirectly activate motor neurons projecting to muscles. The experiments on motor system plasticity in rats have concentrated on the organizational changes in M1 that follow peripheral motor nerve injury. The focus of such studies has been to reveal changes in the functional organization of motor cortex in experiments where microelectrodes are systematically used to stimulate arrays of sites within M1. Such stimulations produce muscle movements, and the part of the body moved depends on the location stimulated in M1.

Somatosensory inputs to M1 of mammals provide an organized representation or map of the cutaneous receptors of the opposite side of the body,[6] and a similarly organized map of the movements of body parts on the opposite side of the body.[7] Neither the sensory nor the motor map in M1 appears to be as precisely or predictably organized as the somatosensory map in primary somatosensory cortex, but the maps have a global order that is consistent across individuals and similar across species. Thus, the movement map in M1 of rats and other mammals proceeds from hindlimb movements medially, to face and tongue movements laterally. In microstimulation experiments, M1 can be divided into territories devoted to movements within major divisions of the body, such as forelimb or face, and the effects of the loss of motor neuron outputs to these body regions on the organization of the evoked movement map in M1 can be studied. However, the consequences of electrically stimulating any site in M1 depend on the parameters of stimulation. High levels of current and long trains of current pulses produce complex sequences of movements often involving a number of body parts,[8] while low levels of current and short sequences of pulses often produce movements or twitches in a few isolated muscles. Movement maps in M1 are typically based on low current levels that produce just noticeable movements of a body part or muscle. Thus, movement maps in mammals with motor nerve injury need to be interpreted concerning both the somatotopic pattern produced, and the levels of current needed to produce that pattern.

The normal organization of M1 in rats is known from a number of studies using microstimulation at threshold levels.[9] An early set of experiments on motor system plasticity,[2,3] described the effects of either a forelimb amputation or a facial motor nerve transection on the organization of M1 in adult rats studied one week to five months after injury. Both of these injuries sectioned the axons of a major motor neuron pool that was targeted directly or indirectly by corticospinal neurons in a well-defined zone in M1. One prediction of post-surgical results would be that microstimulation of neurons throughout the deprived zone of M1 would fail to produce movements, unless perhaps very high levels of stimulating current were used. Instead, stimulations of the deprived M1 zone, at normal threshold levels of current, produced movements of body parts adjacent to the missing or denervated body part. When the motor nerve to the muscles that move the facial vibrissa was cut, stimulation of the deprived vibrissa cortex produced forelimb, eye, and eyelid movements. After forelimb amputation, stimulation of the deprived forelimb cortex moved the shoulder and stump of the missing limb (Figure 9.1).

In a second set of experiments with a chronically implanted stimulating electrode, changes in muscle targets did not appear immediately after the stimulated location was deprived by nerve section, but forelimb activity could be evoked from vibrissa cortex within hours of vibrissa nerve section.[4] Thus, reorganization of motor cortex appeared to occur within hours[10,11,12] and remain stable over months of recovery. Because of the rapid change in the organization of M1, the reorganization was attributed to a rapid strengthening of existing synaptic contacts, rather than the formation of new connections that would likely take longer to grow and become functional.

One mechanism proposed for such reorganization was the long-term potentiation (LTP) of horizontal connections that are intrinsic to M1 and interconnect functionally distinct sectors of M1.[11,13,14] However, it is not known if horizontal connections are potentiated during the course of the rapid reorganization of the motor map in M1, and what factors could be responsible for such potentiation.

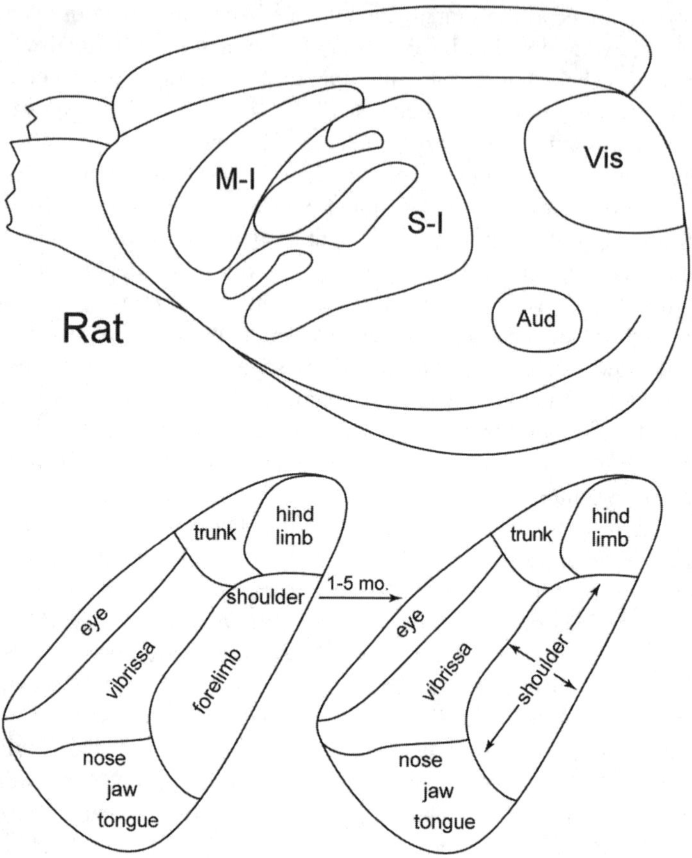

FIGURE 9.1 The reorganization of primary motor cortex, M1, in adult rats after amputation of a forelimb. A dorsolateral view of a brain of a rat (above) shows the location of M1 relative to primary somatosensory cortex (S1) primary visual cortex (Vis.) and auditory cortex (Aud.). The somatotopy of the movement map is shown in enlarged views of M1 below. In normal M1 (left), electrical stimulation of microelectrode site in caudomedial M1 produces hindlimb movements, and more rostrally, trunk movements. A large caudal region is devoted to forelimb movements, including a medial zone for shoulder movements. More rostrally, a large region relates to facial vibrissa movements, and a narrow zone is for eye movements and eye blinks. Rostrolaterally, M1 represents nose, jaw and tongue movements. One to five months (1–5 mos.) after forelimb amputation, the map in M1 has changed (right) so that stimulation throughout the former territory of forelimb cortex evokes shoulder (forelimb stump) movements.[3]

One possibility is that the amputations and peripheral nerve lesions reduce the overall level of activity in M1 as a result of the loss of sensory activation based on movements. Reduced neural activity in M1 would result in activity dependent reductions in the expression of the inhibitory neural transmitter, GABA, and receptors for GABA.[14] GABA and GABA receptors can be rapidly down regulated, even within hours, by the suppression of neural activity.[15–18] Both motor nerve section and amputation would reduce the sensory activation of motor cortex, and this would likely reduce the overall level of inhibition in M1. However, sectioning sensory neurons only, while leaving motor nerves intact, would also reduce the activity in M1, but peripheral sensory nerve section by itself does not appear to produce reorganization in M1 of rats.[19] Thus, there is no direct evidence to support the hypothesis that reduced sensory drive of motor cortex produces reorganization of M1 by reducing inhibition so that horizontal excitatory connections are facilitated to the extent that they are strengthened by LTP.

Reorganization of motor cortex (M1) in rats has also been studied after the descending corticospinal tract in the lower thoracic spinal cord was cut.[20] Such lesions abolished evoked muscle activity in the hindlimb when the hindlimb portion of M1 was microstimulated. In these experiments, there were no rapid reorganizations of the motor cortex, as microstimulation of the hindlimb cortex after two days of recovery continued to produce no motor responses. However, after 4 weeks of recovery, microstimulation of the hindlimb cortex produced whisker, forelimb, and trunk movements, demonstrating a functional reorganization of the motor cortex. Anterograde tracing of corticospinal axons from the reorganized hindlimb portion of M1 indicated that many of these axons had sprouted into the cervical spinal cord. Thus, much of this slowly emerging reorganization of the motor cortex could be attributed to the growth of new connections so that corticospinal axons reached new motor neuron pools.

In related experiments, transected hindlimb corticospinal axons sprouted to contact propriospinal neurons that projected past the lesion to lumbar motor neurons for the hindlimb.[2] After 3 to 12 weeks of recovery, neurons in the hindlimb portion of M1 were transynaptically labeled by the retrograde transport of pseudorabies virus injected into the hindlimb muscles. Thus, neurons in the hindlimb motor cortex had indirect access to hindlimb muscles. The sprouting of cut corticospinal axons to form new connections likely contributed to the considerable behavioral recovery of hindlimb function in these rats.

III. MOTOR SYSTEM PLASTICITY AFTER CORTICAL LESIONS IN RATS

The plasticity of motor cortex in neonatal rats has also been investigated after partial lesion of M1 of one hemisphere. After these lesions, motor performance improved over a period of weeks, and M1 of the intact hemisphere formed new connections with the deafferented striatum, red nucleus, basilar pontine nuclei, and grey matter of the spinal cord.[22] While this extensive growth of new connections did not occur after such lesions in mature rats, new connections did form when anti-Nogo-A antibodies are used to block the Nogo-A protein (a neurite growth inhibitor) on the surface of oligodendrocytes.[22] New axon growth after cortical lesions has not yet been demonstrated without such treatments. However, unilateral lesions of M1 in adult rats may be followed by growth of dendrites in pyramidal cells of the opposite, intact M1,[23,24] possibly as a result of behavioral compensations and greater use of the intact M1.

IV. MOTOR CORTEX REORGANIZATION AFTER THE LOSS OF A LIMB IN MATURE AND DEVELOPING PRIMATES

Because injuries extensive enough to require therapeutic amputation of a limb are fortunately rare, the motor systems of only a few such primates have been available for experimental study. Investigations of the effects of this type of major loss of motor neuron targets and somatosensory feedback on the organization of primary motor cortex, M1, and the connections of the motor system are limited to two galagos (prosimian primates) and three squirrel monkeys,[25] and five macaque monkeys.[26,27] Yet, much has been learned from these few animals where motor cortex organization contralateral to an amputated limb was compared to motor cortex organized in normal primates.

As in other mammals, M1 of primates is located just rostral to somatosensory cortex. This relationship is shown in Figure 9.2 for a squirrel monkey, a small, nocturnal New World monkey that provides the advantage of only having a shallow central dimple or fold rather than a true central sulcus. Thus, all of M1[7] and almost all of somatosensory cortex[28] are exposed on the dorsolateral surface of the brain for direct access with stimulating or recording microelectrodes. As shown, M1 includes large sectors devoted to hindlimb, trunk, forelimb, face, and tongue. At a finer level of analysis, M1 contains a mosaic of efferent zones, each related to a specific movement at near threshold levels of stimulation. Similar movements might be evoked from several nearby, but separate sites, and movements evoked from adjacent sites need not be of the same body part. In the forelimb region of galagos and monkeys, sites producing digit, wrist, forearm, and shoulder movements are mixed with no obvious order. Galagos have few sites for digit movements, and they usually involve several digits, while macaques have many sites for digit movements.

The posterior border of M1 is formed by area 3a, which receives muscle spindle information from the thalamus, and then area 3b (S1 proper), which is activated by slowly adapting and rapidly adapting cutaneous receptors. Microstimulation of sites in area 3a and even 3b often result in evoked movements, sometimes at current thresholds that are comparable to those for M1. The forelimb representations are aligned across M1, 3a, and 3b. Thus, microstimulation results alone might not always identify the border between M1 and area 3a. In galagos, M1 is conveniently located between two shallow frontal sulci, FSa just anterior to M1 and FSp just posterior to M1 along the face-hand representational border of areas 3a and 3b.[29] Squirrel monkeys have a shallow, short central sulcus that disrupts part of the hand, forelimb, and trunk portions of area 3b.[30] However, M1 is uninterrupted. Macaque monkeys, in contrast, have a deep central sulcus enclosing most of area 3b and 3a, and the posterior half of M1. Thus, M1 is more difficult to systematically explore with stimulating microelectrodes, and deep electrode penetrations along the anterior bank of the central sulcus are needed to completely map M1.

Microstimulation motor maps of M1 contralateral to a long-standing amputation were obtained in three squirrel monkeys.[25] One squirrel monkey had its forelimb amputated near the shoulder joint at two months of age, and was studied 8 years later. Another was amputated at the shoulder joint at four months of age and studied

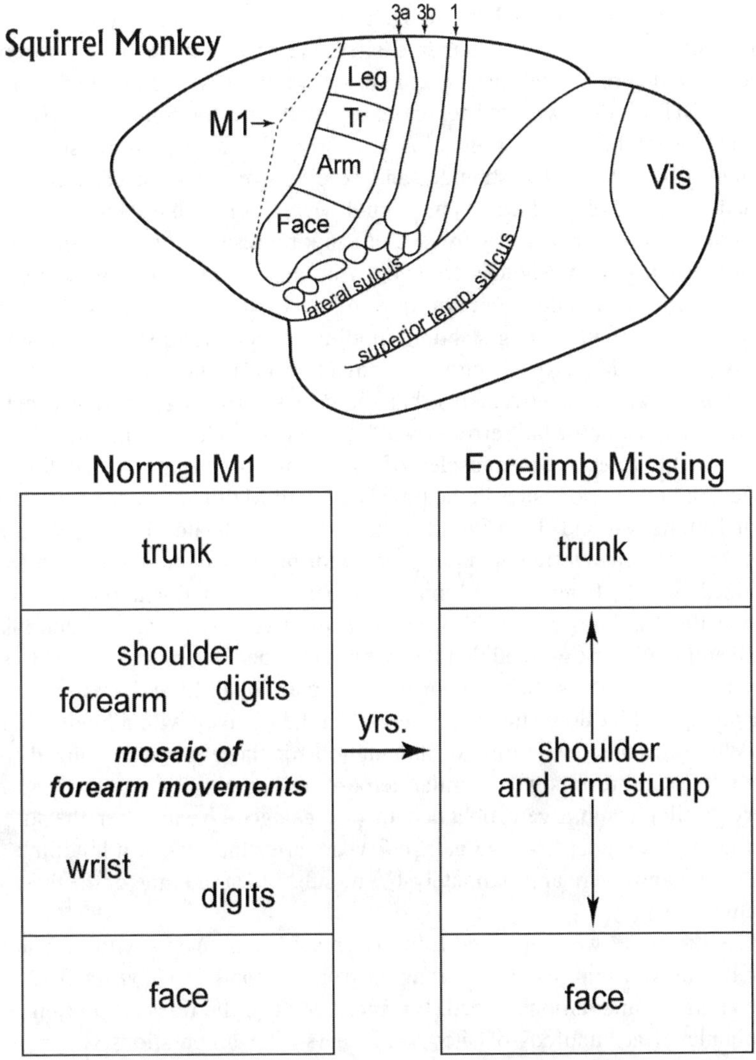

FIGURE 9.2 The reorganization of M1 after the therapeutic amputation of an injured forelimb in squirrel monkeys. Above: a lateral view of a squirrel monkey brain showing the location of primary motor cortex (M1) and the regions of M1 that represent the hindlimb (leg), trunk (tr), forelimb (arm) and face, jaw, and tongue (face). Below left: the normal organization of the midportion of M1. The forelimb region constitutes a mosaic of modules related to movements of digits, hand, forearm, and shoulder. Distal limb movement predominates. Below right: after years of recovery following the amputation of an injured limb above the elbow, the complete forelimb portion of M1 is devoted to shoulder and arm stump movements. Current levels for evoking these movements are in the normal range for arm movements, or slightly higher. Few or no stimulation sites in the region produce face or trunk movements. Based on Wu and Kaas (1999).[38]

5 years later. A third had the hindlimb amputated near the hip joint at 6 years of age and was studied 12 years later. In each monkey, M1 organization was explored by microstimulation at several hundred sites and results were related to cortical architecture to determine the anterior and posterior boundaries of M1. In the forelimb portion of M1 of the two squirrel monkeys with the long-standing loss of the contralateral forelimb, microstimulation at nearly all sites moved the stump of the remaining upper arm or the shoulder and adjoining trunk muscles. Occasionally, movements of parts of the face or upper trunk were elicited, but there was no major invasion of the forelimb territory by face or trunk representations. The shoulder and stump movements at threshold were evoked at current levels that were similar or somewhat higher to shoulder and arm movements in normal monkeys. Similarly, in a squirrel monkey with a long-standing hindlimb amputation, sites throughout the hindlimb region of M1 evoked stump movements at normal or higher current levels. A few sites evoked tail movements, but there was no change in representational boundaries. The trunk or tail territories did not expand into hindlimb M1.

Thus, in all three cases, muscles of the stump of the amputated limb were activated from many more sites in the limb portion of M1 than activate these muscles in normal monkeys, and these muscles were mostly activated by typical levels of current. As in the rats with amputations or motor nerve section, the map in M1 had reorganized. Results from a more limited number of sites in dorsal premotor cortex, PMd, and the forelimb region of area 3a in the two monkeys with contralateral forelimb amputation, suggested that these areas also became more devoted to stump and shoulder movements. In a similar manner, parts of PMd and area 3a appeared to over represent hindlimb stump movements in the monkey with a hindlimb amputation. While the age at the time of amputation for these monkeys ranged from 2 months to 6 years, results were similar across these ages.

Very similar results were obtained in one galago 4 years after the loss of a forelimb at the shoulder and one galago 7 years after the loss of a hindlimb at the hip.[25] Both galagos were approximately $1^1/_2$ months old at the time of the therapeutic amputation.

Related results[27] were obtained from five macaque monkeys with limb amputations that followed injuries at ages ranging from 5 months to 7 years. The extents and locations of amputations varied. For each monkey, the relevant portions of M1 were stimulated at hundreds of sites 5–17 years after amputations. There was no obvious effect of age at the time of amputation or years of recovery.

Results from four macaques with amputations near or above the elbow revealed that nearly all of the forelimb territory of M1 was devoted to movements of the stump and adjoining shoulder. These movements were evoked by current levels that were similar to those for arm movements in normal macaques. Few or no sites in the former forelimb territory evoked face and trunk movements. Results were similar for monkeys injured as adults or 5-month-old infants. In another macaque with a hindlimb amputation at the knee at 4 months of age, stimulations of sites throughout the hindlimb territory evoked movements of the stump at current thresholds in the range of those for normal leg movements. In a sixth monkey with a long-standing loss of digits, much of the hand portion of M1 represented movements of the digit stumps.

Schieber and Deuel (1977)[26] reported similar results from a single macaque monkey that had been studied 15 years after an arm amputation at 2 years of age. Movements of the stump of the amputated limb were evoked at sites throughout the normal territory of the missing limb, and current levels for evoking these movements were about the same or somewhat higher than those for arm movements in cortex contralateral to the normal arm.

Collectively, these results suggest that the extensive losses of motor neuron targets involved in limb amputations are followed by alterations in the motor system that allow the deprived forelimb or hindlimb portion of M1 to become fully devoted to stump movements. This is not an iceberg effect since current levels for evoking movements at most sites are not unusually high. The similar results across prosimian primates, New World monkeys, and Old World monkeys, suggest that primates do not differ in mechanisms of motor system reorganization. Finally, the reorganizations and thus the modes of reorganization seem to be similar in infant and adult primates. However, the effects of such deprivations on newborn or prenatal primates are not known, and the effects of losses at such early stages could have different outcomes.[31]

V. DO NEW CONNECTIONS CONTRIBUTE TO MOTOR CORTEX REORGANIZATION?

The monkeys and prosimian galagos that demonstrate reorganization of motor cortex have been studied long after injury. Thus, there is no information about how long it took cortical reorganization to emerge. Reorganizations of deprived regions of somatosensory cortex in monkeys with nerve injuries, spinal cord damage, or limb amputation depend on the extent of the injury, and involve periods of hours to weeks to over 6 months of recovery.[33] The varying times for reorganization suggests that with major sensory losses and months of recovery, a number of mechanisms are involved. New connections grow over considerable distances.[32] The substitution of stump movements for movements of the distal limb could depend on the potentiation of previously existing connections and/or on the growth of new local connections at several levels of the motor system. Information on the time course of cortical reorganization would provide important clues about possible mechanisms, but this information is missing. However, some of the results from rats and humans suggest that some of the changes in motor cortex could be very rapid.

The potentiation of existing horizontal connections in motor cortex has been suggested as a major source of motor cortex reorganization in rats[14] and humans.[35,36] This seems possible given the extent of normal intrinsic horizontal connections of M1.[11,37] Another possibility, is that amputation induces a denser and more extensive growth of intrinsic horizontal connections in M1, as it does in the somatosensory cortex,[34] but there is yet no evidence for this. A third possibility is that the terminal arbors of corticospinal neurons activating deprived spinal motor pools enlarge to contact motor pools innervating intact muscles. There is some evidence against this possibility in that injections of a retrograde tracer bilaterally into the lower cervical spinal cord of two squirrel monkeys with an amputated forelimb labeled similar distributions of corticospinal neurons in the forelimb portion of M1 of both hemispheres.[38] However, this procedure provides a rather crude measure of connectivity, and the terminal arbors of the corticospinal neurons could have rearranged local contacts in the cervical spinal cord without affecting the uptake of tracers from a large injection in the cervical spinal cord. In rats with a transected corticospinal tract, cut axons sprout to contact local propriospinal neurons to create new and functionally useful circuits.[21]

A fourth possibility is supported by considerable evidence. Motor neurons projecting to distal limb muscles have had their axons severed by limb damage and amputation. Nevertheless, most of these neurons survive and many sprout collaterals to innervate proximal muscles spared by the amputation. Thus, these damaged motor neurons recover and likely contribute to the activation of preserved proximal muscles. As more spinal cord motor neuron pools would be devoted to preserved proximal muscles in animals with amputations, the electrical stimulation of more cortical sites would activate muscles of the stump. Evidence for this possibility comes from comparisons of both sides of the cervical spinal cord in galagos and squirrel monkeys,[25] and in the lumbar spinal cord of a macaque monkey[39] after comparable injections of anterograde tracers into proximal muscles of the intact limb and stump of the amputated forelimb (galagos and squirrel monkeys) or amputated hindlimb (macaque). In all these primates, motor neuron pools normally devoted to the muscles of the amputated limb were present in normal numbers, although these neurons were smaller in size than on the side with the intact limb. In addition, many of these neurons were labeled by the injection in the limb stump. Thus, neurons with severed axons survived for years after amputations, and acquired new or enhanced connections with remaining proximal muscles. Such new patterns of motor neuron connections most likely contribute to the reorganization of the motor map in M1 of all primates with amputations, including humans.

VI. REORGANIZATION AFTER FOCAL LESIONS OF MOTOR CORTEX IN MONKEYS

The primary motor cortex may reorganize after part of the map is damaged. Lesions of M1 result in impaired motor performance in monkeys[40,41] and humans.[42] In squirrel monkeys, a small lesion of part of the hand representation in M1 typically produces a mild impairment in hand use such that the monkeys have more difficulty retrieving small food objects from narrow wells.[43–46] Hand use typically improves over time, and normal hand use in retrieval may return with training on the task over a period of one month. In such cases, the damaged representation reorganizes so that more sites in the remaining hand cortex evoke digit movements, especially after rehabilitative training. These results suggest that the behavioral recoveries are mediated by modifications in primary motor cortex. The changes in the M1 motor map, according to this view, are driven by hand use and training on specific tasks. An extension of this view is that if motor cortex lesions result in reduced use of the opposite hand, M1 reorganizes in a nonproductive way causing further deterioration of skills.[44,47]

Several investigators have studied the effects of motor cortex lesions in infant monkeys. While behavioral recoveries after large lesions may be incomplete,[48] recoveries can be extensive.[49] Such recoveries seem to depend on the reorganization of the remaining portions of the primary motor cortex. Rouiller et al.[50] lesioned the hand portion of M1 in infant monkeys and found that after maturation they had nearly normal hand use. These monkeys had acquired a new, displaced, and relatively complete hand representation in M1 just medial to the site of the old lesion. When this new hand representation was inactivated by cooling, the use of the opposite hand was greatly impaired. Similar results were obtained after lesions in juvenile (2–2.5 yr. old) macaque monkeys.[40] Small lesions of the thumb representation in M1 were followed by a recovery of thumb use and an increased involvement of perilesion cortex in thumb movement as demonstrated by the reappearance of the motor deficit with subsequent damage to the perilesion cortex. As the motor system in 2–2.5 yr. old macaques is likely to be reasonably mature, these results suggest that motor cortex in adult macaque monkeys has the capacity to reorganize after focal lesions, much as in mature squirrel monkeys.

VII. MOTOR CORTEX REORGANIZATION IN HUMANS

Noninvasive methods have been used to reveal changes in motor cortex contralateral to an amputated limb in patients. One procedure has been to measure evoked potentials in muscles in the intact proximal part of the amputated limb and the same muscles in the normal limb, while stimulating contralateral motor cortex with transcranial magnetic pulses. As larger evoked potentials were recorded in muscles of the amputated limbs, the motor cortex contralateral to the amputation was assumed to be more excitable for these muscles.[51,52,53,54] Also, the result could simply reflect an increase in the number of sites in M1 that are devoted to the proximal limb, and as such, transcranial stimulation would then activate more of these sites, thereby producing a greater muscle response. In addition, optimal scalp positions for transcranial stimulation of these muscles were displaced medially toward the missing hand representation in M1 suggesting that M1 had reorganized to over represent the remaining arm muscles in hand cortex. For uncertain reasons, this change in representation seemed to occur only in patients with phantom limb pain. Finally, there was a correlation between motor and somatosensory cortex reorganization in these patients, as face stimulation activated somatosensory hand cortex in the same patients with evidence for motor cortex organization. These results suggest that motor cortex reorganization in humans with amputations can be very similar to that demonstrated with invasive methods in monkeys.

Other results from humans indicate that some plastic changes in motor cortex can be very rapid. In a number of studies, a limb (arm) is made ischemic in order to temporarily and reversibly block nerve conduction. A pressure block has been used to make a subject's arm ischemic, thus providing a lack of effective motor output and sensory input. During the block of nerve conduction in the ischemic limb, transcranial magnetic stimulation of the motor cortex more effectively activates muscles proximal to the nerve block, suggesting that a rapid increase in motor cortex excitability has occurred.[35,36,55–57] A suggested mechanism for this rapid change in motor cortex excitability is a rapid down-regulation of GABA-based inhibitory circuits.[57] It is not yet known if this rapid change in motor cortex is responsible for all of the plasticity that seems to occur in M1 of humans with long-standing limb loss, or if structural changes in the motor system due to new growth of connections are also a factor, as in monkeys.

VIII. PLASTICITY AFTER LESIONS OF MOTOR CORTEX IN HUMANS

Considerable behavioral recovery may follow strokes or tumors that damage parts of primary motor cortex in humans (see below), although full compensation may be unlikely.[58] Such recoveries could be mediated by a number of factors including, the recovery of function in damaged but not destroyed parts of M1, behavioral compensation based on other parts of the motor system, an increased role for the intact M1 in the control of ipsilateral body movements, and the reorganization of preserved parts of the damaged M1 to restore cortical control. There is some evidence from focal transcranial magnetic stimulation that the intact motor cortex becomes more involved in eliciting ipsilateral muscle responses in patients with hemiparesis after ischemic strokes affecting M1 of the other hemisphere.[59] Similar studies suggest that preserved motor areas, possibly including the premotor cortex, may reorganize to compensate for lost parts of representations after strokes or tumors, and contribute to motor control.[60–63] Thus, some of the reorganizational changes seen in M1 of monkeys after focal lesions are likely to occur in humans. As in monkeys, some reorganization may rapidly follow lesions as intra-operative mapping of cortically evoked hand and forearm movements in patients before and after therapeutic lesion demonstrated local changes in the motor map.[64]

IX. SUMMARY AND CONCLUSIONS

Placental mammals have a primary motor area, M1, of the frontal cortex that is identified by containing a characteristic map of movements of muscles of the contralateral body. Normally, M1 represents the hindlimb, trunk, forelimb, and head in a caudomedial to rostrolateral sequence across M1. The organization of M1 is revealed by evoking muscle movements in these body parts with short trains of electrical pulses via microelectrodes positioned in the lower layers of the cortex. At low levels of current, only a few muscles are activated, but higher levels of current stimulate more muscles and evoke more complex movements. In humans, crude but useful motor maps can be produced with noninvasive transcranial magnetic stimulation. Changes in the motor maps are revealed with noninvasive transcranial stimulation by shifts in the location on the scalp that is most effective in eliciting a specific movement, or a lowering of the stimulation threshold for a given movement at a given site. The motor maps are globally similar across individuals of the same species.

In rats, prosimian primates, monkeys, and humans, the motor map in M1 has been altered in internal organization by direct damage to parts of the map, lesions of corticospinal pathways and motor nerves, or by limb amputations. When motor maps in individuals with damage to the motor system are compared to those from normal individuals, reorganizations of motor maps are indicated when cortical locations normally evoking one type of movement evoke another type of movement at normal or nearly normal levels of stimulating current.

Reorganizations of M1 were first demonstrated in rats where cortex normally devoted to facial whisker movements, or forelimb movements, became devoted to other movements after cutting the motor nerve to the whisker muscles or amputating a forelimb. While the mechanism for this change in functional organization is unclear, the reorganization occurred rapidly in a matter of hours. Because of this rapid change, the proposed mechanism of change was that intrinsic horizontal connections in M1 rapidly became potentiated so that effects of stimulations at sites in deprived parts of M1 would relay to other nearby locations in M1 with intact muscle targets. The long-term potentiation (LTP) of horizontal connections in M1 was thought to be a consequence of a loss of sensory feedback from muscle movement that would normally be relayed to M1. According to the hypothesis, this reduction in sensory input would reduce neural activity levels in M1, causing a down regulation of the inhibitory neurotransmitter, GABA, and GABA receptors so that previously inhibited horizontal pathways became effective. Continued activity in these horizontal pathways would strengthen and preserve them via LTP in the face of any return of normal levels of inhibition. A similar explanation has been given for the changes in the motor cortex of humans that follow the ischemic block of nerves in an arm. The difficulty with this otherwise reasonable hypothesis is that a block of sensory input to M1 alone, without a motor loss, does not appear to lead to motor cortex reorganization.

The long term consequences of motor loss have been investigated in monkeys, and prosimian primates with long-standing amputations. Years after the amputations, motor neurons in the spinal cord that normally project to missing muscles are intact but reduced in size. Many or most of these preserved motor neurons have grown new connections to muscles of the stump of the amputated limb where they likely contribute to muscle contractions. Thus, the reorganization of motor cortex in these primates may be largely or at least partly based on these new motor nerve connections. According to this view, there are both rapidly and slowly emerging components to the reorganization of motor maps in M1, with rapid changes based on the disinhibition and potentiation of existing pathways, and the slower changes based on the growth of new connections.

Small lesions of parts of M1 also change M1 organization. Sites around the lesion become more effective in evoking movements that were represented in the lesioned cortex. As use of the impaired limb, and especially training on the tested task appear to be important in the progressive reorganization of cortex around the lesion, experience-driven neural activity in M1 may strengthen connections related to the use of the impaired limb, especially in the training task. Motor cortex organization appears to respond to the requirements of any repeated motor behavior by involving more of motor cortex in the task. Research on the reorganization of damaged motor systems has the potential of usefully guiding the clinical treatment of patients with amputations and motor system injuries. Because motor neurons in the spinal cord with cut axons after amputations regenerate axons to muscles of distal limbs, the outputs of these neurons could be used to guide electromechanical substitutes for missing limbs.

REFERENCES

1. Kaas, J. H., The Mutable Brain, Kaas, J. H. Harwood Academic Publishers, Amsterdam, 2001.
2. Sanes, J. N., Suner, S., Lando, J. F., and Donoghue, J. P., Rapid reorganization of adult rat motor cortex somatic representation patterned after motor nerve injury, *Proc Natl Acad Sci USA,* 85, 2003-2007, 1988.
3. Sanes, J. N., Suner, S., and Donoghue, J. P., Dynamic organization of primary motor cortex output to target muscles in adult rats. I. Long-term pattern of reorganization following motor or mixed peripheral nerve lesion, *Exp Brain Res,* 79, 479-491, 1990.
4. Donoghue, J. P., Suner, S., and Sanes, J. N., Dynamic organization of primary motor cortex output to target muscles in adult rats. II. Rapid reorganization following motor nerve lesion, *Exp Brain Res,* 79, 492-503, 1990.
5. Kaas, J. H., The evolution of the large, complex sensorimotor systems of anthropoid primates, *Int J Comp Psych,* 17, 34-52, 2004.
6. Asanuma, H., Functional role of sensory inputs to the motor cortex, *Prog Neurobiol,* 16, 241-262, 1981.
7. Gould, H. J., Cusick, C. G., Pons, T. P., and Kaas, J. H., The relationship of corpus callosum connections to electrical stimulation maps of motor, supplementary motor, and the frontal eye fields in owl monkeys, *J Comp Neurol,* 247, 297-325, 1986.
8. Graziano, M. S. A., Taylor, C. S. R., and Moore, T., Complex movements evoked by microstimulation of precentral cortex, *Neuron,* 34, 841-851, 2002.
9. Wise, S. P. and Donoghue, J. P., Motor cortex of rodents, in *Cereb Cortex*, Jones, E. G. and Peters, A. Plenum, New York, 1986, 243-270.
10. Toldi, J., Laskawi, R., Landgrebe, M., and Wolff, J. R., Biphasic reorganization of somatotopy in the primary motor cortex follows facial nerve lesions in adult rats, *Neurosci Lett,* 203, 179-182, 1996.
11. Huntley, G. W., Correlation between patterns of horizontal connectivity and the extent of short-term representational plasticity in rat motor cortex, *Cereb Cortex,* 7, 143-156, 1997.
12. Franchi, G., Time course of motor cortex reorganization following botulinum toxin injection into the vibrissal pad of the adult rat, *Eur J Neurosci,* 16, 1333-1348, 2002.
13. Jacobs, K. M. and Donoghue, J. P., Reshaping the cortical motor map by unmasking latent intracortical connections, *Science,* 251, 944-947, 1991.
14. Hess, G. and Donoghue, J. P., Long-term potentiation of horizontal connections provides a mechanism to reorganize cortical motor maps, *J Neurophysiol,* 71, 2543-2547, 1994.
15. Welker, E., Soriano, E., and Van der Loos, H., Plasticity in the barrel cortex of the adult mouse: effects of peripheral deprivation on GAD-immunoreactivity, *Exp Brain Res,* 74, 441-452, 1989.
16. Garraghty, P. E., LaChica, E. A., and Kaas, J. H., Injury-induced reorganization of somatosensory cortex is accompanied by reductions in GABA staining, *Somatosens Mot Res,* 8, 347-354, 1991.
17. Jones, E. G., GABAergic neurons and their role in cortical plasticity in primates, *Cereb Cortex,* 3, 361-372, 1993.
18. Rosier, A. M., Arckens, L., Demeulemeester, H., Orban, G. A., Eysel, U. T., Wu, Y. J., and Vandesande, F., Effect of sensory deafferentation on immunoreactivity of GABAergic cells and on GABA receptors in the adult cat visual cortex, *J Comp Neurol,* 359, 476-489, 1995.

19. Franchi, G., Persistence of vibrissal motor representation following vibrissal pad deafferentation in adult rats, *Exp Brain Res,* 137, 180-189, 2001.
20. Fouad, K., Pedersen, V., Schwab, M. E., and Brösamle, C., Cervical sprouting of corticospinal fibers after thoracic spinal cord injury accompanies shifts in evoked motor responses, *Curr Biol,* 11, 1766-1770, 2001.
21. Bareyre, F. M., Kerschensteiner, M., Raineteau, O., Mettenleiter, T. C., Weinmann, O., and Schwab, M. E., The injured spinal cord spontaneously forms a new intraspinal circuit in adult rats, *Nat Neurosci,* 7, 269-277, 2004.
22. Emerick, A. V., Neafsey, E. J., Schwab, M. E., and Kartje, G. L., Functional reorganization of the motor cortex in adult rats after cortical lesion and treatment with monoclonal antibody, *J Neurosci,* 23, 4820-4830, 2003.
23. Jones, T. A. and Schallert, T., Overgrowth and pruning of dendrites in adult rats recovering from neocortical damage, *Brain Res,* 581, 156-160, 1992.
24. Jones, T. A. and Schallert, T., Use-dependent growth of pyramidal neurons after neocortical damage, *J Neurosci,* 14, 2140-2154, 1994.
25. Wu, C. W. and Kaas, J. H., Spinal cord atrophy and reorganization of motor neuron connections following long-standing limb loss in primates, *Neuron,* 28, 967-78, 2000.
26. Schieber, M. H. and Deuel, R. K., Primary motor cortex reorganization in a long-term monkey amputee, *Somatosens Mot Res,* 14, 157-167, 1997.
27. Qi, H. X., Stepniewska, I., and Kaas, J. H., Reorganization of primary motor cortex in adult macaque monkeys with long-standing amputations, *J Neurophysiol,* 84, 2133-47, 2000.
28. Merzenich, M. M., Kaas, J. H., Wall, J., Nelson, R. J., Sur, M., and Felleman, D., Topographic reorganization of somatosensory cortical areas 3b and 1 in adult monkeys following restricted deafferentation, *Neurosci,* 8, 33-55, 1983.
29. Sur, M., Nelson, R. J., and Kaas, J. H., The representation of the body surface in somatic koniocortex in the prosimian (*Galago senegalensis*), *J Comp Neurol,* 180, 381-402, 1980.
30. Sur, M., Nelson, R. J., and Kaas, J. H., Representations of the body surface in cortical areas 3b and 1 of squirrel monkeys: Comparisons with other primates, *J Comp Neurol,* 211, 177-192, 1982.
31. Florence, S. L., Jain, N., Pospichal, M. W., Beck, P. D., Sly, D. L., and Kaas, J. H., Central reorganization of sensory pathways following peripheral nerve regeneration in fetal monkeys, *Nature,* 381, 69-71, 1996.
32. Jain, N., Florence, S. L., Qi, H.-X., and Kaas, J. H., Growth of new brain stem connections in adult monkeys with massive sensory loss, *Proc Natl Acad Sci USA,* 97, 5546-5550, 2000.
33. Kaas, J. H. and Florence, S. L., Reorganization of sensory and motor systems in adult mammals after injury, in *The Mutable Brain*, Kaas, J. H., 2001.
34. Florence, S. L., Taub, H. B., and Kaas, J. H., Large-scale sprouting of cortical connections after peripheral injury in adult macaque monkeys, *Science,* 282, 1117-1121, 1998.
35. Brasil-Neto, J. P., Cohen, L. G., Pascual-Leone, A., Jabir, F. K., Wall, R. T., and Hallett, M., Rapid reversible modulation of human motor outputs after transient deafferentation of the forearm: a study with transcranial magnetic stimulation, *Neurol,* 42, 1302-1306, 1992.
36. Brasil-Neto, J. P., Valls-Solé, J., Pascual-Leone, A., Cammarota, A., Amassian, V. E., and Cracco, R., Rapid modulation of human cortical motor outputs following ischaemic nerve block, *Brain,* 116, 511-525, 1993.

37. Huntley, G. W. and Jones, E. G., Relationship of intrinsic connections to forelimb movement representations in monkey motor cortex: a correlative anatomic and physiological study, *J Neurophysiol,* 66, 390-413, 1991.
38. Wu, C. W.-H. and Kaas, J. H., Reorganization in primary motor cortex of primates with long-standing therapeutic amputations, *J Neurosci,* 19, 7679-97, 1999.
39. Qi, H.-X., Phillips, W. S., and Kaas, J. H., Connections of neurons in the lumbar ventral horn of the spinal cord are altered after long-standing limb loss in a macaque monkey, *Somatosens Mot Res*, 2004.
40. Glees, P. and Cole, J., Recovery of skilled motor functions after small repeated lesions of motor cortex in macaque, *J Neurophysiol,* 13, 133-148, 1950.
41. Hoffman, D. S. and Strick, P. L., Effects of a primary motor cortex lesion on step-tracking movements of the wrist, *J Neurophysiol,* 73, 891-895, 1995.
42. Porter, R. and Lemond, R., *Corticospinal function and voluntary movement*, Oxford University Press, New York, 1993.
43. Nudo, R. L. and Milliken, G. W., Reorganization of movement representations in primary motor cortex following focal ischemic infarcts in adult squirrel monkeys, *J Neurophysiol,* 75, 2144-2149, 1996.
44. Nudo, R. J., Wise, B. M., SiFuentes, F., and Milliken, G. W., Neural substrates for the effects of rehabilitative training on motor recovery after ischemic infarct, *Science,* 272, 1791-1794, 1996.
45. Friel, K. M. and Nudo, R. J., Recovery of motor function after cortical injury in primates. Compensatory movement patterns used during rehabilitative training, *Somatosens Mot Res,* 15, 173-189, 1998.
46. Plautz, E. J., Milliken, G. W., and Nudo, R. J., Effects of repetitive motor training on movement representations in adult squirrel monkeys: role of use versus learning, *Neurobiol Mem Learn,* 74, 27-55, 2000.
47. Taub, E., Miller, N. E., Novack, E. W., Cook, W. C., Fleming, C. S., Nepomuceno, J. S., Connell, J. S., and Crago, J. E., Technique to improve chronic motor deficit after stroke, *Arch Phys Med Rehabil,* 74, 347-354, 1993.
48. Passingham, R. E., Perry, V. H., and Wilkinson, F., The long-term effects of removal of sensorimotor cortex in infant and adult rhesus monkeys, *Brain,* 106, 675-705, 1983.
49. Kennard, M. A., Cortical reorganization of motor function, *Arch Neurol Psychiat,* 48, 227-240, 1942.
50. Rouiller, E. M., Yu, X. H., Moret, V., Tempini, A., Wiesendanger, M., and Liang, F., Dexerity in adult monkeys following early lesion of the motor cortical hand area: the role of cortex adjacent to the lesion, *Eur J Neurosci,* 10, 729-740, 1998.
51. Hall, E. J., Flament, D., Fraser, C., and Lemon, R. N., Non-invasive brain stimulation reveals reorganized cortical outputs in amputees, *Neurosci Lett,* 116, 379-386, 1990.
52. Cohen, L. G., Bandinelli, S., Findley, T. W., and Hallett, M., Motor reorganization after upper limb amputation in man. A study with focal magnetic stimulation, *Brain,* 114, 615-627, 1991.
53. Fuhr, P., Cohen, L. G., Dang, N., Findley, T. W., Haghighi, S., Oro, J., and Hallett, M., Physiological analysis of motor reorganization following lower limb amputation, *Electroenceph Clin Neurophysiol,* 85, 53-60, 1992.
54. Kew, J. J., Ridding, M. C., Rothwell, J. C., Passingham, R. E., Leigh, P. N., Sooriakumaran, S., Franckowiak, R. S., and Brooks, D. J., Reorganization of cortical blood flow and transcranial magnetic stimulation maps in human subjects after upper limb amputation, *J Neurophysiol,* 72, 2517-2524, 1994.
55. Ridding, M. C. and Rothwell, J. C., Reorganisation in human motor cortex, *Can J Physiol Pharm,* 73, 218-222, 1995.

56. Ziemann, U., Corwell, B., and Cohen, L. G., Modulation of plasticity in human motor cortex after forearm ischemic nerve block, *J Neurosci,* 18, 1115-1123, 1998.
57. Ziemann, U., Hallett, M., and Cohen, L. G., Mechanisms of deafferentation-induced plasticity in human motor cotex, *J Neurosci,* 18, 7000-7007, 1998.
58. Lang, C. E. and Schieber, M. H., Reduced Muscle selectivity during individuated finger movements in humans after damage to the motor cortex or corticospinal tract, *J Neurophysiol,* 91, 1722-1733, 2004.
59. Netz, J., Lammers, T., and Hömberg, V., Reorganization of motor output in the non-affected hemisphere after stroke, *Brain,* 120, 1579-1586, 1997.
60. Rossini, P. M., Tecchino, F., Pizzella, V., Lupoi, D., Cassetta, E., Pasqualetti, P., Pomani, G. L., and Orlacchio, A., On the reoganization of sensory hand areas after mono-hemispheric lesion: a functional (MEG)/anatomical (MRI) integrative study, *Brain Res,* 782, 153-166, 1998.
61. Liepert, J., Miltner, W. H. R., Bauder, H., and Sommer, M., Motor cortex plasticity during constraint-induced movement therapy in stroke patients, *Neurosci Lett,* 250, 5-8, 1998.
62. Traversa, R., Cicinelli, P., Bassi, A., Rossini, P. M., and Bernardi, G., Mapping of motor cortical reorganization after stroke. A brain stimulation study with focal magnetic pulses, *Stroke,* 28, 110-117, 1997.
63. Seitz, R., Huang, Y., Knorr, U., Telimann, L., Herzog, H., and Freund, H.-J., Large-scale plasticity of the human motor cortex, *NeuroReport,* 6, 742-744, 1995.
64. Duffau, H., Acute functional reorganisation of the human motor cortex during resection of central lesions: a study using intraoperative brain mapping, *J Neurol, Neurosurg, Psychiat,* 70, 506-513, 2001.

10 Modulation of Cortical Function and Plasticity in the Human Brain

Friedhelm Hummel, Christian Gerloff, and Leonardo G. Cohen

CONTENTS

Abbreviations 207
I. Introduction 208
II. Somatosensory Input Modulates Motor Cortical Plasticity 208
A. Chronic Limb Deafferentation 208
B. Acute Limb Deafferentation 210
1. Effects on Contralateral Cortical Function 210
2. Effects on Ipsilateral Cortical Function 211
C. Somatosensory Stimulation 214
III. Effects of Cortical Stimulation on Motor Cortical Function and Cortical Plasticity 215
IV. Conclusions 217
Acknowledgments 217
References 218

ABBREVIATIONS

EEG electroencephalogram
fMRI functional magnetic resonance imaging
ICF intracortical facilitation
ICI intracortical inhibition
IHI interhemispheric inhibition
INB ischemic nerve block
PET positron emission tomography
RC recruitment curve
tDCS transcranial direct current stimulation
TMS transcranial magnetic stimulation

I. INTRODUCTION

Plasticity is the ability of the central nervous system (CNS) to adapt to environmental challenges or to compensate for lesions.[29] This definition assumes that the CNS can identify changes in the environment relevant to the individual, process this information, and generate appropriate behavioral responses. Until recently, it was thought that the brain's ability to reorganize could occur only during development. The last two decades have provided ample evidence challenging this concept and demonstrating that the adult brain can experience substantial reorganization.[63,64,87,88,89,105,106,163,164] These plastic changes can occur within the same modality, as in cortical reorganization demonstrated within the primary motor,[55,69,83,134] somatosensory,[10,27,32,122,162] auditory[39,48,60,78,96,117,123] or visual,[8,9,24,53,126] domains, and they can occur in one modality (e.g., in the primary motor cortex) induced by changes in another modality (e.g., the somatosensory system).[11,19,33,68,108,132] Specific forms of cortical reorganization appear to endure if they influence behavior or fade away if behaviorally irrelevant,[2,3,4,89] and operate in both health and disease, as for example after a stroke.[103] Understanding the mechanisms underlying cortical plasticity can provide clues to develop more effective strategies that will enhance neurorehabilitative efforts.

II. SOMATOSENSORY INPUT MODULATES MOTOR CORTICAL PLASTICITY

The somatosensory and the motor cortices are anatomically interconnected (Figure 10.1). Consistently, changes in somatosensory input often influence motor output[12,42,51,66,67,113,132] and contribute to motor learning and skill acquisition.[113,114] The somatosensory cortex has projections to layers II/III of the motor cortex, which are highly interconnected to motor output neurons in layer V,[66,67] providing an anatomical pathway for this communication. Physiological experiments demonstrated that stimulation of the somatosensory cortex can induce long-term potentiation (LTP) in the motor cortex.[132] On the other hand, reduction of somatosensory input by local anesthesia results in impaired motor control,[7,31] as shown in patients with large fiber sensory neuropathy who display characteristically abnormal motor behavior.[49,129] Thus, it is not surprising that changes in somatosensory input could impact cortical organization.

A. Chronic Limb Deafferentation

In humans, limb amputation, a form of chronic deafferentation, leads to extensive cortical reorganization within the somatosensory[10,32,37,68,72,75,141] and the motor cortices.[17,18,42,58,68,73,84,107,108,127] In the motor domain, upper limb amputation (e.g., at the elbow level) results in an increase in the excitability of body part representations in the motor cortex near the deafferented one (i.e., upper arm) in the form of decreased motor thresholds, larger motor maps and a lateral shift of the center of gravity with transcranial magnetic stimulation.[18,58] This increased excitability appears to be predominantly cortical in origin and may well be a consequence of a release of inhibition

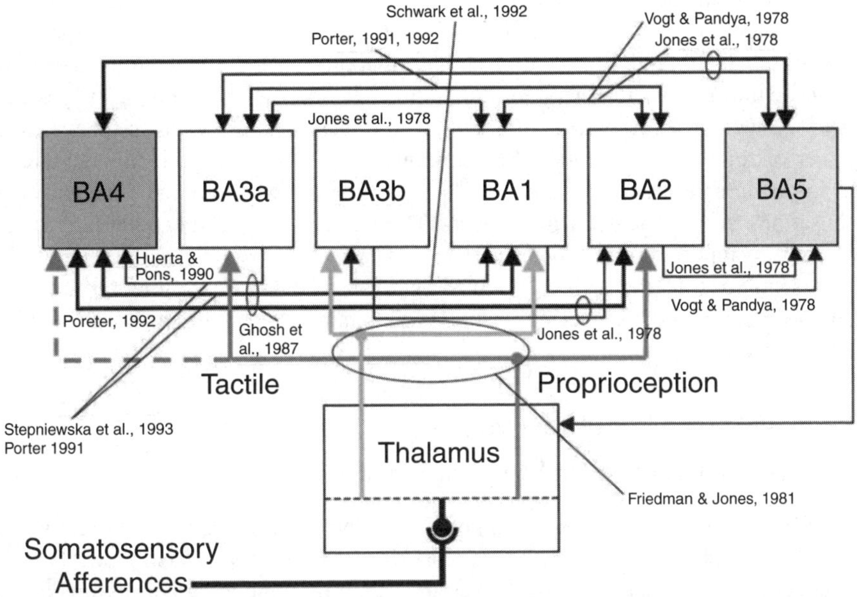

FIGURE 10.1 Connections between somatosensory and motor cortices. BA = Brodman areas; BA1, 2, 3, 5: somatosensory cortex, BA4: primary cortex. Note the highly interconnected somatosensory and motor cortices. The thalamus acts as a relay station for peripheral somatosensory input into the cortex. Thalamo-cortical connections are shown in grey (tactile thalamo-cortical inputs in light grey, deep thalamo-cortical inputs in dark grey). Cortico-cortical connections are shown in black.

originating in the deafferented hand representation.[59] Similar results have been reported with lower limb amputations.[17,42] The mechanisms underlying these reorganizational changes are incompletely understood. However, we now know that intracortical inhibition in the motor cortex contralateral to an amputated limb is decreased relative to healthy subjects[17] favoring the view that reduction in GABA-ergic inhibition might be an operating mechanism.

One of the behavioral consequences of amputations is phantom limb pain, a condition characterized by the presence of painful perceptions referring to the missing limb. Phantom limb pain is associated with profound changes in the cortical,[36,37] as well as the subcortical organization.[25] Flor et al. demonstrated reorganization in the primary somatosensory cortex that strongly correlated with the magnitude of phantom limb pain.[36,37] Interestingly, phantom pain was more prominent in patients in whom the motor representations of face muscles were displaced medially, possibly reflecting an invasion of the face motor representation over the upper limb deafferented representation.[68]

Another form of chronic deafferentation is spinal cord injury. Here, low thoracic spinal cord lesions leading to paraplegia result in interruption of sensory input originating in the lower extremities. Three features characterize this form of plasticity. First, corticomotor excitability targeting muscles immediately proximal to the deafferented level (in this case abdominal muscles immediately above the lesion level) is enhanced.[77,80,157] Second, cortical movement representation of muscles immediately proximal to the lesioned level expand medially toward the representation of the deafferented lower limbs.[85] Third, noninvasive cortical stimulation in the form of transcranial magnetic stimulation (TMS) delivered to mid-scalp positions over the leg cortical representation results in transient paresthesias in the paralyzed lower extremities.[20] A single TMS stimulus in one of these patients resulted, for example, in a surprised exclamation: "I felt my legs", when he had experienced no sensations from the legs for many years. These findings suggest (a) that interruption of sensory input from one body part results in increases in cortical excitability in representations nearby the deafferented one, and (b) that despite long-term deafferentation, the primary sensorimotor cortex maintains at least a memory trace of the deafferented limb.

B. Acute Limb Deafferentation

1. Effects on Contralateral Cortical Function

Deafferentation leads to rapid changes in the CNS. For example, rapid reorganizational changes may be caused by unmasking of previously existing GABAergic connections.[59] Changes in GABAergic inhibition may induce a permissive state in the cortex that facilitate longer-term cortical reorganization[30] such as axonal sprouting and LTP.[23,71] Therefore, modulation of cortical activity detected after long-term deafferentation in amputees could theoretically be influenced by a variety of different mechanisms. One approach to obtain information about mechanisms operating in this form of plasticity is to evaluate patients serially after an amputation.[108] A complementary approach to look more specifically at rapid changes following deafferentation is to study the effects of acute limb anesthesia, as for example in ischemic nerve block (INB). INB represents a strategy to induce acute reversible limb deafferentation in healthy subjects.[11,12,161,162,166,167,168] In this experimental setting, INB is induced by inflation of a blood pressure cuff around the forearm above systolic blood pressure for up to an hour.[12] The INB of one hand results in enlarged motor cortical output targeting upper arm muscles immediately proximal to the deafferented level.[11,12,131] We now know that this form of plasticity is down-regulated by administering a single oral dose of a GABAergic agent,[161,167] suggesting the involvement of GABAergic transmission. Consistent with this proposal, a novel method using J-resolved magnetic resonance spectroscopy (MRS) revealed that GABA levels in the human sensorimotor cortex are quickly reduced within minutes of the INB of the contralateral hand.[79] This finding strongly supports the view that the release of latent cortico-cortical projections from tonic inhibition through decreased GABA availability is a mechanism of rapid cortical plasticity in intact humans. Reduction of brain GABA can play a pivotal role in regulating the extent of rapid cortical reorganization following lesions or changes in sensory input.[79,161,162]

While these studies characterized aspects of cortical reorganization associated with acute limb deafferentation, they did not provide information on the behavioral consequences of these changes. One study recently addressed this issue. The authors demonstrated that hand deafferentation can enhance the beneficial effects of upper arm motor training on motor performance.[168] These findings, in combination with previous results in the basic science domain showing large scale cortical reorganization after deafferentation,[38,44,64,87] raised the exciting hypothesis that deafferentation induced plasticity might contribute to practice-dependent improvements in motor function. This effect may be relevant for patients with brain lesions undergoing rehabilitative treatment. It is conceivable that in these patients, deafferentation of one body part representation could facilitate the ability of a nearby body part representation to learn or to perform a task. This idea led to a small open label study in a group of patients with chronic stroke and hand weakness.[95] Anesthesia was applied to components of the brachial plexus that selectively innervate the upper arm but not the weak hand. It was shown that anesthesia of the upper arm during hand motor practice improved practice effects on hand motor function,[95] presumably by allowing the weak hand representation to expand over the upper arm deafferented representation.

In summary, these results indicate that an acute decrease in sensory input in one body part representation elicits reorganizational changes in the adjacent sensorimotor cortex of the contralateral hemisphere,[12] which can be associated with behavioral gains.[168] This type of information might be useful to develop novel interventional strategies based on principles of neuroplasticity.

2. Effects on Ipsilateral Cortical Function

Acute limb deafferentation leads to reorganizational changes in both cerebral hemispheres.[14,140] In humans, hand anesthesia results in increased bilateral cerebral blood flow in the primary motor cortex,[131] in increased excitability of the homonymous non-deafferented hand motor cortical representation, and decreased interhemispheric inhibition[161] (see Figure 10.2). This effect developed rapidly after the onset of hand anesthesia, disappeared after cuff deflation and consequent restoration of sensation in the hand and was blocked by oral intake of a single dose of a GABAergic agent.[161] Therefore, acute unilateral hand anesthesia can elicit a rapid increase in excitability in the hand motor representation contralateral to the deafferented cortex that is influenced by transcallosal inhibition and GABAergic transmission. Physiological and anatomical substrates for interhemispheric interactions between cortical representations that could explain this result are present in animals[130] and humans[34,43,91,92,93] These interactions are predominantly inhibitory[26,45,137,139] and operate during voluntary movement generation.[97] Indeed, cortical stimulation of one hemisphere can induce remote changes in the sensorimotor cortex of the other hemisphere.[26,34,76,91,92,97,133] For example, down-regulation of excitability in one hand motor representation by 1 Hz TMS, results in enhanced excitability of the opposite hand motor representation,[118,136] a phenomenon that may have direct functional relevance.[70] Overall, these results are consistent with the view that the enhanced motor cortical output to one hand by anesthesia of the other hand is mediated through modulation of interhemispheric interactions.[161]

Modulation of these interhemispheric interactions may also operate in the somatosensory domain. In animal models, acute limb deafferentation results in rapid changes in receptive fields in the somatosensory cortex of both cerebral

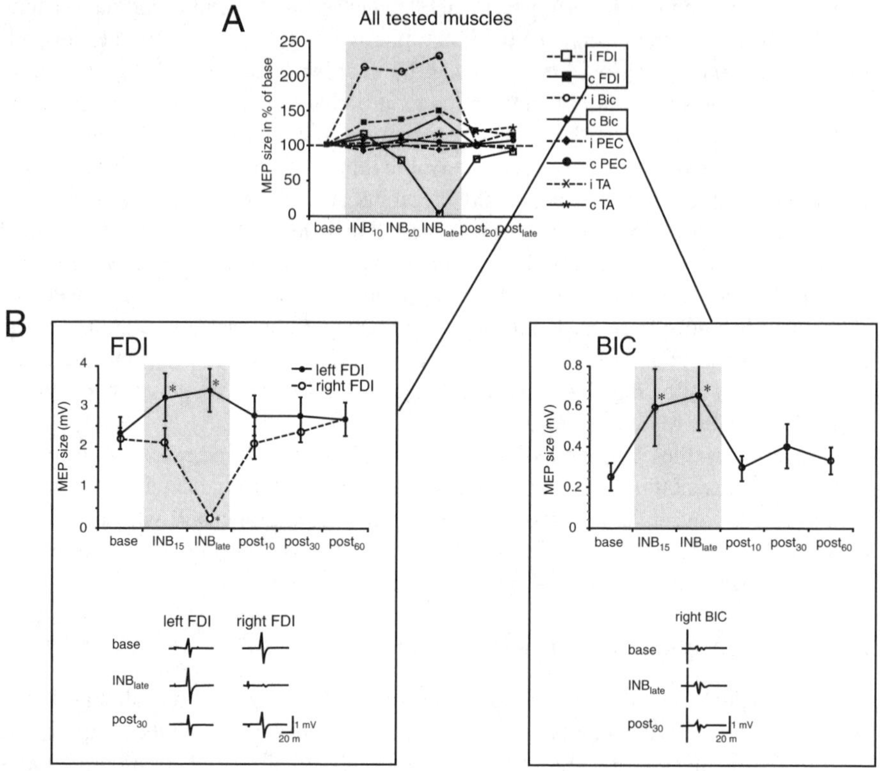

FIGURE 10.2 Effects of unilateral hand deafferentation on motor cortical excitability in healthy volunteers. A. Time course of amplitude changes in eight different muscles (bilateral first dorsal interosseus (FDI), biceps brachii (Bic), pectoralis major (Pec) and tibialis anterior (TA)) before, during and after right hand deafferentation ischemis nerve block (INB). B. Left, effects of right hand deafferentation on MEP amplitudes of bilateral small hand muscles FDI (upper traces show group data, lower traces show example). Right, effects of right hand deafferentation on MEP amplitudes from the right Bic, immediately proximal to the deafferentation level (upper traces show group data, lower traces show example). Shaded areas illustrate the deafferentation period. Note that right hand anesthesia significantly increased MEP amplitudes in the right Bic (a muscle immediately proximal to the tourniquet) and in the FDI contralateral to the tourniquet. Therefore, hand anesthesia elicited plastic changes in cortical representations immediately adjacent to the deafferented one (Bic) and in a representation homologous to the deafferented one in the contralateral hemisphere (left FDI) (modified from Werhahn et al. 2002a).

hemispheres.[14] In humans, acute anesthesia of one hand by INB improved performance of a tactile discriminative task in the other, nondeafferented hand,[162] an effect that lasts for the duration of anesthesia and rapidly returns to baseline when sensation is recovered. This behavioral gain, consistent with that previously reported in the motor domain, is accompanied by increased cortical processing in the primary somatosensory cortex, in the form of enhanced N1-P1 components of the somatosensory evoked potentials. The precise mechanisms mediating this effect remain to be determined. However, the authors suggested that the documentation of this effect is consistent with interhemispheric interaction models of sensory processing.[90] Besides the functional implications of this finding in the intact CNS, these findings raised the possibility of utilizing similar principles to enhance function in, for example, a weak hand after cortical lesions such as a stroke. In one such study, Floel et al. reported improvements in performance of a dynamic finger motor task by the paretic hand of patients with chronic stroke during a period of anesthesia of the intact hand[35] (Figure 10.3, representative subject). The mechanisms underlying this effect in patients with brain lesions remain to be determined but may relate to the transient correction of abnormal interhemispheric interactions in stroke patients[97] elicited by deafferentation of the

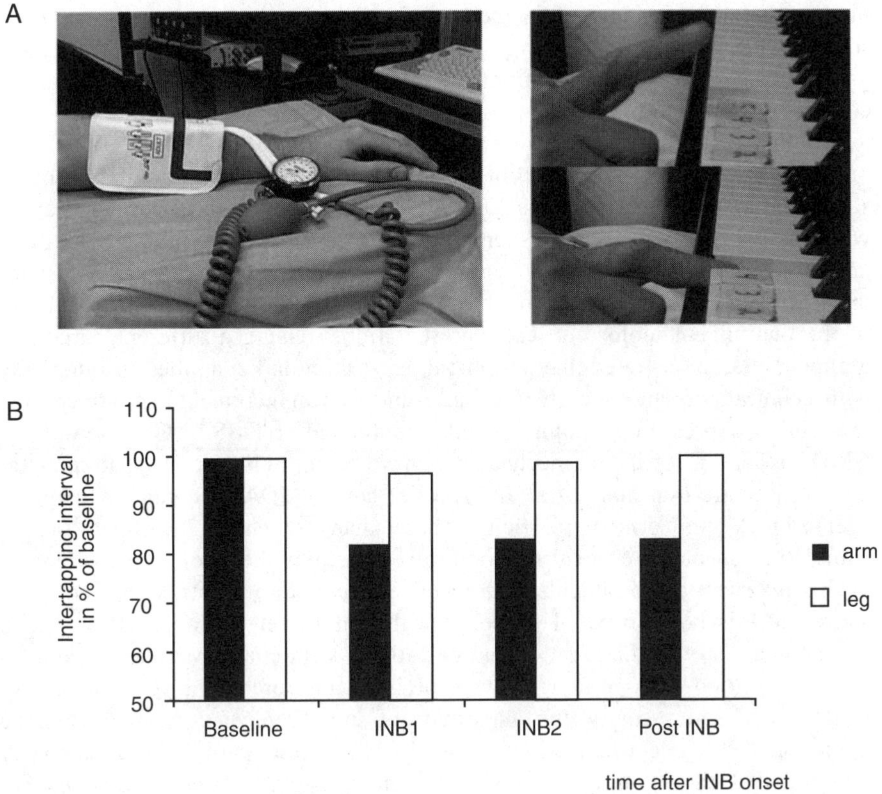

FIGURE 10.3 A. Effect of anesthesia of the intact hand (ischemic nerve block) on motor performance (finger tapping) of the paretic hand in a patient with chronic stroke. Index finger tapping was done with the paretic hand in separate randomized sessions of arm or leg anesthesia. Patients were instructed to tap as fast and regularly as possible. B. Intertapping tapping intervals during arm (black bars) and leg (white bars) anesthesia in a representative patient. Note that finger-tapping intervals were shorter with hand deafferentation, denoting improved tapping speed that outlasted the time of anesthesia (modified from Floel et al. 2004).

intact hand. This type of intervention may have potential implications for the developing rehabilitative treatments for disability after a stroke, the main cause of adult long-term disability.

In summary, deafferentation of one body part representation elicits reorganizational changes in the sensorimotor cortex not only of the contralateral but also the ipsilateral hemisphere[11,161] that are associated with behavioral gains.[162,168] These approaches in combination with training or rehabilitative treatments could conceivably contribute to enhance recovery of function after brain lesions.

C. Somatosensory Stimulation

Somatosensory input is required for proper motor control in healthy volunteers[7,31] and patients with brain lesions.[125] In animal models, peripheral nerve stimulation induced identifiable changes in the stimulated body part representation in the somatosensory cortex.[63,64] In turn, somatosensory stimulation also results in reorganizational changes in the stimulated body part representation in the motor cortex.[5,6,65,86,113,128] In rodents, peripheral nerve stimulation elicited enhanced excitability of the stimulated body part representation in the motor cortex[86] that outlasted the stimulated period. In humans, peripheral nerve stimulation has a similar influence on motor cortical function. Stimulation of the ulnar nerve at the wrist for two hours resulted in increased motor cortical excitability in the ulnar nerve-innervated muscle abductor digiti minimi[65] consistent with previous findings. Similarly, stimulation of pharyngeal areas results in increased excitability of the pharyngeal motor cortical representation.[51] Using a different paradigm, Stefan et al. demonstrated that peripheral nerve stimulation applied in synchrony with a contralateral motor cortical stimulus induced an increment in motor cortical excitability expressed as motor evoked potentials after TMS.[148] GABAergic and NMDA receptor-dependent mechanisms[44] have been implicated since administration of a single oral dose of a GABAergic[65] and NMDA receptor antagonist[147] agents block this form of plasticity. These changes evoked by somatosensory stimulation on motor cortical function raised the hypothesis of the possible use of this intervention in patients with weakness secondary to stroke.[28] One study addressed this possibility.[22] Here, electrical stimulation was applied for 2 h to the median nerve of the paretic hand of patients with chronic stroke. This intervention resulted in a mild improvement in pinch force relative to a control intervention. Interestingly, the magnitude of improvement correlated with the intensity of the electrical stimulation. Two of these patients spontaneously reported that they could hold objects, write and play cards more easily for approximately 24 h. Similar effects were later reported by different laboratories.[150,159] Preliminary experiments in our lab now indicate that peripheral nerve stimulation, in addition to influencing motor cortex function and motor performance, can enhance the beneficial effects of motor training (use-dependent plasticity) in patients with chronic stroke.[135]

In summary, basic science and human studies in healthy volunteers and patients with chronic stroke indicate that somatosensory stimulation may be a useful adjuvant of motor training in neurorehabilitation.

III. EFFECTS OF CORTICAL STIMULATION ON MOTOR CORTICAL FUNCTION AND CORTICAL PLASTICITY

In animal models, cortical stimulation of a body part representation in the motor cortex results in increased motor maps targeting that particular body part.[104] Human studies have provided important clues on these effects. In healthy human volunteers, application of cortical stimulation techniques also modulates cortical excitability.[21,109,111,142] Extradural electrical motor cortex stimulation to relieve chronic pain or movement disorders led to modification in motor symptoms in some patients.[15,70,98,158] Interestingly, human studies performed over the last decade indicate that depending on the way in which stimulation is administered, it can lead to increased[111] or decreased[16] motor cortical excitability. These features make this technique potentially an appealing tool to purposefully modulate cortical function. Three main parameters of cortical stimulation influence the outcome: frequency, intensity, and duration of stimulation.[145] For example, repetitive TMS in the form of trains of 20 Hz for 2 sec delivered at suprathreshold intensity can disrupt function in the primary motor cortex under the stimulating coil[46] whereas, stimulation of the motor cortex at a frequency of 1 Hz for 15 min delivered at subthreshold intensity may induce a long-lasting decrease in cortical excitability.[16] Similarly, the functional effects of application of transcranial direct current stimulation (tDCS) depend on duration, polarity, and the strength of the current.[100,101]

In addition to nonspecific effects on motor cortical excitability, cortical stimulation influences the ability of the CNS to respond to changes in the environment. Ziemann et al demonstrated that low-frequency TMS applied to the motor cortex can enhance deafferentation-induced plasticity in intact humans,[166] expressed as changes in intracortical inhibition and intracortical facilitation. The effects of cortical stimulation have demonstrated topographic specificity. They have been identified with stimulation of the plastic motor cortical representation (upper arm in this study), but not with stimulation of nearby body part representations (hand or face).[169] These results were interpreted as indicative of the intense corticocortical interaction and communication between nearby body part representations in the human motor cortex. The importance of these findings is that they demonstrated, as a proof of principle, that TMS can modulate human cortical plasticity.

These findings led to the investigation of the effects of cortical stimulation on use-dependent plasticity (UDP), fundamentally relevant in neurorehabilitation. A recent study demonstrated for the first time that cortical stimulation can enhance human cortical plasticity elicited by motor training. Buetefisch et al., applied TMS synchronously to a motor cortex engaged in a motor training task and showed that by doing so, UDP became more prominent.[13] Healthy volunteers were studied in different sessions: training alone, training with synchronous application of TMS to the contralateral or ipsilateral motor cortex, and training with TMS delivered asynchronous to the training movement to the motor cortex contralateral to the training hand. It was reported that longevity of UDP was significantly enhanced by TMS. These results demonstrated that UDP can be promoted by synchronous Hebbian stimulation of the motor cortex.[52] Along the same line, tDCS applied to the motor cortex induced

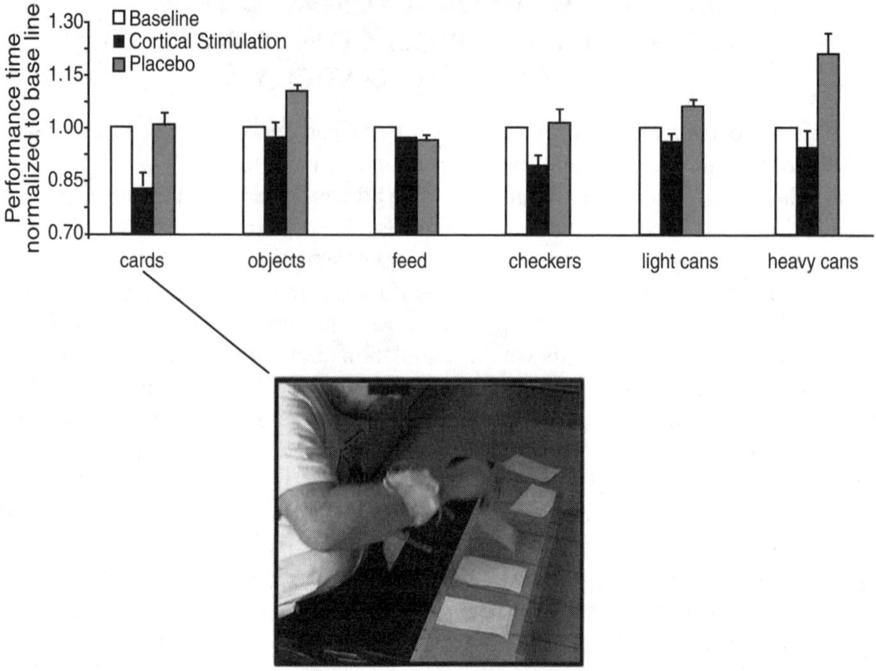

FIGURE 10.4 Effect of cortical stimulation (transcranial direct current stimulation = tDCS) of the affected hemisphere on performance of the Jebsen-Taylor-Hand function-Test (JTT) in chronic stroke patients. The JTT evaluates various functional motor skills. Performance time of each subtest with cortical stimulation (black bars) and placebo stimulation (grey bars) relative to baseline (white bars) is shown in a representative patient. Subtests consist of turning cards (exemplarily shown in the lower row), picking up small objects, mimicking feeding by putting beans with a spoon in a can, stacking checkers, and lifting light and heavy cans is shown. Note the improvement of performance time during tDCS compared to placebo stimulation, indicating that cortical stimulation of the affected hemisphere improved functional motor skills of the paretic hand in this particular patient (modified from Hummel et al.).

improvements in implicit learning in healthy volunteers.[102] Overall, these findings have important implications for neurorehabilitation and are consistent with the view that cortical stimulation could represent an adjuvant to motor training in efforts to recover lost function after cortical lesions like stroke, a proposal consistent with preliminary animal studies.[1,74,104,115,155,156,160] A recent double-blind crossover study showed evidence for this contention. Hummel et al. showed that cortical stimulation in the form of tDCS can transiently improve performance on the JTT functional motor test[50,61,146] relative to placebo brain stimulation[56] (Figure 10.4, see representative data).

Results depicted so far have demonstrated that cortical stimulation applied to one site can enhance excitability or plasticity at that site. Additionally, cortical stimulation applied to one site can induce distant effects on cortical function and behavior.[40,112,118,136,144] TMS applied to one motor cortex elicit blood flow changes in positron emission tomography (PET) brain activation in the opposite motor cortex in healthy volunteers.[40,112,144] Low frequency repetitive TMS (rTMS) applied to one motor cortex influences motor cortical excitability in the homonymous motor representation in the opposite hemisphere[118,136] supporting the concept of a physiological balance of reciprocal inhibitory projections between both hemispheres. It has been proposed that this balance may be disturbed in patients with cortical lesions such as stroke. Indeed, an abnormally high interhemispheric inhibitory drive from M1 in the intact hemisphere to M1 in the affected hemisphere has been documented during generation of a voluntary movement by the paretic hand.[97] The finding that this deficit was more prominent in patients with poorer motor function suggests a contribution to motor performance. Therefore, it is conceivable that one way to improve motor function in the paretic hand is to decrease motor cortical excitability in the ipsilateral, intact motor cortex (with the purpose of reducing abnormal inhibition from the intact to the affected hemisphere), a hypothesis under investigation.

In summary, animal models and human studies in healthy volunteers and stroke patients suggest that cortical stimulation may potentially become an adjuvant to improve motor function and enhance the beneficial effects of motor training in patients with brain lesions.

Improved understanding of the way in which somatosensory input influences motor function led to the development of novel rehabilitative interventions. One example is constraint-induced movement therapy, a strategy consisting of immobilization of the intact hand of stroke patients associated with intensive practice performed with the weak hand. This intervention may enhance functional recovery in patients with motor deficits following a stroke.[94,151–165] Clinical benefits on motor function are associated with documented reduction of cortical excitability of the intact motor cortex and increased excitability of the affected motor cortex.[81,82] Decreased excitability in the intact motor cortex could be the consequence of disuse of the intact hand elicited by immobilization. The combination of constraint and practice in these patients may result in a reduction of exaggerated interhemispheric inhibition from M1 in the intact hemisphere to M1 in the affected hemisphere.[91,161] In turn, such modulation of interhemispheric interactions could result in the documented increase of cortical excitability in the affected motor cortex.[82]

IV. CONCLUSIONS

The somatosensory and motor cortices are highly interconnected, operate in various settings of learning and skill acquisition, and experience constant reorganization in response to environmental challenges or lesions. Acute and chronic deafferentation, somatosensory stimulation, and cortical stimulation can modulate plasticity in both cerebral hemispheres. Improved understanding of these plastic changes has recently raised the exciting hypothesis of utilizing these tools to modify function after brain lesions such as stroke, hopefully evolving to the development of new strategies in neurorehabilitation.

ACKNOWLEDGMENTS

F.H. was supported by a grant from the Alexander v. Humboldt Foundation (Feodor-Lynen).

REFERENCES

1. Adkins-Muir DL, Jones TA (2003) Cortical electrical stimulation combined with rehabilitative training: enhanced functional recovery and dendritic plasticity following focal cortical ischemia in rats. Neurol Res 25: 780-788
2. Ahissar E, Abeles M, Ahissar M, Haidarliu S, Vaadia E (1998) Hebbian-like functional plasticity in the auditory cortex of the behaving monkey. Neuropharmacology 37: 633-655
3. Ahissar E, Ahissar M (1994) Plasticity in auditory cortical circuitry. Curr Opin Neurobiol 4: 580-587
4. Ahissar E, Vaadia E, Ahissar M, Bergman H, Arieli A, Abeles M (1992) Dependence of cortical plasticity on correlated activity of single neurons and on behavioral context. Science 257: 1412-1415
5. Asanuma H, Larsen KD, Zarzecki P (1979) Peripheral input pathways projecting to the motor cortex in the cat. Brain Res 172: 197-208
6. Asanuma H, Rosen I (1972) Functional role of afferent inputs to the monkey motor cortex. Brain Res 40: 3-5
7. Aschersleben G, Gehrke J, Prinz W (2001) Tapping with peripheral nerve block. a role for tactile feedback in the timing of movements. Exp Brain Res 136: 331-339
8. Bartoletti A, Medini P, Berardi N, Maffei L (2004) Environmental enrichment prevents effects of dark-rearing in the rat visual cortex. Nat Neurosci 7: 215-216
9. Bear MF, Rittenhouse CD (1999) Molecular basis for induction of ocular dominance plasticity. J Neurobiol 41: 83-91
10. Birbaumer N, Lutzenberger W, Montoya P, Larbig W, Unertl K, Topfner S, Grodd W, Taub E, Flor H (1997) Effects of regional anesthesia on phantom limb pain are mirrored in changes in cortical reorganization. J Neurosci 17: 5503-5508
11. Brasil-Neto J, Cohen LG, Pascual-Leone A, Jabir FK, Wall RT, Hallett M (1992) Rapid reversible modulation of human motor outputs after transient deafferentation of the forearm: a study with transcranial magnetic stimulation. Neurology 42: 1302-1306
12. Brasil-Neto J, Valls-Sole J, Pascual-Leone A, Cammarota A, Amassian VE, Cracco R, Maccabee P, Cracco J, Hallett M, Cohen LG (1993) Rapid modulation of human cortical motor outputs following ischaemic nerve block. Brain 116: 511-525
13. Buetefisch CM, Khurana V, Kopylev L, Cohen LG (2004) Enhancing encoding of a motor memory in the primary motor cortex by cortical stimulation. J Neurophysiol
14. Calford MB, Tweedale R (1990) Interhemispheric transfer of plasticity in the cerebral cortex. Science 249: 805-807
15. Canavero S, Bonicalzi V, Paolotti R, Castellano G, Greco-Crasto S, Rizzo L, Davini O, Maina R (2003) Therapeutic extradural cortical stimulation for movement disorders: a review. Neurol Res 25: 118-122
16. Chen R, Classen J, Gerloff C, Celnik P, Wassermann EM, Hallett M, Cohen LG (1997) Depression of motor cortex excitability by low-frequency transcranial magnetic stimulation. Neurology 48: 1398-1403
17. Chen R, Corwell B, Yaseen Z, Hallett M, Cohen LG (1998) Mechanisms of cortical reorganization in lower-limb amputees. J Neurosci 18: 3443-3450
18. Cohen LG, Bandinelli S, Findley TW, Hallett M (1991a) Motor reorganization after upper limb amputation in man. A study with focal magnetic stimulation. Brain 114: 615-627
19. Cohen LG, Celnik P, Pascual-Leone A, Corwell B, Falz L, Dambrosia J, Honda M, Sadato N, Gerloff C, Catala MD, Hallett M (1997) Functional relevance of cross-modal plasticity in blind humans. Nature 389: 180-183

20. Cohen LG, Topka H, Cole RA, Hallett M (1991b) Leg paresthesias induced by magnetic brain stimulation in patients with thoracic spinal cord injury. Neurology 41: 1283-1288
21. Cohen LG, Ziemann U, Chen R, Classen J, Hallett M, Gerloff C, Butefisch C (1998) Studies of neuroplasticity with transcranial magnetic stimulation. J Clin Neurophysiol 15: 305-324.
22. Conforto AB, Kaelin-Lang A, Cohen LG (2002) Increase in hand muscle strength of stroke patients after somatosensory stimulation. Ann Neurol 51: 122-125
23. Darian-Smith C, Gilbert CD (1994) Axonal sprouting accompanies functional reorganization in adult cat striate cortex. Nature 368: 737-740
24. Darian-Smith C, Gilbert CD (1995) Topographic reorganization in the striate cortex of the adult cat and monkey is cortically mediated. J Neurosci 15: 1631-1647
25. Davis KD, Kiss ZH, Luo L, Tasker RR, Lozano AM, Dostrovsky JO (1998) Phantom sensations generated by thalamic microstimulation. Nature 391: 385-387
26. Di Lazzaro V, Oliviero A, Profice P, Insola A, Mazzone P, Tonali P, Rothwell JC (1999) Direct demonstration of interhemispheric inhibition of the human motor cortex produced by transcranial magnetic stimulation. Exp Brain Res 124: 520-524
27. Dinse HR, Ragert P, Pleger B, Schwenkreis P, Tegenthoff M (2003) Pharmacological modulation of perceptual learning and associated cortical reorganization. Science 301: 91-94
28. Dobkin BH (2003) Editorial Review: Do electrically stimulated sensory inputs and movements lead to long-term plasticity and rehabilitation gains? Curr Opin Neurol 16: 685-691
29. Donoghue JP (1995) Plasticity of adult sensorimotor representations. Curr Opin Neurobiol 5: 749-754
30. Dykes RW (1997) Mechanisms controlling neuronal plasticity in somatosensory cortex. Can J Physiol Pharmacol 75: 535-545
31. Edin BB, Johansson N (1995) Skin strain patterns provide kinaesthetic information to the human central nervous system. J Physiol 487 (Pt 1): 243-251
32. Elbert T, Flor H, Birbaumer N, Knecht S, Hampson S, Larbig W, Taub E (1994) Extensive reorganization of the somatosensory cortex in adult humans after nervous system injury. Neuroreport 5: 2593-2597
33. Farkas T, Kis Z, Toldi J, Wolff JR (1999) Activation of the primary motor cortex by somatosensory stimulation in adult rats is mediated mainly by associational connections from the somatosensory cortex. Neuroscience 90: 353-361
34. Ferbert A, Priori A, Rothwell JC, Day BL, Colebatch JG, Marsden CD (1992) Interhemispheric inhibition of the human motor cortex. J Physiol 453: 525-546
35. Floel A, Nagorsen U, Werhahn C, Birbaumer N, Knecht S, Cohen LG (2004)
36. Flor H (2002) Phantom-limb pain: characteristics, causes, and treatment. Lancet Neurol 1: 182-189
37. Flor H, Elbert T, Knecht S, Wienbruch C, Pantev C, Birbaumer N, Larbig W, Taub E (1995) Phantom-limb pain as a perceptual correlate of cortical reorganization following arm amputation. Nature 375: 482-484
38. Florence SL, Taub HB, Kaas JH (1998) Large-scale sprouting of cortical connections after peripheral injury in adult macaque monkeys. Science 282: 1117-1121
39. Folmer RL, Griest SE, Martin WH (2001) Chronic tinnitus as phantom auditory pain. Otolaryngol Head Neck Surg 124: 394-400
40. Fox P, Ingham R, George MS, Mayberg H, Ingham J, Roby J, Martin C, Jerabek P (1997) Imaging human intra-cerebral connectivity by PET during TMS. Neuroreport 8: 2787-2791

41. Friedman DP, Jones EG (1981) Thalamic input to areas 3a and 2 in monkeys. J Neurophysiol 45: 59-85
42. Fuhr P, Cohen LG, Dang N, Findley TW, Haghighi S, Oro J, Hallett M (1992) Physiological analysis of motor reorganization following lower limb amputation. Electroencephalogr Clin Neurophysiol 85: 53-60
43. Funnell MG, Corballis PM, Gazzaniga MS (2000) Cortical and subcortical interhemispheric interactions following partial and complete callosotomy. Arch Neurol 57: 185-189
44. Garraghty PE, Kaas JH (1991) Large-scale functional reorganization in adult monkey cortex after peripheral nerve injury. Proc Natl Acad Sci U S A 88: 6976-6980
45. Gerloff C, Cohen LG, Floeter MK, Chen R, Corwell B, Hallett M (1998a) Inhibitory influence of the ipsilateral motor cortex on responses to stimulation of the human cortex and pyramidal tract. J Physiol (Lond) 510: 249-259
46. Gerloff C, Corwell B, Chen R, Hallett M, Cohen LG (1998b) The role of the human motor cortex in the control of complex and simple finger movement sequences. Brain 121: 1695-1709
47. Ghosh S, Brinkman C, Porter R (1987) A quantitative study of the distribution of neurons projecting to the precentral motor cortex in the monkey (M. fascicularis). J Comp Neurol 259: 424-444
48. Giraud AL, Chery-Croze S, Fischer G, Fischer C, Vighetto A, Gregoire MC, Lavenne F, Collet L (1999) A selective imaging of tinnitus. Neuroreport 10: 1-5
49. Gordon J, Ghilardi MF, Ghez C (1995) Impairments of reaching movements in patients without proprioception. I. Spatial errors. J Neurophysiol 73: 347-360
50. Hackel ME, Wolfe GA, Bang SM, Canfield JS (1992) Changes in hand function in the aging adult as determined by the Jebsen Test of Hand Function. Phys Ther 72: 373-377
51. Hamdy S, Rothwell JC, Aziz Q, Singh KD, Thompson DG (1998) Long-term reorganization of human motor cortex driven by short-term sensory stimulation. Nat Neurosci 1: 64-68
52. Hebb DO (1949) The organization of behavior: A neurophysiological theory. Wiley, New York
53. Heynen AJ, Yoon BJ, Liu CH, Chung HJ, Huganir RL, Bear MF (2003) Molecular mechanism for loss of visual cortical responsiveness following brief monocular deprivation. Nat Neurosci 6: 854-862
54. Huerta MF, Pons TP (1990) Primary motor cortex receives input from area 3a in macaques. Brain Res 537: 367-371
55. Hummel F, Andres F, Altenmuller E, Dichgans J, Gerloff C (2002) Inhibitory control of acquired motor programmes in the human brain. Brain 125: 404-420.
56. Hummel F, A, Gerloff C, Cohen LG
57. Iriki A, Pavlides C, Keller A, Asanuma H (1989) Long-term potentiation in the motor cortex. Science 245: 1385-1387
58. Irlbacher K, Meyer BU, Voss M, Brandt SA, Roricht S (2002) Spatial reorganization of cortical motor output maps of stump muscles in human upper-limb amputees. Neurosci Lett 321: 129-132
59. Jacobs KM, Donoghue JP (1991) Reshaping the cortical motor map by unmasking latent intracortical connections. Science 251: 944-947
60. Jastreboff PJ (1990) Phantom auditory perception (tinnitus): mechanisms of generation and perception. Neurosci Res 8: 221-254
61. Jebsen RH, Taylor N, Trieschmann RB, Trotter MJ, Howard LA (1969) An objective and standardized test of hand function. Arch Phys Med Rehabil 50: 311-319

62. Jones EG, Coulter JD, Hendry SH (1978) Intracortical connectivity of architectonic fields in the somatic sensory, motor and parietal cortex of monkeys. J Comp Neurol 181: 291-347
63. Kaas JH (1991) Plasticity of sensory and motor maps in adult mammals. Annu Rev Neurosci 14: 137-167
64. Kaas JH (2000) The reorganization of somatosensory and motor cortex after peripheral nerve or spinal cord injury in primates. Prog Brain Res 128: 173-179
65. Kaelin-Lang A, Luft AR, Sawaki L, Burstein AH, Sohn YH, Cohen LG (2002) Modulation of human corticomotor excitability by somatosensory input. J Physiol 540: 623-633
66. Kaneko T, Caria MA, Asanuma H (1994a) Information processing within the motor cortex. I. Responses of morphologically identified motor cortical cells to stimulation of the somatosensory cortex. J Comp Neurol 345: 161-171
67. Kaneko T, Caria MA, Asanuma H (1994b) Information processing within the motor cortex. II. Intracortical connections between neurons receiving somatosensory cortical input and motor output neurons of the cortex. J Comp Neurol 345: 172-184
68. Karl A, Birbaumer N, Lutzenberger W, Cohen LG, Flor H (2001) Reorganization of motor and somatosensory cortex in upper extremity amputees with phantom limb pain. J Neurosci 21: 3609-3618
69. Karni A, Meyer G, Jezzard P, Adams MM, Turner R, Ungerleider LG (1995) Functional MRI evidence for adult motor cortex plasticity during motor skill learning. Nature 377: 155-158
70. Katayama Y, Fukaya C, Yamamoto T (1997) Control of poststroke involuntary and voluntary movement disorders with deep brain or epidural cortical stimulation. Stereotact Funct Neurosurg 69: 73-79
71. Keller A, Arissian K, Asanuma H (1992) Synaptic proliferation in the motor cortex of adult cats after long-term thalamic stimulation. J Neurophysiol 68: 295-308
72. Kew JJ, Halligan PW, Marshall JC, Passingham RE, Rothwell JC, Ridding MC, Marsden CD, Brooks DJ (1997) Abnormal access of axial vibrotactile input to deafferented somatosensory cortex in human upper limb amputees. J Neurophysiol 77: 2753-2764
73. Kew JJ, Ridding MC, Rothwell JC, Passingham RE, Leigh PN, Sooriakumaran S, Frackowiak RS, Brooks DJ (1994) Reorganization of cortical blood flow and transcranial magnetic stimulation maps in human subjects after upper limb amputation. J Neurophysiol 72: 2517-2524
74. Kleim JA, Jones TA, Schallert T (2003) Motor enrichment and the induction of plasticity before or after brain injury. Neurochem Res 28: 1757-1769
75. Knecht S, Henningsen H, Elbert T, Flor H, Hohling C, Pantev C, Taub E (1996) Reorganizational and perceptional changes after amputation. Brain 119: 1213-1219
76. Kobayashi M, Hutchinson S, Theoret H, Schlaug G, Pascual-Leone A (2004) Repetitive TMS of the motor cortex improves ipsilateral sequential simple finger movements. Neurology 62: 91-98
77. Laubis-Herrmann U, Dichgans J, Bilow H, Topka H (2000) Motor reorganization after spinal cord injury: evidence of adaptive changes in remote muscles. Restor Neurol Neurosci 17: 175-181
78. Lenarz T, Schreiner C, Snyder RL, Ernst A (1993) Neural mechanisms of tinnitus. Eur Arch Otorhinolaryngol 249: 441-446
79. Levy LM, Ziemann U, Chen R, Cohen LG (2002) Rapid modulation of GABA in sensorimotor cortex induced by acute deafferentation. Ann Neurol 52: 755-761

80. Levy WJ, Jr., Amassian VE, Traad M, Cadwell J (1990) Focal magnetic coil stimulation reveals motor cortical system reorganized in humans after traumatic quadriplegia. Brain Res 510: 130-134
81. 81. Liepert J, Bauder H, Wolfgang HR, Miltner WH, Taub E, Weiller C (2000) Treatment-induced cortical reorganization after stroke in humans. Stroke 31: 1210-1216
82. Liepert J, Miltner WH, Bauder H, Sommer M, Dettmers C, Taub E, Weiller C (1998) Motor cortex plasticity during constraint-induced movement therapy in stroke patients. Neurosci Lett 250: 5-8
83. Lotze M, Braun C, Birbaumer N, Anders S, Cohen LG (2003) Motor learning elicited by voluntary drive. Brain 126: 866-872
84. Lotze M, Grodd W, Birbaumer N, Erb M, Huse E, Flor H (1999a) Does use of a myoelectric prosthesis prevent cortical reorganization and phantom limb pain? Nat Neurosci 2: 501-502
85. Lotze M, Laubis-Herrmann U, Topka H, Erb M, Grodd W (1999b) Reorganization in the primary motor cortex after spinal cord injury — A functional Magnetic Resonance (fMRI) study. Restor Neurol Neurosci 14: 183-187
86. Luft AR, Kaelin-Lang A, Hauser TK, Buitrago MM, Thakor NV, Hanley DF, Cohen LG (2002) Modulation of rodent cortical motor excitability by somatosensory input. Exp Brain Res 142: 562-569.
87. Merzenich MM, Kaas JH, Wall JT, Sur M, Nelson RJ, Felleman DJ (1983) Progression of change following median nerve section in the cortical representation of the hand in areas 3b and 1 in adult owl and squirrel monkeys. Neuroscience 10: 639-665
88. Merzenich MM, Nelson RJ, Stryker MP, Cynader MS, Schoppmann A, Zook JM (1984) Somatosensory cortical map changes following digit amputation in adult monkeys. J Comp Neurol 224: 591-605
89. Merzenich MM, Recanzone GH, Jenkins WM, Grajski KA (1990) Adaptive mechanisms in cortical networks underlying cortical contributions to learning and non-declarative memory. Cold Spring Harb Symp Quant Biol 55: 873-887
90. Mesulam MM (1999) Spatial attention and neglect: parietal, frontal and cingulate contributions to the mental representation and attentional targeting of salient extrapersonal events. Philos Trans R Soc Lond B Biol Sci 354: 1325-1346
91. Meyer B-U, Roericht S, Woiciechowsky C, Brandt SA (1995a) Interhemispheric inhibition induced by transcranial magnetic stimulation: localisation of involved fibres studied in patients after partial callosotomy. *J. Physiol.* 487: 68P
92. Meyer BU, Roricht S, Graefin von Einsiedel H, Kruggel F, Weindl A (1995b) Inhibitory and excitatory interhemispheric transfers between motor cortical areas in normal humans and patients with abnormalities of the corpus callosum. Brain 118: 429-440
93. Meyer BU, Roricht S, Woiciechowsky C (1998) Topography of fibers in the human corpus callosum mediating interhemispheric inhibition between the motor cortices. Ann Neurol 43: 360-369
94. Miltner WH, Bauder H, Sommer M, Dettmers C, Taub E (1999) Effects of constraint-induced movement therapy on patients with chronic motor deficits after stroke: a replication. Stroke 30: 586-592
95. Muellbacher W, Richards C, Ziemann U, Wittenberg G, Weltz D, Boroojerdi B, Cohen L, Hallett M (2002) Improving hand function in chronic stroke. Arch Neurol 59: 1278-1282
96. Muhlnickel W, Elbert T, Taub E, Flor H (1998) Reorganization of auditory cortex in tinnitus. Proc Natl Acad Sci U S A 95: 10340-10343

97. Murase N, Duque J, Mazzocchio R, Cohen LG (2004) Influence of interhemispheric interactions on motor function in chronic stroke. Ann Neurol 55: 400-409
98. Nguyen JP, Pollin B, Feve A, Geny C, Cesaro P (1998) Improvement of action tremor by chronic cortical stimulation. Mov Disord 13: 84-88
99. Nitsche MA, Nitsche MS, Klein CC, Tergau F, Rothwell JC, Paulus W (2003a) Level of action of cathodal DC polarisation induced inhibition of the human motor cortex. Clin Neurophysiol 114: 600-604
100. Nitsche MA, Paulus W (2000) Excitability changes induced in the human motor cortex by weak transcranial direct current stimulation. J Physiol 527 Pt 3: 633-639
101. Nitsche MA, Paulus W (2001) Sustained excitability elevations induced by transcranial DC motor cortex stimulation in humans. Neurology 57: 1899-1901
102. Nitsche MA, Schauenburg A, Lang N, Liebetanz D, Exner C, Paulus W, Tergau F (2003b) Facilitation of Implicit Motor Learning by Weak Transcranial Direct Current Stimulation of the Primary Motor Cortex in the Human. J. Cogn. Neurosci. 15: 619-626
103. Nudo RJ (2003) Functional and structural plasticity in motor cortex: implications for stroke recovery. Phys Med Rehabil Clin N Am 14: S57-76
104. Nudo RJ, Jenkins WM, Merzenich MM (1990) Repetitive microstimulation alters the cortical representation of movements in adult rats. Somatosens Mot Res 7: 463-483
105. Nudo RJ, Milliken GW, Jenkins WM, Merzenich MM (1996a) Use-dependent alterations of movement representations in primary motor cortex of adult squirrel monkeys. J Neurosci 16: 785-807
106. 106. Nudo RJ, Wise BM, SiFuentes F, Milliken GW (1996b) Neural substrates for the effects of rehabilitative training on motor recovery after ischemic infarct [see comments]. Science 272: 1791-1794
107. Ojemann JG, Silbergeld DL (1995) Cortical stimulation mapping of phantom limb rolandic cortex. Case report. J Neurosurg 82: 641-644
108. Pascual-Leone A, Peris M, Tormos JM, Pascual AP, Catala MD (1996) Reorganization of human cortical motor output maps following traumatic forearm amputation. Neuroreport 7: 2068-2070
109. Pascual-Leone A, Tormos JM, Keenan J, Tarazona F, Canete C, Catala MD (1998) Study and modulation of human cortical excitability with transcranial magnetic stimulation. J Clin Neurophysiol 15: 333-343
110. Pascual-Leone A, Valls-Sole J, Brasil-Neto JP, Cammarota A, Grafman J, Hallett M (1994a) Akinesia in Parkinson's disease. II. Effects of subthreshold repetitive transcranial motor cortex stimulation. Neurology 44: 892-898
111. Pascual-Leone A, Valls-Solé J, Wassermann EM, Hallett M (1994b) Responses to rapid-rate transcranial stimulation of the human motor cortex. Brain 117: 847-858
112. Paus T, Jech R, Thompson CJ, Comeau R, Peters T, Evans AC (1998) Dose-dependent reduction of cerebral blood flow during rapid-rate transcranial magnetic stimulation of the human sensorimotor cortex. J Neurophysiol 79: 1102-1107
113. Pavlides C, Miyashita E, Asanuma H (1993) Projection from the sensory to the motor cortex is important in learning motor skills in the monkey. *J. Neurophysiol.* 70: 733-741
114. Pearson K (2000) Motor systems. Curr Opin Neurobiol 10: 649-654
115. Plautz EJ, Barbay S, Frost SB, Friel KM, Dancause N, Zoubina EV, Stowe AM, Quaney BM, Nudo RJ (2003) Post-infarct cortical plasticity and behavioral recovery using concurrent cortical stimulation and rehabilitative training: a feasibility study in primates. Neurol Res 25: 801-810

116. Pleger B, Dinse HR, Ragert P, Schwenkreis P, Malin JP, Tegenthoff M (2001) Shifts in cortical representations predict human discrimination improvement. Proc Natl Acad Sci U S A 98: 12255-12260
117. Plewnia C, Bartels M, Gerloff C (2003a) Transient suppression of tinnitus by transcranial magnetic stimulation. Ann Neurol 53: 263-266
118. Plewnia C, Lotze M, Gerloff C (2003b) Disinhibition of the contralateral motor cortex by low-frequency rTMS. Neuroreport 14: 609-612
119. Porter LL (1991) Patterns of connectivity in the cat sensory-motor cortex: a light and electron microscope analysis of the projection arising from area 3a. J Comp Neurol 312: 404-414
120. Porter LL (1992) Patterns of projections from area 2 of the sensory cortex to area 3a and to the motor cortex in cats. Exp Brain Res 91: 85-93
121. Priori A, Berardelli A, Rona S, Accornero N, Manfredi M (1998) Polarization of the human motor cortex through the scalp. Neuroreport 9: 2257-2260
122. Ragert P, Dinse HR, Pleger B, Wilimzig C, Frombach E, Schwenkreis P, Tegenthoff M (2003) Combination of 5 Hz repetitive transcranial magnetic stimulation (rTMS) and tactile coactivation boosts tactile discrimination in humans. Neurosci Lett 348: 105-108
123. Rauschecker JP (1999) Auditory cortical plasticity: a comparison with other sensory systems. Trends Neurosci 22: 74-80
124. Recanzone GH, Allard TT, Jenkins WM, Merzenich MM (1990) Receptive-field changes induced by peripheral nerve stimulation in SI of adult cats. J Neurophysiol 63: 1213-1225
125. Reding MJ, Potes E (1988) Rehabilitation outcome following initial unilateral hemispheric stroke. Life table analysis approach. Stroke 19: 1354-1358
126. Rittenhouse CD, Shouval HZ, Paradiso MA, Bear MF (1999) Monocular deprivation induces homosynaptic long-term depression in visual cortex. Nature 397: 347-350
127. Roricht S, Meyer BU, Niehaus L, Brandt SA (1999) Long-term reorganization of motor cortex outputs after arm amputation. Neurology 53: 106-111
128. Rosen I, Asanuma H (1972) Peripheral afferent inputs to the forelimb area of the monkey motor cortex: input-output relations. Exp Brain Res 14: 257-273
129. Rothwell JC, Traub MM, Day BL, Obeso JA, Thomas PK, Marsden CD (1982) Manual motor performance in a deafferented man. Brain 105 (Pt 3): 515-542
130. Rouiller EM, Babalian A, Kazennikov O, Moret V, Yu XH, Wiesendanger M (1994) Transcallosal connections of the distal forelimb representations of the primary and supplementary motor cortical areas in macaque monkeys. Exp Brain Res 102: 227-243
131. Sadato N, Zeffiro TA, Campbell G, Konishi J, Shibasaki H, Hallett M (1995) Regional cerebral blood flow changes in motor cortical areas after transient anesthesia of the forearm. Ann. Neurol. 37: 74-81
132. Sakamoto T, Porter LL, Asanuma H (1987) Long-lasting potentiation of synaptic potentials in the motor cortex produced by stimulation of the sensory cortex in the cat: a basis of motor learning. Brain Res 413: 360-364
133. Salerno A, Georgesco M (1996) Interhemispheric facilitation and inhibition studied in man with double magnetic stimulation. Electroencephalogr Clin Neurophysiol 101: 395-403
134. Sanes JN, Donoghue JP (2000) Plasticity and primary motor cortex. Annu Rev Neurosci 23: 393-415
135. Sawaki L, Wu CW, Cohen LG (2004) Enhancement of use-dependent plasticity by peripheral nerve stimulation in patients with chronic stroke. In: 5th World Stroke Congress, Vancouver, B.C., Canada, June 23-26, 2004

136. Schambra HM, Sawaki L, Cohen LG (2003) Modulation of excitability of human motor cortex (M1) by 1 Hz transcranial magnetic stimulation of the contralateral M1. Clin Neurophysiol 114: 130-133
137. Schnitzler A, Kessler KR, Benecke R (1996) Transcallosally mediated inhibition of interneurons within human primary motor cortex. Exp Brain Res 112: 381-391
138. Schwark HD, Esteky H, Jones EG (1992) Corticocortical connections of cat primary somatosensory cortex. Exp Brain Res 91: 425-434
139. Shibata T, Shimoyama I, Ito T, Abla D, Iwasa H, Koseki K, Yamanouchi N, Sato T, Nakajima Y (1997) The time course of interhemispheric EEG coherence during a GO/NO-GO task in humans. Neurosci Lett 233: 117-120
140. 140. Shin HC, Won CK, Jung SC, Oh S, Park S, Sohn JH (1997) Interhemispheric modulation of sensory transmission in the primary somatosensory cortex of rats. Neurosci Lett 230: 137-139
141. Sica RE, Sanz OP, Cohen LG, Freyre JD, Panizza M (1984) Changes in the N1-P1 component of the somatosensory cortical evoked response in patients with partial limb amputation. Electromyogr Clin Neurophysiol 24: 415-427
142. Siebner HR, Filipovic SR, Rowe JB, Cordivari C, Gerschlager W, Rothwell JC, Frackowiak RS, Bhatia KP (2003) Patients with focal arm dystonia have increased sensitivity to slow-frequency repetitive TMS of the dorsal premotor cortex. Brain 126: 2710-2725
143. Siebner HR, Mentschel C, Auer C, Conrad B (1999) Repetitive transcranial magnetic stimulation has a beneficial effect on bradykinesia in Parkinson's disease. Neuroreport 10: 589-594
144. Siebner HR, Peller M, Willoch F, Minoshima S, Boecker H, Auer C, Drzezga A, Conrad B, Bartenstein P (2000) Lasting cortical activation after repetitive TMS of the motor cortex: a glucose metabolic study. Neurology 54: 956-963
145. Siebner HR, Rothwell J (2003) Transcranial magnetic stimulation: new insights into representational cortical plasticity. Exp Brain Res 148: 1-16
146. Spaulding SJ, McPherson JJ, Strachota E, Kuphal M, Ramponi M (1988) Jebsen Hand Function Test: performance of the uninvolved hand in hemiplegia and of right-handed, right and left hemiplegic persons. Arch Phys Med Rehabil 69: 419-422
147. Stefan K, Kunesch E, Benecke R, Cohen LG, Classen J (2002) Mechanisms of enhancement of human motor cortex excitability induced by interventional paired associative stimulation. J Physiol 543: 699-708
148. Stefan K, Kunesch E, Cohen LG, Benecke R, Classen J (2000) Induction of plasticity in the human motor cortex by paired associative stimulation. Brain 123 Pt 3: 572-584
149. Stepniewska I, Preuss TM, Kaas JH (1993) Architectonics, somatotopic organization, and ipsilateral cortical connections of the primary motor area (M1) of owl monkeys. J Comp Neurol 330: 238-271
150. Struppler A, Havel P, Muller-Barna P (2003) Facilitation of skilled finger movements by repetitive peripheral magnetic stimulation (RPMS) — a new approach in central paresis. NeuroRehabilitation 18: 69-82
151. Taub E, Morris DM (2001) Constraint-induced movement therapy to enhance recovery after stroke. Curr Atheroscler Rep 3: 279-286
152. Taub E, Ramey SL, DeLuca S, Echols K (2004) Efficacy of constraint-induced movement therapy for children with cerebral palsy with asymmetric motor impairment. Pediatrics 113: 305-312
153. Taub E, Uswatte G, Morris DM (2003) Improved motor recovery after stroke and massive cortical reorganization following Constraint-Induced Movement therapy. Phys Med Rehabil Clin N Am 14: S77-91, ix

154. Taub E, Uswatte G, Pidikiti R (1999) Constraint-Induced Movement Therapy: a new family of techniques with broad application to physical rehabilitation — a clinical review. J Rehabil Res Dev 36: 237-251
155. Teskey GC, Flynn C, Goertzen CD, Monfils MH, Young NA (2003) Cortical stimulation improves skilled forelimb use following a focal ischemic infarct in the rat. Neurol Res 25: 794-800
156. Teskey GC, Monfils MH, VandenBerg PM, Kleim JA (2002) Motor map expansion following repeated cortical and limbic seizures is related to synaptic potentiation. Cereb Cortex 12: 98-105
157. Topka H, Cohen LG, Cole RA, Hallett M (1991) Reorganization of corticospinal pathways following spinal cord injury. Neurology 41: 1276-1283
158. Tsubokawa T, Katayama Y, Yamamoto T, Hirayama T, Koyama S (1993) Chronic motor cortex stimulation in patients with thalamic pain. J Neurosurg 78: 393-401
159. Uy J, Ridding MC, Hillier S, Thompson PD, Miles TS (2003) Does induction of plastic change in motor cortex improve leg function after stroke? Neurology 61: 982-984
160. VandenBerg PM, Hogg TM, Kleim JA, Whishaw IQ (2002) Long-Evans rats have a larger cortical topographic representation of movement than Fischer-344 rats: a microstimulation study of motor cortex in naive and skilled reaching-trained rats. Brain Res Bull 59: 197-203
161. Werhahn KJ, Mortensen J, Kaelin-Lang A, Boroojerdi B, Cohen LG (2002a) Cortical excitability changes induced by deafferentation of the contralateral hemisphere. Brain 125: 1402-1413
162. Werhahn KJ, Mortensen J, Van Boven RW, Zeuner KE, Cohen LG (2002b) Enhanced tactile spatial acuity and cortical processing during acute hand deafferentation. Nat Neurosci 5: 936-938
163. Wiesel TN, Hubel DH (1963) Single cell responses in striate cortex of kittens deprived of vision in one eye. *J. Neurophysiol.* 26: 1003-1017
164. Wiesel TN, Hubel DH (1965) Comparison of the effects of unilateral and bilateral eye closure on cortical unit responses in kittens. J Neurophysiol 28: 1029-1040
165. Wittenberg GF, Chen R, Ishii K, Bushara KO, Eckloff S, Croarkin E, Taub E, Gerber LH, Hallett M, Cohen LG (2003) Constraint-induced therapy in stroke: magnetic-stimulation motor maps and cerebral activation. Neurorehabil Neural Repair 17: 48-57
166. Ziemann U, Corwell B, Cohen LG (1998a) Modulation of plasticity in human motor cortex after forearm ischemic nerve block. J Neurosci 18: 1115-1123
167. Ziemann U, Hallett M, Cohen LG (1998b) Mechanisms of deafferentation-induced plasticity in human motor cortex. J Neurosci 18: 7000-7007
168. Ziemann U, Muellbacher W, Hallett M, Cohen LG (2001) Modulation of practice-dependent plasticity in human motor cortex. Brain 124: 1171-1181.
169. Ziemann U, Wittenberg GF, Cohen LG (2002) Stimulation-induced within-representation and across-representation plasticity in human motor cortex. J Neurosci 22: 5563-5571

11 Behavioral Basis of Focal Hand Dystonia: Aberrant Learning in the Somatosensory Cortex

Nancy N. Byl

CONTENTS

I. Abstract 227
II. Introduction 228
  A. Focal Hand Dystonia (FHd) 229
  B. The Evidence for Aberrant Learning 231
  C. Primate Studies 231
  D. Rodent Animal Models 238
  E. Summary of Animal Studies 239
  F. Human Research Models 240
    1. Experiment I: Relationship of Clinical Performance and Neural Structure 240
    2. Experiment II: Intervention (12 Subjects) 247
    3. Experiment III: Three Case Studies 250
III. Summary of Intervention Strategies 252
References 253

I. ABSTRACT

Focal hand dystonia (FHd) is a disabling movement disorder of unknown etiology that can develop in productive, motivated individuals who perform highly repetitive, intensive hand tasks. This chapter summarizes our research supporting aberrant learning as one origin of FHd. Our studies document degradation of the somatosensory representation of the hand in animals and patients with dystonic hand movements as characterized by large receptive fields (rfs), overlapped across adjacent digits and glabrous-dorsal surfaces, persistence of digital receptive fields across broad cortical distances, high ratio of amplitude to latency in somatosensory evoked field responses, and decreased spread and abnormal sequencing of digit representation.. Challenging, rewarded, repetitive behavioral tasks that require high speed, high force, precision and intense work cycles with minimal breaks can accelerate the expression of FHd, particularly when paired with physical characteristics such as joint and soft tissue inflexibility, poor posture, slow sensory processing with inaccurate sensory discrimination, imbalance of extrinsic and intrinsic muscles, and rapid reversals of digital flexion and extension. The development of FHd may be minimized if individuals use the hands in a functional, mid-range position, take frequent breaks, work at variable speeds for short durations, attend to sensormotor feedback, and initiate digital movements with the intrinsic muscles. Attended, progressive, rewarded, learning based sensorimotor training consistent with the principles of neuroplasticity can facilitate recovery of task specific motor control in patients with focal hand dystonia.

II. INTRODUCTION

We perform delicate, complex, individuated, fine motor movements with our hands.[45,66,122] This is reflected by large, orderly, somatotopic, highly differentiated representations of the hand in the thalamus, basal ganglia, and cortex.[58,59,63,68,106,139,140] Topographical representations are complemented by integrative functional representations of well learned tasks like instrumental play or writing.[118] Topographical representations can be modified across the lifetime by environmental enrichment or deprivation, drugs and attended, spaced, repetitive, learning based, goal-directed, rewarded, nonstereotypical, progressive practice[51,52,54,64,139,140] (Byl and Merzenich, 2000e). Task-specific, learning-based repetitive behaviors, drive selective changes in cell assemblies that differentiate and selectively specialize representations. These changes occur simultaneously with the emergence of more efficient, accurate, and differentiated behaviors[35,87,114,115,116] (Merzenich et al. 1996).

Learning based activities, that are progressed in difficulty, rewarded, and attended, up-regulate neurotransmitters like dopamine and acetylcholine.[46,67,71] In human and animal studies, these behavioral activities are associated with changes in neural structure. For example, expansion of cortical representations, reduction in the size of receptive fields, narrowed columnar spread, co-selection of complementary inputs, increased excitable neurons, enhanced salience and specificity of feedback, increased myelination, strengthened synapses between coincident inputs, shortened integration time, and increased complexity of dendritic branching[1,41,53,62,63,68,83–86,88–90,92,96–99,106,107,114–116,120,123,139,140,145] (Jenkins et al., 1990; Xerri et al., 1996;1999[64]). Practice (mental or physical) not only modifies neural structure, but enhances the learning of new tasks, improves task proficiency, and increases recovery following neural insults.[40,41,62,65,88–90,92,93,120,123] Metabolic state, emotions, sleep, and natural endorphins can improve and refresh good networks as well as erase poor connections[65,41] (Jenkins et al., 1987[62]).

However, there are inherent limits to neural plasticity due to physiological time constants, inhibition, and integration time[121] (Byl and Merzenich, 2000e). When inputs occur within the inhibitory or integration period, they may no longer be registered as temporally distinct,[3,31,45,90,105,124,139,140,144] with the stimulated skin surfaces forming a unified rather than a unique spatial and temporal representation in the cerebral cortex.[14,15] Specificity of digital representation is critical to the maintenance of the normal sensory organization, sensorimotor feedback and fine motor control of the digits.[45,104] Is it possible that rapid, rewarded, repetitive, stereotypic, near simultaneous alternation of flexors and extensors could degrade representational specificity, disrupt sensorimotor feedback and cause involuntary, task specific dystonic movements of the hand?[3,14,15,18,99,120,139,140]

A. Focal Hand Dystonia (FHd)

FHd is a disabling condition that can bring an early end to a successful career. This is one of the limb dystonias. There is a loss of inhibition of agonists and antagonists that is associated with involuntary, writhing, twisting end range movements that interfere with normal hand specific fine motor control.[2,7,33,34,43,55,60,81,95,119,133,135,138] The etiology of FHd is still considered idiopathic. The hypotheses range from a genetic origin,[44,57,76,102] an imbalance in the globus pallidus/substantia nigra (lack of inhibitory and excitatory pathways),[8,37,38,67,94,103,117] cortical dysfunction,[29,36,39,47,117,127,131] degradation of receptive fields in the sensory thalamus,[75,135,148] disruption in cortical sensory activation, somatosensory representation and spatial perception,[3,4,11,12–19,22,31,40,56,75,101,122,128,129] (McKenzie et al., 2003) abnormal presynaptic desynchronization of movement or abnormal muscle spindle afferent firing,[49,131] abnormal gating of somatosensory inputs,[91] disruption of inhibition in the spinal cord,[30,70,93,94,102,103] or secondary consequences of peripheral trauma, nerve root irritation, peripheral nerve entrapment, or anatomic restrictions.[27,28,72,74,113,130,141–143] (Jankovic, 2001) Aberrant learning may also be an etiology of FHd (Byl, 2003).

Sanger and Merzenich (2000) proposed the computational model as one origin of FHd. This is consistent with the somatosensory learning hypothesis proposed by Byl et al in 1996.[14,15] As a result of repetitive use, simultaneous firing, coupling of multiple sensory signals, and voluntary coactivation of muscles, there could be a degradation of the sensory cortical representation of the hand and disruption in sensorimotor feedback.[15,35,85,87,96,97,114–116,139,140,148] (Merzenich et al., 1983, 1990ab; Byl et al., 1996). This model could explain why: 1. symptoms develop in otherwise healthy individuals in response to highly attended repetitive movements; 2. evolution of symptoms is variable in terms of time; 3. symptoms appear only during the performance of a target specific task; 4. dystonic movements persist despite stopping the task; 5. symptoms decrease, but may not be remediated with dopamine-depleting drugs or botulinum toxin; and 6. there are abnormalities in motor and sensory cortical representations of the dystonic limb. The loop through the deep nuclei to the cortex, basal ganglia, and thalamus combines with the sensorimotor loop gain and contributes to instability.[48,120,121] If only certain mechanical models of the sensorimotor loop are unstable, then a focal dystonia rather than a generalized dystonia develops.

The computational model further suggests that the appropriate treatment needs to decrease the imbalance in the loop gain. Intervention should be based on the principles of plasticity to redifferentiate cortical and subcortical representations. If the dystonia is severe, it may be necessary to temporarily break the cycle before retraining can be effectively implemented (e.g., botulinum toxin injections). This retraining requires variability of practiced movements to increase uncorrelated movement components, each engaging specific, relevant sensory neurons. This is comparable to behaviors directed to uncoupling pathologically coupled modes. The diagnosis of FHd is made by a careful history. In some individuals, the history may reveal a strong temporal-anatomical relationship between trauma and the onset of FHd. Jankovic (2001)[61] suggests that peripherally induced movement disorders should be considered if: 1. trauma is severe enough to cause persistent local symptoms (e.g., symptoms persist at least 2 weeks and cause a person to seek late medical attention); 2. the anatomical site of the original injury is the same site as the initial manifestation of the movement disorder; 3. the movement disorder develops within days or months, (up to a year) post injury and; 4. there are preexisting or trauma-related contractures and limitations of passive movement. In contrast, Weiner (2001)[141] argues that there is little evidence in support of peripheral trauma as a likely cause of focal dystonia since only a very small proportion of a large population of individuals suffering traumatic injuries develop a movement disorder, and only a few of the millions of people performing repetitive work-related hand tasks develop a focal hand dystonia. On the other hand, the work history frequently reveals a stressful period of repetitive hand use, job stress or instability, application of a new technique, change in equipment, and increased time on task to improve quality or quantity of work (Kolle, 2000)[73]. Initially, patients report an earlier history of pain that was diagnosed as repetitive strain injury (cumulative trauma) and effectively treated as a condition of acute inflammation from microtrauma.[5,6,125,137] (Millender et al., 1992).[91] Some individuals continue the repetitive movements and either develop chronic pain or degenerative, painful conditions like tendinonsis.[69] Others report fatigue, incoordination, and ultimately deterioration of performance with the fingers developing a life of their own (curling or extending involuntarily when attempting to perform a familiar task).[33] Personality characteristics such as perfectionism, anxiety, stress, phobias, and emotional instability may also be abstracted from the history.[2,59]

FHd is described as painless and the neurological examination is reported as normal even though some patients complain of a physiological tremor, uncontrollable excitability, numbness or dullness of the hand when simply placing the finger pads on the target surface. The critical factor in the examination is the observation of selective, painless, involuntary movements during the performance of a target-specific task. While not commonly ordered in the clinic, functional magnetic resource imaging and magnetoencephalography can document differences in neural firing patterns, blood flow patterns with task performance, and representational topography (e.g., representational size, location, digit spread and order).

To date, there are no intervention strategies that are 100% effective for restoring normal motor control in patients with FHd. While botulinum toxin injections or Baclophen can decrease dystonic cramping,[10,26,34,42,71,112,132,136] the medications do not generally improve somatosensory differentiation and rarely enable musicians to return to high levels of performance. Conservative intervention strategies based on the principles of neuroplasticity have been applied as alternate intervention strategies to drugs and surgery. These promising paradigms include constraint-induced therapy now called sensory motor retuning,[24,25] sensitivity training,[133] conditioning techniques,[77,78] kinematic training,[79] immobilization, (Priori et al., 2001)[110] and learning based sensorimotor training (Byl et al 1998, 2000a, 2000c, 2000d).[16,17,19,20] None of these strategies have been confirmed by randomized clinical trials.

B. The Evidence for Aberrant Learning

When movements become excessively repetitive, deterioration in motor speed and accuracy can develop.[5] With highly technical, repetitive, stereotypical, and near simultaneous movements, abnormal, involuntary, dystonic movements can develop.[14] The question is whether there is sufficient evidence to support the hypothesis that focal hand dystonia is a consequence of abnormal learning. The purpose of this chapter is to summarize the evidence from our animal and human studies to support aberrant somatosensory learning as one valid etiology of FHd and clarify the factors that contribute to the risk for developing this condition.

C. Primate Studies

The objective of the primate studies[9,14,15,130] was to carry out a series of experiments in a species of animals that had no history of dystonia and determine if it was possible: l. to induce a focal hand dystonia by having the animals perform repetitive hand tasks; 2. to correlate the dystonia with the presence of inflammatory cells and fibrosis; and 3. to correlate deterioration of clinical performance with change in the cortical hand representation. Seven adult Aotus nancimae owl monkeys and historic reference controls from other research studies were included[114–116] (Jenkins et al., 1990).[64]

The monkeys were trained to do one of three tasks: 1. place the hand on a hand-piece that passively opened and closed the hand (2 monkeys); 2. place the hand on a hand-piece and actively close and open the hand (4 monkeys); or 3. place the index finger and the thumb onto two marked, indented, areas requiring a wide spread of the two digits (1 monkey). Once the behavior was shaped, the monkeys were brought to the Keck Center for Neuroscience for training. Training was carried out in a cage mounted in a sound-isolated test chamber. A video system outside the cage monitored the monkeys' behavior. A short cylinder mounted on the cage front guided the monkey to reach to a hand grip molded to fit the monkey's hand in a vertically oriented position. A pellet feeder or a juice feeder was attached to the sidewall of the cage. The monkeys were deprived of food for 20–22 h before beginning each training week (with weight maintained at 80–90% of normal). Nutritionally complete, whole-grain, banana-flavored pellets of 45 mg (Bio-serve, Frenchtown, NH) or Tang served as behavioral rewards. Monkeys received water ad libitum and food supplements after training.

The monkeys engaged in behavioral training 5 d a week over 2–12 months. In the passive task, the digits had to make contact with the detectors to activate the spring-loaded device that opened and closed the hand-piece. A second spring-loaded solenoid was mounted on the thumb pad that opened and closed the thumb. The excursion of the fingers was 6.44 mm and the thumb plate 1.5 mm. The openings occurred quickly within 16 msec and closure required approximately 50 msec. In the *active* opening and closing paradigm, contact detectors were mounted on the thumb

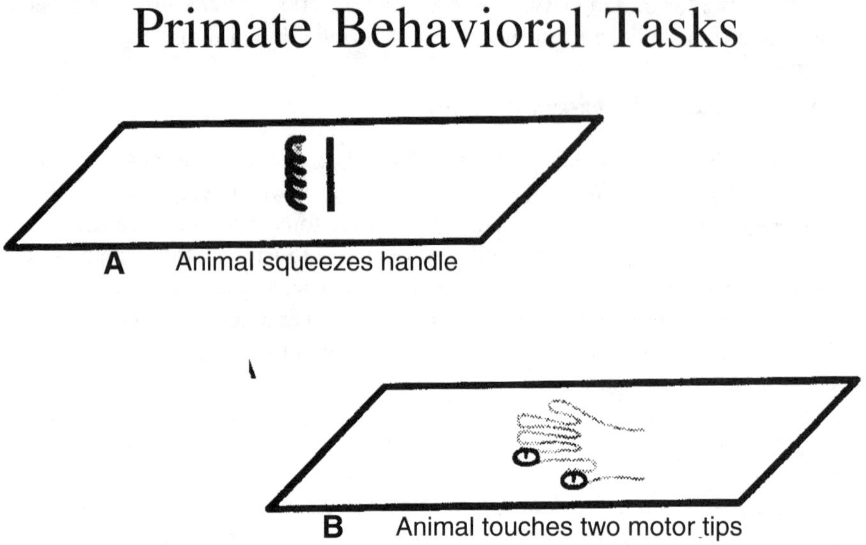

FIGURE 11.1 Behavioral Tasks. On the model, 1A the monkey had to open and close the hand piece for a variable number of trials. On the model 1B the money had to accurately place the tips of D1 and D2 onto the target and leave the digits in place during a random series of taps to the digits. From Blake, D.T., Byl, N.N., Cheung, S., Bedenbaugh, P., Nagarajan, S., Lamb, M., Merzenich, M. 2002. Sensory representation abnormalities that parallel focal hand dystonia in primate model. *Somatosens Mot Res* 19: 347–357. With permissionn.

piece and each finger groove. The hand-piece was driven by a spring-loaded solenoid to provide a known force (80 g). The monkey closed the hand-piece over a distance of approximately 7 mm. The hand-piece vibrated when the closure was complete. When the vibration stopped, the monkey had to quickly release finger contact by extending the fingers; then the hand piece automatically reopened. In the reaching task, the animal positioned its right, dominant hand in a hand mold, with the first and second digits on two metal contacts with the forearm pronated. Each contact was 1 mm in diameter and positioned the two digits in an unnatural position. To receive an award, the animal was required to hold the hand in place for several hundred milliseconds before releasing. The animal was trained on a task that delivered 1000–200 μm taps to its fingertips. This task was a cross-digit interval discrimination task. When the animal placed the hand in the mold making electrical contact with the tips of the two motors on the pads of the thumb and index finger, then a series of stimuli were delivered to the index finger and the thumb with a change in the inter-tap interval time for a pair of taps, decreasing from 500 to 100 msec (See Figure 11.1).

All three hand tasks were controlled by LabView® virtual instruments software.[14,15] All tasks had the following characteristics: 1. attended; 2. rapid; 3. rewarded with food; 4. stereotypical, near coincident in time; 5. repetitive (@2 h/d, 3–5 d a week; and 6. spaced over time (5 weeks to 6 months). Speed of repetitions, number of repetitions, time of training, and accuracy of task performance (videotaped) and monitored with Labview®. After motor performance deteriorated by 50% in speed and accuracy, training continued at least another 2 weeks. These protocols were approved by the Committee on Animal Research.

The details of anesthesia, surgery and electrophysiological monitoring have been detailed in a variety of other studies and have been determined to meet the criteria for safe, animal care protocols for research.[15,85,108,139,140] (Blake et al., 2002; Byl et al.,[14] 1996; Jenkins et al.,[65] 1990; Merzenich et al., 1996a) Monkeys were either mapped for 15 h or mapped for 5 d and not recovered. MAP 50 software[109] was used to construct and measure the cortical representation and to measure the size of the cutaneous receptive fields. The clinical dependent variables included motor performance at the target task. A food retrieval task (picking foot out of trays of graded size) was also rated for quality to confirm that the movement disorder was confined primarily to the target task. Following cortical mapping, anatomic dissections were performed in the monkeys with analysis of the tissues for inflammatory cells, fibroblasts and macrophages.

In the monkey performing the reaching task, a dense microelectrode array was implanted in the left hemisphere, shortly before the full blown focal hand dystonia developed (right hand, D1, D2). The techniques for implantation have been previously described.[35] There were 49 high impedance parylene-iridium microelectrodes implanted into a 2 by 2 mm cortical area.

These experiments were based on a post test experimental design. Although the number of subjects was small, well over 100 d of data were gathered on motor accuracy and frequency of task performance, and 300–400 receptive fields were mapped. The clinical dependent variables included accuracy, speed and quality of task performance and food retrieval. For the electrophysiological data, the area of the topographical field was mapped, the total area was calculated, the cortical distances between separate receptive fields was measured, the number of receptive fields were plotted per electrode penetration, the number of overlaps across adjacent digits and across glabrous and dorsal receptive fields were counted, and the circumference of the receptive fields were calculated.

The Student t test was used to determine the significance of differences between the trained animals and the controls. The decline in speed and accuracy of performance over time was analyzed using the Page Test for Trends. ($p<0.05$).[80] Each dependent variable was considered an independent family. The presence of inflammatory cells and fibroblasts were described post anatomical dissection and immunochemical analysis but were not tested for significance.

The normal topography of the hand is characterized with one receptive field per electrode penetration, small receptive fields (8.0 ± 3.0 mm^2) unique to each digit, orderly sequencing of digits from inferior to superior and segments from proximal to distal, distinct differentiation of the digits at 100–600 μm, and an area of representation of 3.2 to 5.1 mm^2.[114,115,116,126] (Jenkins et al., 1986). With training, the area of the representation increases in size while the receptive fields decrease in size and increase in specificity and density (Figure 11.2).

Two owl monkeys performed the attended, repetitive, passive hand opening and closing task (1.5–2 h/d) 5 d a week for 12–25 weeks. Initial task performance was greater than 90% accurate. Between 5–8 weeks of training, both monkeys sponta-

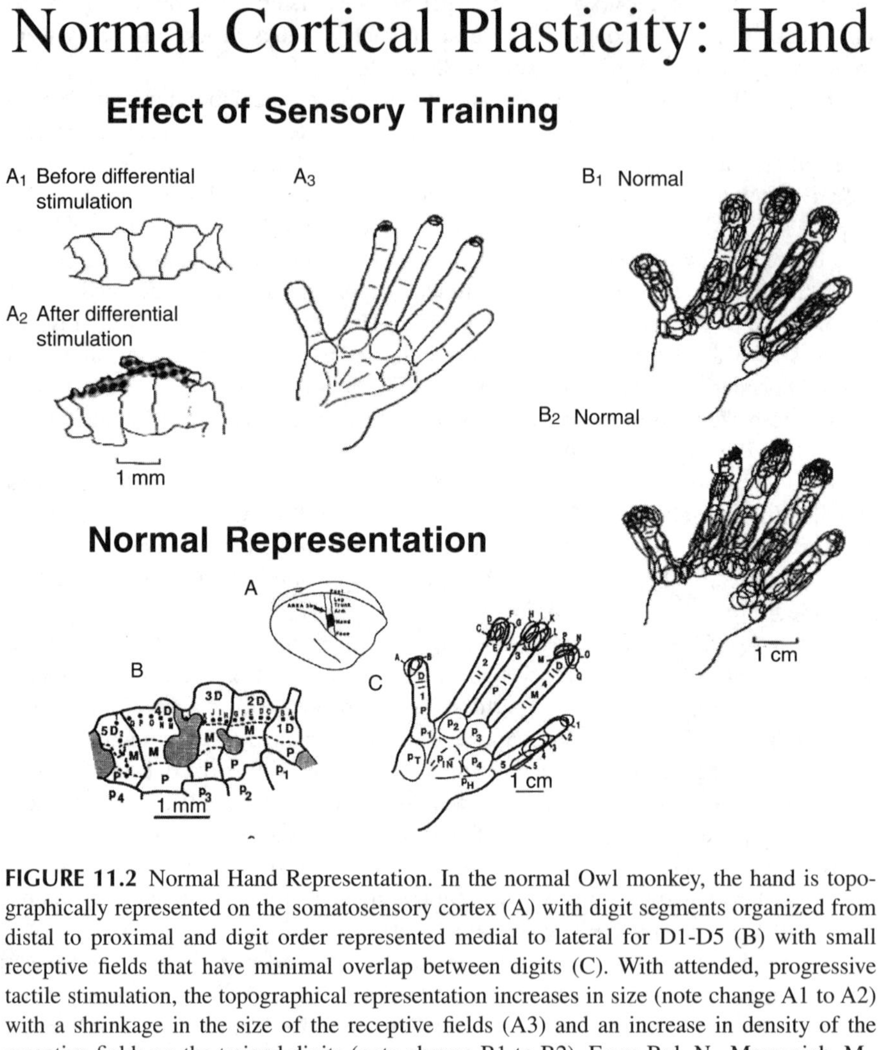

FIGURE 11.2 Normal Hand Representation. In the normal Owl monkey, the hand is topographically represented on the somatosensory cortex (A) with digit segments organized from distal to proximal and digit order represented medial to lateral for D1-D5 (B) with small receptive fields that have minimal overlap between digits (C). With attended, progressive tactile stimulation, the topographical representation increases in size (note change A1 to A2) with a shrinkage in the size of the receptive fields (A3) and an increase in density of the receptive fields on the trained digits (note change B1 to B2). From Byl, N., Merzenich, M., and Jenkins, W. 1996c. A primate genesis model of focal dystonia and repetitive strain injury: I. Learning-induced dedifferentiation of the representation of the hand in the primary somatosensory cortex in adult monkeys. *Neurology* 47: 508–520. Lippiincott, Williams and Wilkins. With permission.

neously decreased their repetition rates. Both resumed training. In 12–25 weeks, the monkeys decreased task frequency decreased from 15-16 trials/minute to 8-9 trials per minute ($p<0.001$) and accuracy dropped below 50% ($p<0.001$). The monkeys continued to perform the task for another 2 weeks with unusual posturing of the hand.

The two passively trained monkeys (OM175 and 281) both showed a significant de-differentiation of the somatosensory hand representation on the trained side (Figure 11.3) and mild de-differentiation on the untrained side. Multiple receptive

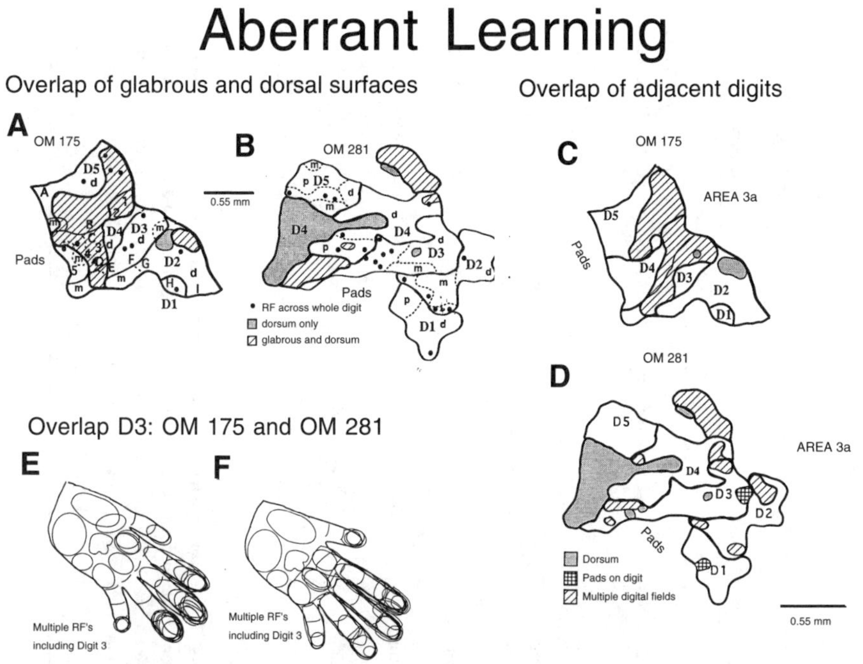

FIGURE 11.3 Abnormal Hand Representation: Aberrant Learning. Following excessive, rapid repetitive hand opening and closing, the Owl monkey was unable to complete the task. The size of the topographical hand representation decreased (A and B and C and D) and there were large cortical areas with overlapping receive fields. The receptive fields were larger than normal with a single cortical penetration representing receptive fields across multiple digits (E and F). From Byl, N., Merzenich, M., and Jenkins, W. 1996c. A primate genesis model of focal dystonia and repetitive strain injury: I. Learning-induced dedifferentiation of the representation of the hand in the primary somatosensory cortex in adult monkeys. *Neurology* 47: 508–520. Lippiincott, Williams and Wilkins. With permission.

fields were recorded per cortical penetration with frequent overlap of receptive fields across adjacent digits and across glabrous and dorsal surfaces. The receptive fields were significantly larger than normal ($p<0.0001$ compared to normal) and the areas of the cortical hand representations were significantly reduced on the contralateral hemisphere of the trained monkeys ($p<0.001$, respectively, compared to normal). There was also a breakdown in the normally separated cortical representations of different digits, with overlapping of receptive fields ranging up to a cortical distance of 2000 microns on the trained side. (See Figure 11.3 and Summary Table 11.1.)

After shaping to perform the active hand opening/closing task, three of the four monkeys had a temporary decrease in the rate of task repetition in the 3rd week of training while one continued to work at a moderate pace but taking a lot of breaks (OM 624). After a week, the three monkeys (OM177,574,311) returned to a rapid rate of squeezing, using a power type grip to open and close the hand-piece followed by rapid opening. However, one monkey (410) resumed task training using a prox-

TABLE 11.1
Summary of Cortical Changes Across Multiple Primate Studies. Summary of Primate Studies: Aberrant Learning and FHD

	Trained				Untrained		
	Dystonia			No Dystonia			
Dependent Variables	**Passive Grip**	**Active Grip**	**D1/D2 Contact**	**Active Grip**	**Passive Grip**	**Active Grip**	**D1/D2 Contact**
Number of animals	2	3	1	2	2	5	1
Speed	20–30x /min	16–50x / min	Variable by task discrimin	10–20x/ min	NA	NA	NA
Task Description	Rapid, alternating flex/ext no breaks	Rapid, alternating flex/ext; no breaks	Difficult end range rapid, no breaks	Slow, short work period, or lots of breaks	NA	NA	NA
Size of hand Representation	1.3 mm^2, 2.0 mm^2	So much overlap Could not measure	Could not measure	4.93 mm^2	WNL	WNL	WNL
Average size receptive fields	78–150 mm^2	21–110 mm^2	88–115 mm^2	17–40 mm^2	60 mm^2	20–42 mm^2	37.1 mm^2
Percent overlap adjacent digits	70%	50%	Replaced with face	11%	NA	12%	NA
Percent overlap glabrous/ dorsal surface	53–90%	33–70%	NA	17–20%	NA	33%	NA
Cortical distance	> 2 mm	>2 mm	NA	<1 mm	<1.6 mm	<1.6 mm	NA

Data was not available on all characteristics for each monkey experiment.

The monkeys who developed a hand dystonia worked at high speeds with minimal breaks. These monkeys had a reduction in the area of the representation of the digits, receptive field size 10–100 times normal, overlap of receptive fields between adjacent digits and across glabrous and dorsal surfaces, and digit representations were twice the normal cortical distance. On the untrained side, no dystonic movements were observed. The receptive fields were 2–6x normal, but the cortical distances between the digits were only 50% longer than normal. In the two monkeys who performed the target task at slow speeds and took a lot of breaks, there were no signs of dystonia. The receptive fields were 3 times larger than normal and the columnar distances 50% longer than normal.

imal arm-trunk pulling strategy (keeping the hand on the hand-piece, but leaning backwards or extending the shoulder to close the hand-piece and bending forward or releasing shoulder extension to open the hand-piece).

The three monkeys using rapid, stressful, articulated digital strategy to open and close the hand, slowed down their repetition rate by 50% ($p<0.001$) and decreased their accuracy by 50% ($p<0.001$) or more in 4–40 weeks. In 4 weeks, monkey 311 developed an unusual, uncontrollable extension posturing of D4 and the rate of trials per minute dropped from 15 to 7. At 20 weeks, monkey 177 had difficulty opening and closing the hand on the hand piece with trials per minute decreasing from 15 to 6 ($p<0.001$). In 40 weeks, monkey 574 began to use only D3 and D4 to squeeze down on the handle while D1, 2 and 5 hyperextended at the metacarpophalangeal joint and flexed at the IP joints, decreasing the trials from 44–50 trials/minute to 13 ($p<0.001$).

The somatosensory organization of the hand for monkeys 177, 574 and 311 was seriously degraded on the trained side. The mean size of the digital receptive fields was significantly larger than controls on the trained side. The majority of the cortical penetrations had multiple receptive fields and the receptive fields frequently overlapped the segments on a single digit, adjacent digits, or dorsal and glabrous surfaces (respectively different from controls $p<0.0001$). For OM 311, only the dystonic finger (D4) showed a dense mixing of hairy and glabrous surfaces. When the receptive field overlap was plotted as a function of cortical distance, normal monkeys had minimal overall over lap across 600 *u*m while those with FHd overlapped up to 2 mm, (whether performing the active or passive task). The hand representation was mildly degraded on the untrained side as well with minimal overlap of receptive fields with adjacent digits or glabrous and dorsal surfaces (See Table 11.1).

Two monkeys did not develop motor dysfunction. OM 624 trained for over a year at a good speed (20 repetitions/minute). Characteristically, they trained in bursts taking frequent breaks, requiring about 2 h/d for training instead of 1.5 hours. Over 6 months, OM 410 trained at a slower pace using a proximal arm/trunk pulling strategy (10-13 repetitions/minute) and did not develop dystonia. This monkey would not train for more than 30 min per session and had to be trained twice a day. The receptive fields for these monkeys were larger than normal on the trained and untrained sides, but the size of the area of representation was maintained and there were minimal overlapping receptive fields (similar to the untrained side of the monkeys who developed dystonia) (Table 11.1).

Post mapping, the anatomical dissections of the dystonic and nondystonic hands showed no signs of acute inflammation (Topp et al., 1999)[130]. However, in one monkey there was a congenital defect of the flexor superficialis and flexor profundus noted on the 4th digit on the trained side and the 3rd finger on the untrained side. This monkey developed uncontrollable extension of D4 after 4 weeks of training. There were no signs of movement dysfunction on the left, untrained side. The receptive fields on this digit were larger than normal but the receptive fields did not overlap across adjacent or dorsal and glabrous surfaces.

Within a few weeks of training, the animal training on the reaching task (OM 592) developed an intense tremor while trying to place the digits on the targets. The number of trials/minute decreased from 19 to 9 after 7 weeks of training, with the percent performed correctly dropping to 16%. The tremor began as the hand approached the target. Sometimes the monkey used the unaffected side to retract the training hand. Over 4 months of training, this condition worsened until digits 3, 4, and 5 assumed an abnormal posture (extension of the MP joints and flexion at the interphalangeal joints).

Daily receptive field mapping based on stationary electrodes showed an increase in the size of receptive fields (significantly larger than normal controls *på*<0.0001). The receptive field sizes of D1 and D2 increased over time with increasing overlap. As training continued and movement dysfunction worsened, there was an expansion of the medial face into the hand representation of D1 and D2. This area of the face was consistent with a cortical columnar substitution.

D. Rodent Animal Models

The objective of these series of experiments[5,6,32] was to train Sprague Dawley rats to perform an attended, repetitive, voluntary forelimb reachng task to document: 1. the change in immunohistochemistry of the tissues of the upper limb; 2. the change in motor performance; 3. the changes in both immunohistochemistry and motor performance following reaching at high versus a low rate; and 4. characterize tissue responses relative to time of training.

Fifty-seven adult, female Sprague Dawley rats (age 12–14 weeks) were included in the studies. The rats were trained to reach and grasp foot pellets out of a cylinder. Once the task was shaped for each animal, the animals trained 3 times per week at the task with operant test chambers. Body weight was maintained at 80–90%. The trained animals were food deprived prior to training. A tube of 1.5 cm in length was placed at shoulder height, with a distance set to force the rat to fully extend the elbow to reach the pellets. Pellets (45 mg) were dispensed every 15 or 30 sec (low versus high rate). The delivery of the pellet was signaled with an auditory indicator. After training to the task at 4 reaches per minute, each rat then set their own self-paced rate. The rats trained for 3–8 weeks with the daily task divided into four, 0.5-h training sessions separated by 1.5 h. This kept the reach frequency high during each training period. A reach was defined as a trial where the rat reached the forepaw beyond a line drawn 0.5 cm within the tube. Reaching rate involved minimal force. Reaching rate was monitored throughout training, and movements were videotaped. Two distinct reaching and grasping patterns were noted: 1. scooping: semi-open forepaw placed over the food pellet and then dragged along the tube and scooped into the mouth; or 2. raking: an inefficient extreme of scooping where repeated, unsuccessful attempts were made to contact the food pellet by moving the paw back and forth like a rake to bring the pellet to the mouth. The number of minutes the rat participated in the task was monitored.

The animals were euthanized at weekly endpoints using Nembutol (120 mg/kg body weight with Serum IL-1 α and IL-l β levels examined (3–8/group). Blood samples were collected from the heart, centrifuged, serum aspirated and total protein determined (BCA-200 protein assay).

This was a controlled research design with random assignment to: 1. controls (no shaping); 2. shaping, and 3. shaping plus training. Gross movement patterns were noted and recorded as present >1/min or absent (<1/min). Cytokines were measured (IL-1 α and IL –1 β, inflammatory cells; macrophages including infiltrating macrophages [ED1] and resident tissue macrophages [ED2]). The contralateral, nonreach limb and hindlimb were examined for increased numbers of ED1-1R macrophages and serum elevation of IL-1 α proinflammatory cytokine. Differences in reach rate, movement pattern, task duration and numbers of macrophages by week and by tissue were analyzed using a mixed model ANOVA ($p<0.05$) with post hoc analyses carried out using the Bonferroni method for multiple comparisons ($p<0.0167$).

The mean reach rate was highest at baseline (8.27 reaches/min ± .66 SEM). A significant decrease in reach rate was measured after week 5 (6.82, ± .66 SEM reaches per minute, $p<0.0028$); 6.12 reaches per minute at week 6 (± .52 SEM; <0.0070). For the low repetition group, there were no significant difference in reach rates across weeks with the mean reach rate 3.01 reaches per minute (± 1.03 SEM). The rats in the high repetition group decreased scooping and by week five; 47% were raking. By week 8, raking was present in 100% of the trained animals. In the low repetition group, the scooping remained consistent weeks 1–5. While raking increased across weeks, only 60% of the animals used the scooping strategy by week 7–8.

Cytokine changes were measured. Increases in serum levels of IL-1 alpha were measured in the high repetition group (increased 27%) with no significant change in IL-1 ß. In the low rate group, there were no significant changes in IL-1α or IL-1 β, but there was a trend for a decrease in IL-1 α at 8 weeks, suggesting a dose-response relationship between reach rate and behavioral and physiological responses to repetitive reaching with increased injury with higher rates of repetition.

With high rates of repetitive reaching, there was a decrease in motor performance and measurable signs of tissue injury (cellular and tissue responses associated with inflammation). Reach rate decreased after week 5 as did task performance time and accuracy. Decrease in motor efficiency was followed with the emergence of a clumsy, raking movement pattern instead of scooping. The animals could no longer close the digits to lift the food pellets. These researchers observed discrete sites of disruption in tendon fibers and infiltration of phagocytic macrophages. The number of macrophages remained high at 8 weeks, but there was some rebound in the reach rate.

Collaboration is currently underway between these investigators and researchers at UCSF. We are performing electrophysiological mapping experiments of the motor and somatosensory cortices to determine if the decrease in motor performance (scooping to raking) might be associated with measurable de-differentiation of the topographic representation of the hand in the somatosensory and motor cortices similar to that reported in the primate studies.

E. Summary of Animal Studies

These animal studies provide evidence that stressful, excessive over use of the hand can be associated with early tissue trauma as measured by the presence of inflammatory cells, fibroblasts and macrophages. With persistent repetition (>5 weeks in Sprague Dawley rats and >24 weeks in primates), motor performance was seriously impaired in both the primates and the rats who were using rapid alternating opening and closing of the digits. In the primates, electrophysiological mapping showed clear signs of de-differentiation of the hand on the somatosensory cortex. The tissue histological changes as well as cortical changes were measured on the trained and untrained sides.

The critical training characteristics associated with motor dysfunction were high rates of repetition, use of an alternating flexion/extension strategy or forced end range movements, near simultaneous movements (challenging integration time) and persistence at the task with minimal breaks. Animals that worked slowly, took frequent breaks, or used proximal arm/trunk strategies rather than digital strategies did not develop movement dysfunction. In addition, the presence of a soft tissue restriction may shorten the time to the development of motor dysfunction when the hand is aggressively overused. The critical changes in cortical structure associated with dystonic movement included multiple receptive fields per cortical penetration, large receptive fields overlapping segments, adjacent digits and glabrous and dorsal surfaces which also persisted across large columnar distances.

Animal models may have utility for studying questions of etiology. In these, behavioral models both the primate and rat studies produced models where the animals were normal healthy adults without the gene for dystonia. In addition, the animals trained at a repetitive task with high compliance under supervision over a period of 8 weeks or more. The animals were driven to participate in the tasks and they were rewarded. However, each animal, to some extent selected a training strategy and decided on when to take breaks within the constraints of the training paradigm. Animal models of FHd have limitations for the study of complex relationships such as repetitive behaviors and pain, sensory discrimination, motor reaction time, fine motor perceptual skills and progressive learning based sensorimotor retraining strategies for the remediation of focal hand dystonia.

F. Human Research Models

1. Experiment I: Relationship of Clinical Performance and Neural Structure

The purpose of these series of experiments was: 1. to describe the neurophysiological and clinical correlates of patients with focal hand dystonia (17 subjects); 2. evaluate the effectiveness of learning based sensorimotor training on the remediation of focal hand dystonia (12 subjects); and 3. relate change in clinical function with change in cortical structure (3 case studies).

Subjects were recruited from the Peter Ostwald Health Program for Performing Artists and the Physical Therapy Faculty Practice at the University of California, San Francisco (males or females, betweeen 20 and 60 years of age). The subjects :1. were referred by a neurologist for a diagnosis of FHd; 2. demonstrated observable involuntary twisting movements of the digits and wrist when performing the target or similar tasks; 3. had normal reflexes and no evidence of objective peripheral neuropathy or central nervous system pathology; 4. had the diagnosis of FHd for more than six months; 5. worked in an occupation that demanded repetitive use of the hands (e.g., computer use, card dealing, playing a musical instrument, writing, court reporting); and 6. had not received an injected or systemic drug to control the dystonia for more than 6 months prior to study admission. Subjects lived in the San Francisco Bay Area or were willing to stay in the bay area for several days to complete the testing. From a previous data base of healthy controls, two groups of healthy subjects were age and sex matched to serve as historic reference controls. The purpose and the procedures of the study were explained to each subject and signed consent was obtained prior to testing. The studies were approved by the UCSF Committee on Human Research. In study I, 17 patients with FHd and 15 controls were included, in Experiment II[19,20,22] 12 subjects were included and in. Experiment III, 3 case studies were included.

Each FHd subject and all of the historic and matched controls were administered a broad battery of standardized clinical tests. The test procedures and the reliability of testing have been described in previous studies and are summarized in Table 11.2.[12,13,17,18] Specific subtests were summed into seven dependent variables: 1. physical musculoskeletal performance (selected range of motion, strength, neural tension); 2. sensory discrimination (graphesthesia, localization, kinesthesia, stereognosis); 3. fine motor efficiency (Purdue Test-time), fine motor skill (line tracing accuracy and time) and digital reaction time (averaged across the 5 digits for each hand); 4. motor control at the target task; 5. posture and balance; 6. functional independence; and 7. pain. The subtests allowed for comprehensive measurement of clinical performance. The intercorrelations between the summated dependent variables were low ($r<0.1$), suggesting the dependent variables were measuring unique characteristics[12,13,17] (See Table 11.2).

Subjects in Experiment I were classified with simple or dystonic dystonia: 1. simple = dystonia limited to one target task; and 2. dystonic = dystonia occurred with tasks similar to the target task or surface contact of the hand. Severity was established according to the Tubiana Dystonia Scale for Musicians. 0 = unable to do the target task; 1 = able to do limited aspects of the task or perform the task for very short periods; 2 = able to do the task with modification of technique; 3 = able to do the task but not efficiently or with normal quality.[134] For Experiment I, FHd were classified into one of two severity categories: mild or severe. Those with simple dystonia and rated as 0 or 1 were classified as severe dystonia while those with dystonia at a single task were rated 2 or 3 and classified with mild dystonia.

Somatosensory testing was performed by trained research assistants according to standard protocols. All of the testing in the Biomagnetic Imaging Laboratory at UCSF were performed by trained staff (Roberts et al., 1998; Rowley et al., 1995).[119,120] The test-retest values for the MSI testing established in this lab are high (>0.9).[123] A 37 channel biomagnetometer (Magnes II, 4D Neuroimaging, 1.5 fT, San Diego, CA) placed in a magnetically shielded room with two circular sensors (14.4 cm) was used to create a magnetic source image of the hand. Two hundred and fifty air puffs were delivered within 1 cm 2 sacs, for 30 msec, at 17–20 lb/in^2 (psi), with a pseudorandom inter-stimulus interval 450–500 msec. The stimulus was a super threshold force designed to indent the skin 400 microns. Each digit was stimulated on the distal pad, middle and proximal segment on each digit on each hand. In addition, a similar stimulus was delivered to each side of the upper lip.

TABLE 11.2
Summary of Clinical Tests

Measurement Tool	Dependent Variable	Scoring System	Directions	Reliability	Equipment
Graphesthesia (Modified Subtest of Sensory Integration Praxis Test [SIPT])	Sensory Performance	2 = correct, 1 = partially correct, 0 = incorrect; % error calculated.	Tip of a paperclip used to draw designs on subject's fingers while eyes closed (EC). Subject recreates design with pen with eyes open (EO). 2 designs per finger pad.	Interrater = 0.95, Test-retest r = 0.91.	Paper-clip and design sheet.
Kinesthesia (Subtest of SIPT)	Sensory Performance	Average error (distance from target) in mm.	Subject's hand is moved to target and back to start position; subject attempts to relocate digit, EC. 5 trials per hand.	Interrater = 0.95,[45] Test-retest r = 0.90.	Target sheet and ruler.
Byl-Cheney- Boczai Test (BCB) for Stereognosis	Sensory Performance	2 = correct, 1 = partially correct, 0 = incorrect; % error calculated.	Subject's finger is drawn across the shape twice, EC. Subject attempts to pick correct shape. 10 trials for 2nd and 4th finger pads.	Interrater/ intrarater = 0.995 (ICC), correlation of r = .60 b/w BCB and PurdueTest	20 designs and test sheet of designs
Digital Reaction Time	Fine Motor Performance	Time in msec, average of all trails.	Subject turns stopwatch on/off as quickly as possible. 3 trials per finger.	Intrasession reliability ranges from 0.975–0.99.	Stopwatch.
Purdue Test	Fine Motor Performance	Total time to put pegs in and out.	Subject puts 25 pegs into a board and then removes.	0.60–0.63.	Watch, peg-board.
Manual Muscle Test	Musculo-skeletal Performance	Kilograms of force: UE and LE Scores total all scores.[a]	Performed per procedures defined by Kendall using Microfet dynamometer. Jamar dynamometers used for power, key and pinch grip.	R = 0.887 multiple correlation with MMT.	Jamar Microfet and Baseline dynamometers
Pain	Pain	Visual Analog Scale: Ordinal scale, 0 = no pain; 10 = worse pain every had	Patient circles the ordinal scale that best reflects their pain	R = 0.9	none

Range of Motion	Musculo-skeletal Performance	Degrees; sum of active and passive.	Performed per Norkin. Upper and Lower Extremity Scores totaled.[b]	Intratester: r = 0.91–0.99.	Goniometer
Balance	Ability to maintain uprightness in different sensory conditions	Ordinal scale following performance of balance strategies on different surfaces: eyes open, eyes, closed	Stand feet together EO, EC 20 sec Stand one foot EO, EC 10 sec Stand in tandem Romberg EO, EC 10 sec	ICC = .95	Stop watch
Posture	Postural alignment (side and posterior)	Ordinal scales	Based on ordinal scale for how far off alignment for Kendall's alignment markers for gravity.	Intertester r = .91	none
Motor Control at Target Task	Functional Performance at target task	Ordinal scale for performance characteristics: normal and abnormal patterns 0–3 per criteria and summed to total score	Patient performs target task while being videotaped. Criteria have been defined	ICC = .97–.99 for test retest	Need test sheet with criteria and ordinal scale
CAFÉ 40	Functional Performance	7 point Likert Scale (1 = least independent; 7 = most independent	Self-scoring of ability to perform functional activities. Scores inverted for data analysis.	Test-Retest: r = 0.971.[59]	Written questionnaire.

The clinical testing sampled a broad range of sensory and motor skills in addition to performance on the target task and functional independence

[a] Muscle groups tested: hip flexors and extensors, knee flexors and extensors, ankle dorsiflexors, elbow flexors, shoulder flexors, wrist extensors, lumbricals, grip and pinch (3-jaw chuck and key grip) strength.

[b] Joint motions tested: shoulder flexion, abduction, and external rotation.

From Byl, N.N., Nagarajan, S.S., Merzenich, M.M., Roberts, T., McKenzie. 2002. Correlation of clinical neuromusculoskeletal and central somatosensory performance: variability in controls and patients with severe and mild focal hand dystonia. *Neural Plasticity* 9:177–203. With permission.

A normal, cutaneous, somatosensory evoked field response (SEF) is characterized with a peak amplitude at a latency between 30 and 70 msec, subject to a signal to noise ratio greater than 4, goodness of fit (model/data) greater than 0.95 with a minimal confidence volume less than 3000 mm^3 (Roberts, et al., 1998).[119,120] The dependent variables recorded for each SEF response included latency (msec), root mean square (rms) amplitude across sensor channels (fT), ratio of amplitude to latency, location of the digits on the x-, y-, and z-axes (cm), spread between digits, order of the digits on the z-axis and volume of the hand representation (4/3 times the radius of the spread on x-, y-, and z-axes).

Experiment I was a descriptive study. All dependent variables were described by mean and standard deviation. Each dependent variable for each limb was considered independent and tested for significance at $p \leq 0.05$ (two tailed). Where multiple subtests were combined or multiple trials were combined to create a dependent variable, the number of measurements was based on the number of subjects times the number of test components/trials.

Based on the somatosensory and clinical dependent variables, differences between controls and FHd subjects and within subjects with FHd were analyzed using the Student t Test or Analysis of Variance for the dependent variables measured on ratio scales and the Ranked Sum Wilcoxon or Two Sample Wilcoxon Test for the dependent variables measured on ordinal scales. The severity groupings for the FHd subjects were correlated with the clinical performance parameters and the somatosensory and tested for significance using the z Test for Correlation Coefficients.[80]

There were 9 males and 8 females with FHd. (23 to 55 years, mean of 39.9 years [±11.1 years]). All worked in jobs requiring repetitive hand movements (10 musicians; 11 with simple dystonia and 6 with dystonic dystonia). Ten subjects could no longer practically perform the task (severe dysonia) and seven could perform the task for short periods of time with modification of technique (mild dystonia). The control subjects ranged in age from 23 to 57 years with a mean age of 37.4 years (±9.7 years). There were 8 males and 7 females. Two were musicians and the other subjects worked in jobs requiring repetitive hand use on a computer keyboard. Of the 15 reference controls selected for comparing clinical performance, there were 5 males and 10 females with an average age of 30.2 years (±3.6 years). The majority of control subjects were graduate students, faculty, or friends of students or faculty who had a history of repetitive hand use (e.g., intensive note taking and computer use).

Patients with FHd performed significantly worse than healthy controls when using either the affected or unaffected side on musculoskeletal tasks, balance activities, postural alignment, fine motor control, and sensory discrimination. Using the affected limb, those with severe dystonia demonstrated greater restrictions on musculoskeletal skills and target specific motor control. Although the overall sensory discrimination accuracy was low for all FHd subjects, those with *severe* dystonia performed faster than those with mild dystonia. On the unaffected side, those with mild dystonia demonstrated greater inaccuracy when performing the target specific task (See Table 11.3).

There were no significant differences between mean SEF latency or mean SEF amplitude for FHd subjects and reference controls, but the location of the

TABLE 11.3
Clinical Performance Parameters in Controls and Patients with FHD: FHD (n=17)

	Severe FHD		Mild FHD		Controls (15)	
	Affected	Unaffected	Affected	Unaffected		
Physical Performance	235.9 (36.6) *1	261.0 (35.8) *1	273.7 (32.5) *2	275.7 (42.3) *2	292.5 *1 (25.1)	289.4 *2 (24.2)
Balance Posture	87.1 (12.2)		90.9 (4.5)		96.6 (3.6)	
Motor Skills (n=50 and 35)	189.4 (34.4) *3	198.1 (5.2) *3	186.7 (48.2) *4	191.4 (42.3) *4	139.03 (25.5)	139.04 (25.6)
Sensory Discrimination	276.3 (67.0) *5	274.8 (12.9) *5	211.8 (65.9) *6	231.8 (66.1) *6	175.0 *5 (29.6)	172.1 *6 (16.5)
Functional Independence	91.72 (5.98)		93.5 (4.8)		91.9 (5.6)	
Pain	88.7 (9.5)	96.4 (5.0)	99.0 (3.2)	100.0 (0)	97.2 (6.8)	97.4 (6.1)

Severe Dystonia: included Dystonic Dystonia (0-3) Simple (0-1) Mild Dystonia: Simple Dystonia (2-3)

Differences between Controls and FHD subjects (combined across groups): Critical Values: *1, t = 4.6 ($p<0.001$); *2, t = 3.12 ($p<0.002$); *3, t = 6.09 ($p<0.001$); *4, t = 13.6 ($p<0.001$); *5, t = 8.66 ($p<0.001$); *6, t = 8.87, ($p<0.001$).

Differences between FHD Severe and Mild: Patients with severe dystonia had lower scores for physical performance, sensory discrimination, and were more likely to complain of pain, ($p<0.05$, respectively).

Patients with focal hand dystonia were noted to have decreased strength and range of motion (physical performance) compared to controls as well as poor posture, decreased motor skills, and decreased sensory discrimination accuracy. However, those with hand dystonia self-rated their functional independence to be similar to controls and only a few with severe dystonia complained about pain. Compared to those with severe hand dystonia, those with mild dystonia demonstrated better physical performance, were slower but more accurate in sensory and fine motor performance and demonstrated bilateral problems.

From Byl, N.N., Nagarajan, S.S., Merzenich, M.M., Roberts, T., McKenzie, 2002. Correlation of clinical neuromusculoskeletal and central somatosensory performance: variability in controls with patients with severe and mild focal hand dystonia. *Neural Plasticity* 9:177–203. With permission.

digits on the x (bilateral) and y axes (affected) were significantly different ($p<0.0001$, respectively) and the ratio of SEF mean amplitude to latency was higher for FHd subjects compared to controls ($p\leq0.05$). On the unaffected side, the volume of the hand representation was significantly larger for FHd subjects compared to controls ($p<0.05$).

The ratio of SEF amplitude plotted by response latency was significantly lower in the early phase (<100 msec) for the FHd subjects compared to controls. The amplitude was similar for the control subjects and the FHd subjects for the unaffected digits on the affected limb and the digits on the unaffected limb. For FHd subjects, there was a bimodal distribution of mean SEF amplitude plotted by mean latency

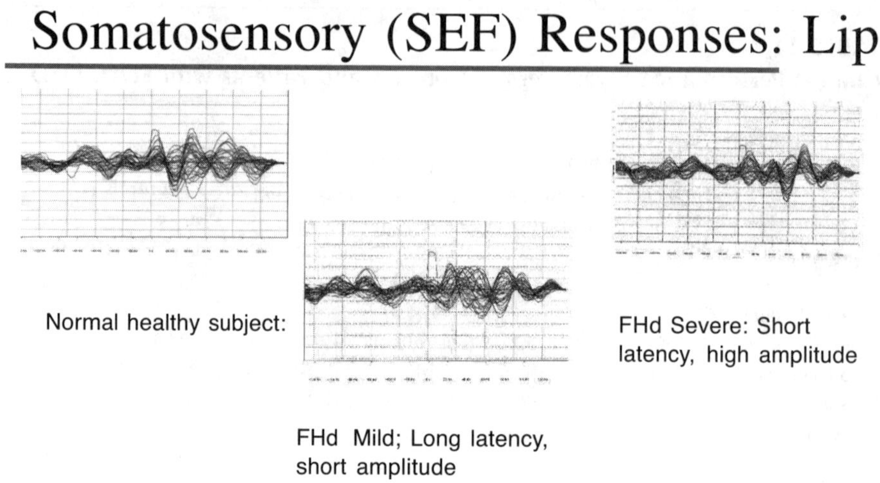

FIGURE 11.4 Somatosensory Evoked Responses Using MEG: The Lip. There were no differences in latency or amplitude of somatosensory evoked responses for normal subjects and subjects with focal hand dystonia based on mapping an uninvolved area. From Byl, N.N., Nagarajan, S.S., Merzenich, M.M., Roberts, T., McKenzie, 2002. Correlation of clinical neuromusculoskeletal and central somatosensory performance: variability in controls with patients with severe and mild focal hand dystonia. *Neural Plasticity* 9:177–203. With permission.

on the affected side (mean latency ranging from 30 to 60 msec and the mean amplitude ranging from 20 to 119 fT). There was a negative linear trend of amplitude by latency for the digits on the unaffected side for FHd subjects and all of the controls (as the latency increased, the amplitude decreased.

The field evoked firing patterns for controls and those with dystonia (mild and severe) were similar when measured on an unaffected part, the lip. (Figure 11.4) However, integrating amplitude by latency, those with severe dystonia had a significantly higher amplitude than those with mild dystonia. Those with severe dystonia had a short latency and a high amplitude and those with mild dystonia had a long latency and a low amplitude. (Figure 11.5) Bilaterally, the volume of the representation of the hand for those with mild dystonia was larger than the volume for subjects with *severe* dystonia.

There were high, significant correlations (0.9029 affected and 0.8477 unaffected; $p<0.001$, respectively) between dystonia severity and the SEF ratio of amplitude to latency. On the affected side, there were negative correlations between SEF ratio and dystonia severity with musculoskeletal performance, motor control on the target task and fine motor skills. FHd subjects with mild dystonia tended to have a low SEF ratio and demonstrated higher performance on these tasks than those with severe dystonia. There was a significantly negative correlation between fine motor skills and SEF ratio on the affected side; those with a high SEF ratio of amplitude to latency demonstrated greater inaccuracy. On the unaffected side, there was a significant, moderately positive correlation between the severity of dystonia performance on the target task; with mild dystonia having lower performance scores on the target task.

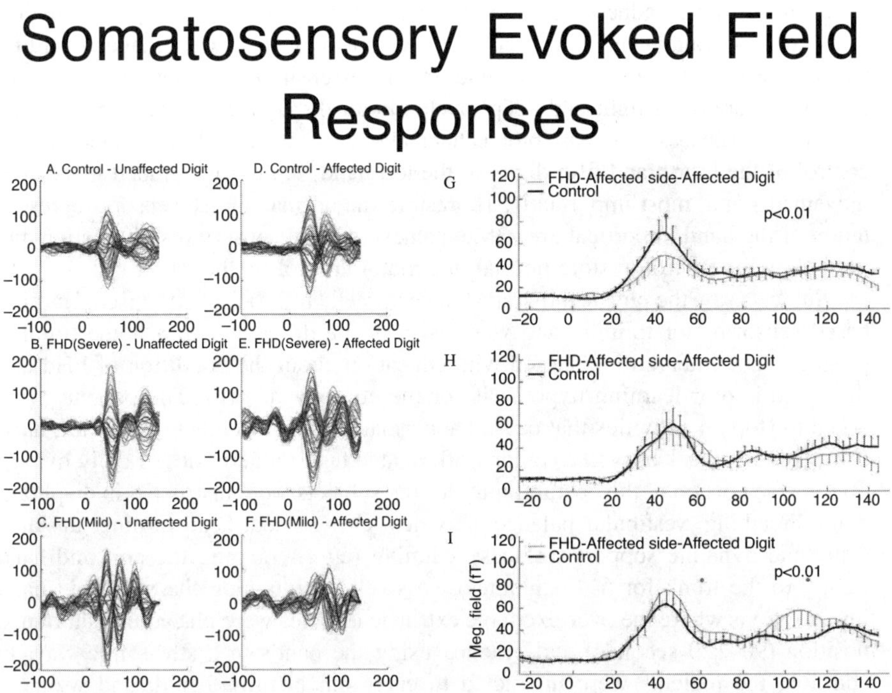

FIGURE 11.5 Somatosensory Evoked Responses Using MEG: The Hand. The subjects with severe focal hand dystonia had a shorter latency and a higher amplitude on the involved digits compared to normal subjects (B E compared to A and D). In addition, those with mild hand dystonia had a longer latency and a lower amplitude than normal subjects (C and F compared to A and D). Compared to controls, on the somatosensory evoked response, the affected digits had a lower amplitude (G) but the somatosensory evoked response on the unaffected digits on the affected side and the digits on the unaffected side of subjects with FHd was similar to normal controls. From Byl, N.N., Nagarajan, S.S., Merzenich, M.M., Roberts, T., McKenzie, 2002. Correlation of clinical neuromusculoskeletal and central somatosensory performance: variability in controls with patients with severe and mild focal hand dystonia. *Neural Plasticity* 9:177–203. With permission.

2. Experiment II: Intervention (12 Subjects)

The purpose of this study was to assess the effectiveness of learning based sensorimotor training and recovery of task specific and sensory motor function in patients with focal hand dystonia. Twelve subjects met the same criteria summarized in Experiment I. A broad range of clinical tests were administered as in Experiment I.

The goals of treatment were to: 1. stop the learning of the abnormal movements (stop performance on the target task); 2. use the hands in a stress-free way for all activities (e.g., maintain carpal and oblique arches, initiate movement at MP joints with lumbricals and interossei, avoid the lateral key pinch grip and forceful gripping, let the sensory information from objects open the hand); 3. improve postural alignment with gravity and reduce adverse neural tension; 4. maximize flexibility in finger spread, forearm rotation and shoulder rotation by decreasing muscle tension and releasing soft tissue adhesions; 5. facilitate positive health and healing (e.g., hydration, nutrition, and general exercise); 6. restore a positve self image regarding performance on the target task; 7. quiet the hypersensitivity of muscle co-contractions when applying light and deep touch stimuli through the skin; 8. modify the instrument as necessary to ease biomechanical stresses and facilitate the independent control of the fourth and fifth digits of the left hand; 9. improve graded fine motor movements and, most importantly, 10. restore the normal somatosensory representation of the hand in cortical area 3b to achieve normal sensory discrimination and normal motor control. restore normal fine motor control on the target task

Subjects saw the physical therapist once a week (1–2 h) for supervised, learning based sensorimotor training and were asked to be diligent with a home training program. The intervention started with education about the condition of FHd and the sensorimotor learning hypothesis for the etiology of FHd. The patients were asked to stop all activities that caused abnormal finger movements of the left hand (e.g., the target task as well as other work related tasks or activities of daily living). To ease the tension in the postural muscles the subjects were instructed in diaphragmatic breathing, vestibular balance activities (eyes closed, head turning, unstable static and dynamic support surfaces), calming (e.g., wrapping the arm and hand tightly to the trunk for 5–15 minutes), passively maintaining the wrist and hand into positions where the over-excitable extrinsic muscles were placed in a shortened position (90–120 seconds) and practice using the hands in a stress free way. In addition, the patients were instructed to carry out positive health and wellness activities (good hydration, regular exercise, balanced diet). If the patient had musculoskeletal problems, then the home program included activities to improve flexibility, strength in the intrinsic hand muscles, and postural alignment in addition to sensory retraining.

In each of the supervised sessions (1.5 h), 5–10 minutes were spent on soft tissue mobilization and inhibitory positioning as needed to relieve the tension in the affected limb, then 30–45 minutes were spent on supervised learning based sensory motor training progressing from sensory discrimination activities, to sensory motor activities and then 10–20 minutes focused on selective fine motor training on nontarget and target tasks using biofeedback (auditory or tape). The sensory discrimination training focused initially on the involved fingers, with each finger individually challenged on the distal pads as well as the dorsum and sides of the fingers. Sensory discrimination activities were done with the patient in different positions (supine, sitting or standing). Both passive and active stimulation was used. The sensory motor activities included graded movements where sensory information was used to control the hand (e.g., eyes closed, objects with different surfaces [rough and smooth] introduced to the hand, explored and then used appropriately). The fine motor tasks included instruction in stress free use of the hand in common tasks such as picking up objects, doing activities of daily living (ADLs), using the computer and ultimately playing the target instrument.

As part of the sensory retraining at home (at least 1 h/d), the patients were asked to physically carry out sensory discrimination tasks, use a mirror of the unaffected side to provide an image of the affected side which was out of sight behind the mirror (Figure 11.6), use biofeedback to minimize cocontractions, and mentally

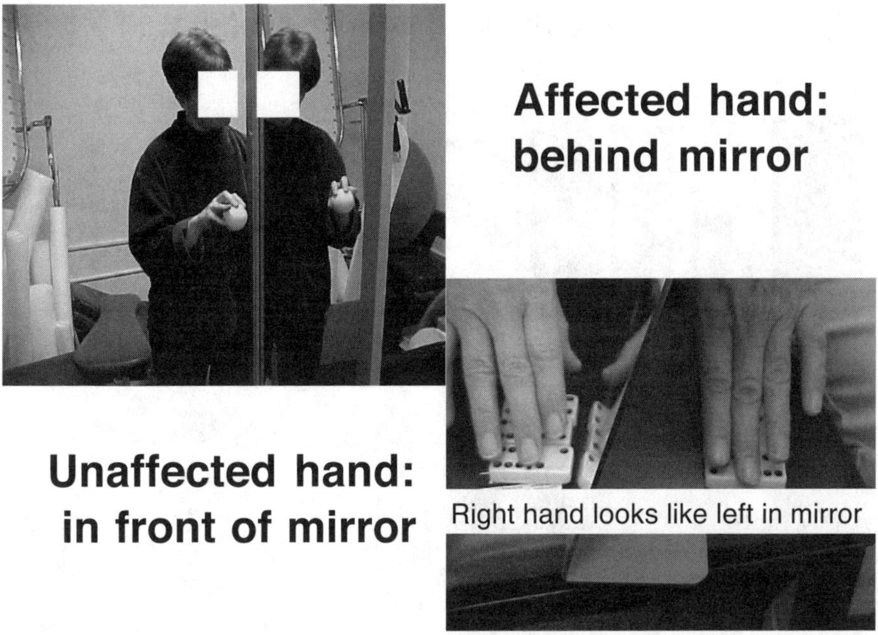

FIGURE 11.6 Sensorimotor Mirror Training. While the subject looked at the mirror image of the affected side, sensory and motor tasks were formed to provide positive feedback and facilitate normal performance.

practice tasks such as: 1. the hand feeling normal; 2. the hand playing the instrument normally and easily with appropriate speed and accuracy; 3. using eating utensils normally; 4. writing without excessive gripping on the pen and moving the pen from the elbow-shoulder; 5. doing detailed hand work; and 6. completing fine motor tasks. Each subject was also encouraged to make a video of someone playing their instrument or doing tasks that they could view and imagine themselves doing the task. As sensory processing skills improved, they were also asked to practice, small, independent, isolated movements of the uninvolved and involved digits. If they returned to instrumental play, they were asked to begin with new music.

This was a pre-experimental single group, prepost test study design with 12 subjects with FHd that participated in a controlled sensorimotor training program for 6 months. All scores were reported descriptively and prepost test differences were tested for significance using the paired Wilcoxon Test or the Paired t Test depending on whether the dependent variables were ordinal or ratio scales.

a. *Study Findings*

All patients improved significantly on all parameters of clinical performance (25% to 80%), bringing the performance of musculoskeletal parameters, sensory discrimination and fine motor control to the level of normal subjects. Task specific motor control increased to 94%. All but two subjects returned to their previous work. However, none gained 100% control of the hand. Rather, they still had to be careful

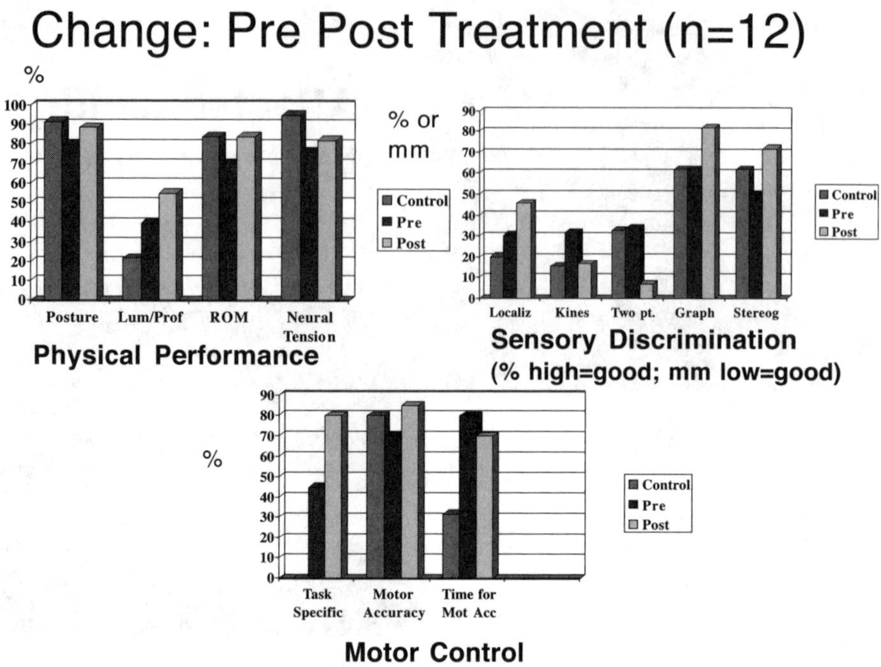

FIGURE 11.7 Summary of Change in Clinical Performance Following Sensorimotor Training for Patients with Focal Hand Dystonia. Post training, the subjects with FHd improved their performance in all sensory and motor areas, matching their performance to controls in all performance areas except accuracy on the Motor Accuracy Test where they required twice a much time as normal subjects. Motor control on the target task improved to 85% of normal.

how they performed the target task Some no longer were playing professionally and others were able to return to performance. (Figure 11.7).

3. Experiment III: Three Case Studies

The purpose of this study was to determine the effect of learning based sensorimotor training on change in structure and clinical function in patients with FHd. Three musicians were referred from the Peter Ostwald Health Program for Performing Artists, University of California, San Francisco to participate in the study. Ten healthy age matched controls served as reference norms for magnetoencephalography and 30 additional healthy subjects served as reference norms for the clinical performance parameters.

Two subjects lived outside the United States (#1 and #2) and the third was from the San Francisco Bay Area (#3). All of the subjects agreed to participate in at least 8 weeks of physical therapy. All of the subjects had been diagnosed with FHd by a neurologist approximately one year prior to this current intervention study.

All of the patients were otherwise healthy except for the complaints of painless, uncontrollable curling of digits four and five (D4–D5) on the left hand when they played their instrument. All indicated that the fifth digit excessively curled or extended and it was hard to lift D4. All three subjects noticed that it was more difficult to control D4 and D5 when D3 was pressing down. All of the subjects were completely independent in personal care and household management, and were well integrated into the community. They all participated to some extent in fitness programs. One subject played for the symphony and was out on medical disability, one subject played for a travelling performance group but was working at a desk job when physical therapy was initiated, and the third subject was a full time music student who was home for two quarters and was working part time as a waitress.

All subjects participated in measurements pre and post treatment including magnetoencephalography and clinical testing as described in Experiments I and II[22] (Byl et al., 2000c).[19]

Subject # 1 participated in supervised treatment for 12 weeks [two 6 week sessions], subject #2 participated daily for two weeks and Subject # 3 participated for 17 weeks). Consequently, the total period of treatment as well as the number of visits with a physical therapist varied across subjects (23 visits for subject #1, 19 visits for subject # 2, and 23 visits for subject #3).

At baseline, somatosensory evoked responses were similar on the right and left sides for controls except the spread of the digits on the dominant hand were greater than the nondominant hand on the z-axis. On both hands, the order and location of the digits on the z-axes followed a predictable pattern with D2-D5 progressing from inferior to superior. For the subjects with FHd, both the amplitude and the spread of the digits on the x,y, and z axes were reduced on the affected side compared to the unaffected side and the digits were not sequentially organized from inferior to superior for D1-D5 on the z axis on either side. Compared to controls, the FHd subjects had a shorter SEF latency, the neuronal burst was higher on the affected and unaffected sides for subjects #1 and #3, and the amplitude was lower in the early phase (30–70 msec) for subjects #2 and # 3. The location of the hand representation on the x, y, and z axes were different for FHd subjects and controls. Bilaterally, the spread of the digits on the x, y, and z-axes was greater for the subjects with FHd (who were all musicians) than the controls.

In general, the reference controls achieved comparable clinical performance bilaterally and across digits except motor reaction time was slower for digits 4 and 5. The controls did have some postural asymmetry and indicated their health sometimes interfered with daily activities (scoring 89.6% out of a maximum score of 100% for functional independence). On the other hand, at baseline, the subjects with FHd demonstrated reduced accuracy and slowing in sensory processing compared to controls on both the affected and unaffected sides. On the motor performance tests, subjects #1 and #3 performed with reduced motor accuracy on both sides with prolonged processing time. On the affected side, Task Specific Motor Control Scores were approximately 50% of that measured on the unaffected side. Subjects #2 and #3 had limited finger spread between D3–D4 and D4–D5 on the affected side (25 degrees on the affected side compared to 35–45 degrees on the unaffected side). Compared to controls, the subjects with FHd were more likely to have poor posture, positive signs of neurovascular entrapment and decreased strength in the lumbricals (on both sides). Two of the subjects with FHd also had limited shoulder internal rotation bilaterally (45–55°). The subjects with FHd were not working at their usual jobs but they were independent in activities of daily living. They reported difficulty with functional activities (ranging from 63–90% of maximum performance on the functional independence test).

Based on magnetic source imaging, the controls did not change with retesting. However, for the three subjects with FHd, there was a general increase in the spread of the digits and the area of representation on the cortex on the trained side (larger than control subjects). There was a decrease in the area of representation on the unaffected side. The order of the digits (D1–D5) on the affected side approximated an inferior to superior progression from D1 to D5, but they were still less orderly than controls. The amplitude of the evoked somatosensory potential, integrated over time, was increased and similar to controls on the affected side.

On the clinical tests, the subjects with FHd performed between 80–90% on the target task. Motor reaction time did not change significantly on either the affected or unaffected side but was similar to controls. The subjects with FHd improved in motor accuracy 27–42%, performing at similar accuracy as controls, however, the time needed to complete the task was still longer than controls. There were measurable improvements in accuracy on all of the sensory tests (25–50%), performing similarly or better than controls. However, the time required to perform the tests remained longer than controls for two of the subjects. (subjects #2 and # 3 required 66–197 seconds compared to 37 seconds for controls). The subjects also improved their range of motion, strength, and posture, raising performance to the level of controls. The FHd subjects also showed improvement in functional independence, similar to controls.

III. SUMMARY OF INTERVENTION STRATEGIES

The consistency of the findings of somatosensory hand degradation with clear objective improvement in clinical function and neural structure following learning based sensory retraining strengthens the evidence in support of aberrant learning as one etiology of FHd. It is important to examine the involved and uninvolved limbs.[12,13,16,31,60,95,113,142] (Charness and Hallett 1992,1993;[27,28] Fry 1986;[43] Jankovic and Shale 1993;[60] Chen and Hallett, 1998; McKenzie et al., 1997,2003;[82] Norkin 1995;[262] Leijinse 1991; Lockwood 2003;[82] Sanger et al., 2000ab,[120] 2001;[121] Tinazzi et al., 2002; Wagner et al., 1974) In addition, voluntary fine motor control at the target task should be videotaped and scored for quality and severity (Tubiana and Chamagne, 1998)[134] Some clinicians may be able to use computer technology (e.g. MIDI) to objectively document abnormalities of timing and force[45,59,105,142,143] (Pascual-Leone et al., 1995).[59]

The clinical studies suggest that the development of FHd is multifactorial. The development of involuntary task-specific dystonic movements can develop under conditions of aggressive, stressful, stereotypical, rapid, repetitive hand use interacting with anxiety, perfection, previous trauma, joint inflexibility or hypermobility, imbalance of extrinsics intrinsics, poor posture, neurovascular entrapment, quick motor reaction time, but slow and inaccurate sensory discrimination. Each individual may present with unique physical characteristics, however for those with a history of overuse, an etiology of aberrant learning should be considered. Thus, intervention strategies should be based on the principles of learning. A learning based sensorimotor strategy was associated with improvement in, physical performance, posture, sensory discrimination, task specific motor control, and somatosensory organization of the hand. However, performance was not 100% on the target task. Within the system of health care constraints, intervention was only once a week, reinforced with a self-guided home program. This may not be sufficiently intense to completely normalize somatosensory structure and task specific performance.

Patients successfully rehabilitated confirm the necessity to stop the abnormal movements which usually means not performing the target task. These individuals also express the need for mentoring and guidance to maintain self esteem and stay focused on sensorimotor retraining, while also integrating biomechanically safe hand techniques, avoiding stereotypical, near simultaneous, alternating contractions of agonists and antagonists or end-range motions. The potential for rewiring the brain will necessitate the incorporation of new computerized learning models that are fun, rewarded, repetitive, engaging and self initiated at home. Randomized clinical trials across multiple centers are needed to continue to identify the risk factors for developing FHd but also the most efficient, effective learning-based retraining strategies.

REFERENCES

1. Allard T, Clark SA, Jenkins WM, Merzenich MM. 1991. Reorganization of somatosensory area 3b representations in adult owl monkeys after digit syndactyly. J Neurophysiol 66: 1048–1058.
2. Altenmueller E. 1998. Causes and cures of focal limb dystonia in musicians. In: Scott R, Black J. (Eds), Proc Intl Congress of Musicians: Health and the Musician (York, 1997). London, UK: Br Assoc Perf Art Med; G-1-1-12.
3. Bara-Jiminez W, Catalan M, Hallett M. 1998. Abnormal somatosensory homunculus in dystonia of the hand. Ann Neurol 44: 828–831.
4. Bara-Jimenez W, Shelton P, Sanger TD, Hallett M. 2000. Sensory discrimination capabilities in patients with focal hand dystonia. Ann Neurol 47: 377–380.
5. Barbe MF, Barr AE, Gorzelany I, Amin M, Gaughan MP, Safadi FF. (2003) Chronic repetitive reaching and grasping results in decreased motor performance and widespread tissue responses in a rat model of MSD. J of Orthopaedic Res 32:167-176
6. Barr AE, Barbe MF. 2002 Pathophysiological tissue changes associated with repetitive movement: A review of the Evidence. Phy Ther 82:173-187.
7. Bell C. 1883. The Hand: Its Mechanism and Vital Endowments as Evincing Design. London, UK: William Pickering; 23 pp.
8. Black KJ, Ongur D, Perlmutter JS. 1998. Putamen volume in idiopathic focal dystonia. Neurology 51: 819–824.
9. Blake DT, Byl NN, Cheung S, Bedenbaugh P, Nagarajan S, Lamb M, Merzenich M. 2002. Sensory representation abnormalities that parallel focal hand dystonia in a primate model. Somatosens Mot Res 19: 347–357.
10. Brin MF, Fahn S, Moskowitz C, Friedman A, Shale HM, Greene PE, et al. 1987. Localized injections of botulinum toxin for the treatment of focal dystonia and hemifacial spasm. Mov Disord 2: 237–254.
11. Butterworth S, Francis, S, Kelly E, McGlone F, Bowtell R, Sawle GV 2003 Abnormal cortical sensory activation in dystonia: An fMRI study. Movement Disorders, 18:623-633

12. Byl N, Wilson F, Merzenich M, Melnick M, Scott P, Oakes A, McKenzie A. 1996a. Sensory dysfunction associated with repetitive strain injuries of tendinitis and focal hand dystonia: a comparative study. J Orthoped Sports Phys Ther 23: 234–244.
13. Byl N, Hamati D, Melnick M, Wilson F, McKenzie A. 1996b. The sensory consequences of repetitive strain injury in musicians: Focal dystonia of the hand. J Back Musculoskel Rehab 7: 27–39.
14. Byl N, Merzenich M, Jenkins W. 1996c. A primate genesis model of focal dystonia and repetitive strain injury: I. Learning-induced dedifferentiation of the representation of the hand in the primary somatosensory cortex in adult monkeys. Neurology 47: 508–520.
15. Byl NN, Merzenich MM, Cheung S, Bedenbaugh P, Nagarajan SS, Jenkins WM. 1997. A primate model for studying focal dystonia and repetitive strain injury: effects on the primary somatosensory cortex. Phys Ther 77: 269–284.
16. Byl N, Topp KS. 1998. Focal hand dystonia. Phys Ther Case Rep 1: 39–52.
17. Byl NN, Nagarajan SS, Newton N, McKenzie AL. 2000a. Effect of sensory discrimination training on structure and function in a musician with focal hand dystonia. Phys Ther Case Rep 3: 94–113.
18. Byl N, McKenzie A, Nagarajan SS. 2000b. Differences in somatosensory hand organization: healthy flutist and flutist with focal hand dystonia. J Hand Ther 13: 301–309.
19. Byl N, Nagarajan SS, McKenzie A. 2000c. Effectiveness of sensory retraining: three case studies of patients with focal hand dystonia. [Abstract] Society for Neuroscience Meeting, New Orleans, Louisiana.
20. Byl N, McKenzie A. 2000d. Treatment effectiveness of patients with a history of repetitive hand use and focal hand dystonia: A planned prospective follow up study. J Hand Ther 13: 289–301.
21. Byl, NN, Merzenich, MM. 2000e. Chapter 26 Principles of Neuroplasticity: Implications for Neurorehabilitation and Learning, in Gonzalez ES, Myers S, Edelstein J, Lieberman JS, Downey JA (Eds), Downey and Darling, Physiological Basis of Rehabilitation Medicine.
22. Byl NN, Nagarajan SS, Merzenich MM, Roberts T, McKenzie. 2002. Correlation of clinical neuromusculoskeletal and central somatosensory performance: variability in controls and patients with severe and mild focal hand dystonia. Neural Plasticity 9:177-203
23. Byl, NN 2004 Focal hand dystonia may result from aberrant neuroplasticity. Dystonia 4: Adv Neurol 94, 19-27.
24. Candia V, Elbert T, Altenmuller E, Rau H, Schafer T, Taub E. 1999. Constraint-induced movement therapy for focal hand dystonia in musicians. [Letter] Lancet 353: 42–43.
25. Candia V, Schafer, T, Taub E, Rau, H, Alternmuller, E, Rockstroh, B, Elbert, T. Sensory motor returning:2002 A behavioral treatment for foal hand dystonia of pianists and guitarists. Arch Phys Med Rehabil 83: 1342-1349
26. Ceballos-Baumann AO, Sheean G, Pasingham RE, Marsden CD, Brooks DJ. 1995. Cerebral activation with stereotyped writing in patients with writer's cramp before and after botulinum toxin treatment: a PET study. Neurology Suppl 45: 393–396.
27. Charness ME. 1992. Unique upper extremity disorders of musicians. In: Millender LH, Louis DS, Simmons BP (Eds) Occupational Disorders of the Upper Extremity. New York, NY: Churchill Livingstone. 117–151.
28. Charness M. 1993. The relationship between peripheral nerve injury and focal dystonia in musicians. Am Acad Neurol 162: 21–27.

29. Chase T, Tamminga C. and Burrows H. 1988. Positron emission tomographic studies of regional cerebral glucose metabolism in idiopathic dystonia. Adv Neurol 50: 237–241.
30. Chen RS, Tsai CH, Lu CS. 1995. Reciprocal inhibition in writer's cramp. Mov Disord 10: 556–561.
31. Chen R, Hallett M. 1998. Focal dystonia and repetitive motion disorders. Clin Ortho 351: 102–106.
32. Clark, BD, Barr, AE, Safadi, FF, Beitman, L, Al-Shatti, T, Amin, M, Gaughan, JP, Barbe, MF.(2003) Median nerve trauma in a rat model of work-related musculoskeletal dirorder. J Neurotrauma, accepted 2003.
33. Cohen LG, Hallett M. 1988. Hand cramps: clinical features and electromyographic patterns in a focal dystonia. Neurology 38: 1005–1012.
34. Cole R, Hallett M, Cohen LG. 1995. Double-blind trial of botulinum toxin for treatment of focal hand dystonia. Mov Disord 10: 466–471.
35. deCharms, RC, Blake, DT, Merzenich MM 1999 A multi-electrode implant device for the cerebral cortex. J Neurosci Methods 93: 27-35
36. Defendini R, Fahn S. 1988. Magnetic resonance imaging of dystonic states. Adv Neurol 50: 265–275.
37. DeLong MR, Crutcher MD, Georgopoulis AP. 1985. Primate globus pallidus and subthalamic nucleus: functional organization. J Neurophysiol 53: 530–543.
38. DeLong MR. 1990. Primate models of movement disorders of basal ganglia origin. Trends Neurosci 13: 281–285.
39. Deuschl G, Toro C, Matsumoto J, Hallett M. 1995. Movement-related cortical potentials in writer's cramp. Ann Neurol 38: 862–868.
40. Elbert T, Candia V, Altenmuller E, Rau H, Sterr A, Rockstroh B, et al. 1998. Alteration of digital representations in somatosensory cortex in focal hand dystonia. Neuroreport 9: 3571–3575.
41. Elbert T, Pantev C, Wienbruach C, Rockstroh B, Taub E. 1995. Increased cortical representation of the fingers of the left hand in string players; Science 370:305-307
42. Fahn S, Marsden CD, Calne D. 1987. The treatment of dystonia. In: Marsden CD, Fahn S, (Eds) Movement Disorders. London, UK: Butterworths 359–382.
43. Fry H. 1986. Overuse syndromes in musicians 100 years ago: An historical review. Med J Aust 146: 620–625.
44. Gasser T, Bove C, Ozelius L, Hallett M, Charness M, Hochberg F, et al. 1996. Haplotype analysis at the DYT1 locus in Ashkenazi Jewish patients with occupational hand dystonia. Mov Disord 11: 163–166.
45. Gerloff CH, Corwell B, Chen R, Hallett M, Cohen LG. 1998. The role of the human motor cortex in the control of complex and simple finger movement sequences. Brain 121: 1695–1709.
46. Gil, Ziv, Connors, BW, Amitai, Y. 1997 Differential regulation of neocortical synapses by neuromodulators and activity. Neuron 19: 679-686
47. Gilman S, Junck L, Young A. 1988. Cerebral metabolic activity in idiopathic dystonia studied with positron emission tomography. Adv Neurol 50: 231–236.
48. Groenwegen HJ, Berendse HW, Wolters JG, Lohman AH. 1990. The anatomical relationship of the prefontral cortex with the striatopallidal system, the thalamus and the amygdala: evidence for parallel organization. Prog in Brain Res 85:95-116
49. Grunewald RA, Yoneda Y, Shipman JM, Sagar HJ. 1997. Idiopathic focal dystonia: a disorder of muscle spindle afferent processing? Brain 120: 2179–2185.
50. Hallett M. 1998. Physiology of dystonia. Adv Neurol 78: 11–18.

51. Hasselmo ME. 1996. Neuromodulation and cortical function; modeling the physiological basis of behavior. Behav Brain Res 57 1-27
52. Hasselmo ME, Bower J. 1993. Acetylcholine and memory. Trends Neurosci 16: 218-222.
53. Hebb DO. 1949. The Organization of Behavior. New York, NY: Wiley.
54. Hikosaka O, Tanaka M, Sakamoto M, Iwamura Y. 1984. Deficits in manipulative behaviors induced by local injections of muscimol in the first somatosensory cortex of the conscious monkey. Brain Res 325:375-380.
55. Hochberg F, Harris S, Blartert T. 1990. Occupational hand cramps: Professional disorders of motor control. Hand Injury Sports Perform Arts 6: 427–428.
56. Ikeda A, Shibasaki H, Kaji R, Terado K, Nugamine T, Honda M, et al. 1999. Abnormal sensorimotor integration in writer's cramp: Study of contingent negative variation. Mov Disord 17: 683–690.
57. Illarioshkin SN, Markova ED, Slominsky PA, Miklena NI, Popovov SN, Limborska SA, et al. 1988. The GTP cyclohydrolase I gene in Russian families with dopa responsive dystonia. Arch Neurol 55: 789–792.
58. Iwamura Y. 1992. Dynamic and hierarchical processing in the monkey somatosensory cortex. Biomed Res 14: 107–111.
59. Jabusch HC, Mueller SV, Altenmueller E. 2002. High levels of perfectionisms and anxiety in musicians with focal dystonia. International Meeting, Dystonia Foundation, Atlanta, Georgia, July
60. Jankovic J, Shale H. 1989. Dystonia in musicians. Sem Neurol 9: 131–135.
61. Jancovic J. Can peripheral trauma induce dystonia and other movement disorders? Yes. Mov Disord 2001. 16:7–11.
62. Jenkins WM, Merzenich MM. 1987 Reorganization of neocortical representations after brain injury: a neurophysiological model of the bases of recovery from stroke. In FJ Seil, Herbert E, Caroson BM. (Eds). Prog Brain Res 71:249–266
63. Jenkins W, Allard T, Nudo R. 1988. Cortical representational plasticity. In: Raskic P, Singer W. (Eds). Neurobiology of the Neocortex. New York, NY: John Wiley and Sons; 41–67.
64. Jenkins WM, Merzenich MM, Ochs MT, Allard T, Guic-Robles E. 1990a. Functional reorganization of primary somatosensory cortex in adult owl monkeys after behaviorally controlled tactile stimulation. J Neurophysiol 53: 82–104.
65. Jenkins, WM, Merzenich MM, Recanzone G.1990b. Neocortical representational dynamics in adult primates: implications for neuropsychology. Neuropsychologia 28:573-584.
66. Johansson RS. 1996. Sensory control of dexterous manipulation in humans. In: Wing AM, Haggard P, Flanagan JR. (Eds) Hand and Brain. The Neuro-physiology and Psychology of Hand Movements, San Diego, CA: Academic Press, 381–414.
67. Juliano SL, Ma W, Eslin D. 1991. Cholinergic depletion prevents expansion of topographic maps in soma-tosensory cortex. Proc Natl Acad Sci 88: 780–784.
68. Kaas JH, Merzenich MM, Killackey HP. 1983. The reorganization of somatosensory cortex following peripheral nerve damage in adult and developing mammals. Annu Rev Neurosci 6: 325–356.
69. Kahn KM, Cook, JL, Bonar, F, Harcourt, P, Astrom, M. 1999 Histopathology of common tendinopathies: update and implications for clinical management. Sports Med 27: 393-408
70. Kaji R, Rothwell J, Katayama M, Ikeda T, Kubori, Kohara N, et al. 1995. Tonic vibration reflex and muscle afferent block in writer's cramp. Ann Neurol 38: 155–162.

71. Karp B, Cole R, Cohen L, Grill L, Lou J, Hallet M. 1994. Long-term botulinum toxin treatment of focal hand dystonia. Neurology 44: 70–76.
72. Katz R, Williams C. 1990. Focal dystonia following soft tissue injury: three case reports with long-term outcome. Arch Phys Med Rehab 71: 345–349.
73. Kolle B. Psychological approach to focal dystonia in musicians, chapter 18, 363-368 in Tubiana R, Amadio PC. (Eds). Medical Problems of the INstrumentalists Musician, Martin Dunitz Pub, 2000, 626 pp.
74. Leijnse JNAL. 1997. Anatomical factors predisposing to focal dystonia in the musician's hand—principles, theoretical examples, clinical significance. J Biomech 30: 659–669.
75. Lenz FA, Byl NN. 1999. Reorganization in the cutaneous core of the human thalamic principal somatic sensory nucleus (Ventral caudal) in patients with dystonia. J Neurophysiol 82: 3204–3212.
76. Leube B, Rudnicki D, Ratzlaff T, Kessler K, Benecke R, Auburger G. 1996. Idiopathic torsion dystonia: Assignment of a gene to chromosome 18p in a German family with adult onset, autosomal dominant inheritance and purely focal distribution. Hum Molec Genet 5: 1673–1677.
77. Liversedge LA, Sylvester JD.1955 Conditioning techniques in the treatment of writer's cramp. Lancet 1: 1147–1149.
78. Liversedge LA, Sylvester JD. 1960. Conditioning techniques in the treatment of writer´s cramp. In: Eysenck HJ. (Ed). Behaviour Therapy and the Neuroses. Pergamon Press: Oxford, UK.
79. Mai N, Marguardt C. 1994. Treatment of writer's cramp: Kinematic measures as assessment tools for planning and evaluating handwriting training procedures. In: Faure C, Keuss PJG, Lorette G, Vinter A. (Eds). Advances in Handwriting and Drawing: A Multidisciplinary Approach, Paris, France: Europia.
80. Marachuilo LA, McSweeney M. 1977 Nonparametric and Distribution-Free Methods for the Social Sciences. Monterey, CA: Brooks/Cole Publishing.
81. Marsden CD, Sheehy MP. 1990. Writer's cramp. Trends Neurosci. 13: 148–153.
82. McKenzie AL, Nagarajan SS, Roberts TP, Merzenich MM, Byl NN. Somatosensory representation of the digits and clinical performance in patients with focal hand dystonia, Am J Phys Med Rehabil, 2003, 82(10) 737-749.
83. Merzenich MM, Kaas JH, Wall J, Nelson RJ, Sur M, Felleman D. 1983a. Topographic reorganization of somatosensory cortical areas 3b and 1 in adult monkeys following restricted deafferentation. Neuroscience 8: 33–55.
84. Merzenich MM, Kaas JH, Wall JT, Sur M, Nelson RJ, Felleman DJ. 1983b. Progression of change following median nerve section in the cortical representation of the hand in areas 3b and 1 in adult owl and squirrel monkeys. Neuroscience 10: 639–665.
85. Merzenich MM, Nelson RJ, Stryker MP, Cynader MS, Schoppmann A, Zook JM. 1984. Somatosensory cortical map changes following digit amputation in adult monkeys. J Comp Neurol 224: 591–605.
86. Merzenich MM. 1990. Development and maintenance of cortical somatosensory representations: Functional "maps" and neuroanatomical repertoires. In Barnard KE, Brazelton TB. (Eds). Touch: The Foundation of Experience. Madison, CT: International Universities Press; 47–71.
87. Merzenich MM, Jenkins WM. 1995. Cortical plasticity, learning, and learning dysfunction. In: Julesz B, Kovacs I. (Eds). Maturational Windows and Adult Cortical Plasticity. New York, NY: Addison-Wesley; 247–272.

88. Merzenich M, Wright B, Jenkins W, Xerri C, Byl N, Miller S, Tallal P. 1996a. Cortical plasticity underlying perceptual, motor, and cognitive skill development: implications for neurorehabilitation: Implications for neurorehabilitation. Cold Spring Harbor Symp Quant Biol 61: 1–8.
89. Merzenich M, deCharms, C. 1996b. Neural representations, experience, and change. In: Llinas, R, Churchland PS. (Eds). Mind Brain Continuum. Cambridge, MA: MIT Press; 61–81.
90. Merzenich MM, Tallal P, Peterson B, Miller SL, Jenkins WM. 1999. Some neurological principles relevant to the origins of—and the cortical plasticity based remediation of—language learning impairments. In: Grafman J, Christen Y, (Eds) Neuroplasticity: Building a Bridge from the Laboratory to the Clinic. Berlin, Germany: Springer Verlag; 169–187
91. Millender LH, Louis DS, Simmons BP. 1992. Occupational Disorders of the Upper Extremity. New York, NY: Churchill Livingstone.
92. Murase N, Kaji R, Shimazu, H, Katayama-Hirota, M, Ikeda, A, Kohara, N, Kimura, J, Shibasaki, H, Rothwell, JD. 2000. Abnormal premovement gating of somatosensory input in writer's cramp. Brain 123:1813-1829.
93. Nagarajan SS, Blake DT, Wright BA, Byl N, Merzenich MM. 1998. Practice-related improvements in somatosensory interval discrimination are temporally specific but generalize across skin location, hemi-sphere, and modality. J Neurosci 18: 1559–1570.
94. Nakashima K, Rothwell JC, Day BL, Thompson PD, Shannon K. 1989. Reciprocal inhibition between forearm muscles in patients with writer's cramp and other occupational cramps, symptomatic hemidystonia and hemiparesis due to stroke. Brain 112: 681–697.
95. 95. Naumann M, Reiners K. 1997. Long-latency reflexes of hand muscles in idiopathic focal dystonia and their modification by botulinum toxin. Brain 120: 409–416.
96. Newmark J, Hochberg F. 1987. Isolated painless manual coordination in 57 musicians. Br J Neurol Neurosurg Psychiat 50: 291–295.
97. Nudo RJ, Wise BM, SiFuentes F, Milliken GW. 1996a. Neural substrates for the effects of rehabilitative training on motor recovery after ischemic infarct. Science 272(5269): 1791–1794.
98. Nudo RJ, Millikin GW. 1996b. Reorganization of movement representations in primary motor cortex following focal ischemic infarcts in adult squirrel monkeys. J Neurophysiol 75: 2144–2149.
99. Nudo RJ. 1999. Recovery after damage to motor cortical areas. Curr Opin Neurobiol 9: 740–747.
100. Nudo RJ, Friel KM, Delia SW. 2000. Role of sensory deficits in motor impairments after injury to primary motor cortex. Neuropharmacology 39: 733–742.
101. Odergren T, Iwasaki N, Borg J, Forssberg H. 1996. Impaired sensory-motor integration during grasping in writer's cramp. Brain 119: 569–583.
102. Ozelius LJ, Hewett JW, Page CE, Bressman SB, Kramer PL, Shalish C, et al. 1997. The early-onset torsion dystonia gene (DYT1) encodes an ATP-binding protein. Nat Genet 17: 40–48.
103. Panizza ME, Hallett M, Nilsson J. 1989. Reciprocal inhibition in patients with hand cramps. Neurology 39: 85–89.
104. Panizza M, Lelli S, Nilsson J, Hallett M. 1990. H-reflex recovery curve and reciprocal inhibition of H-reflex in different kinds of dystonia. Neurology 40: 824–828.

105. Pascual-Leone, A, Dang, N, Cohen, LG, Brasil-Neto, JP, Cammarota, A, Hallett, M. 1995a. Modulation of muscle responses evoked by transcranial magnetic stimulation during the acquisition of new fine motor skills. J Neurophys 74:1037-1045.
106. Pascual-Leone, A, Wassermann, FM, Sadaio N, Hallet M. 1995b The role of reading activity on the modulation of motor cortical outputs to the reading hand in Braille readers. Annal of Neurol 38:910-915.
107. Penfield W, Rasmussen, T. 1950. The cerebral cortex of man: a clinical study of localization of function. New York, NY\: MacMillan Publishing.
108. Perlmutter JS, Stambuk MK, Markham J, Black KJ, McGee-Minnich L, Jankovic J, Moerlein SM. 1997. Decreased [^{18}F]spiperone binding in putamen in idiopathic focal dystonia. J Neurosci 17: 843–850.
109. Peterson BE, Merzenich MM. 1995 MAP: a Macintosh program for generating categorical maps applied to cortical mapping. J Neurosci Methods. 57:133-144
110. Priori A, Pesenti A, Cappellari A, Scarlato G, Barbieri, S. 2001. Limb immobilization for the treatment of focal occupational dystonia. Neurology 57: 405–409
111. Pujol, J, Roset-Llobet, J, Rosines-Cubells, D, Deus, J, Narberhaus, B, Valls-Sole, J, Capdevila, A, Pascual-Leone, 2000 A. Brain cortical activation during guitar-induced hand dystonia studied by functional MRI. NeuroImage, 12:257-267.
112. Pullman SL, Greene P, Fahn S, Pedersen SF. 1996. Approach to the treatment of limb disorders with botulinum toxin A. Experience with 187 patients. Arch Neurol 53: 617–624.
113. Quartarone A, Girlanda P, Risitano G, Picciolo G, Sinicropi S, Nicolosi C, et al. 1998. Focal hand dystonia in a patient with thoracic outlet syndrome. J Neurol Neurosurg Psychiat 65: 272–274.
114. Recanzone GH, Jenkins WM, Hradek GT, Merzenich MM. 1992a. Progressive improvement in discriminative abilities in adult owl monkeys performing a tactile frequency discrimination task. J Neurophysiol 67: 1015–1030.
115. 115. Recanzone GH, Merzenich MM, Jenkins WM, Grajski KA, Dinse HR. 1992b. Topographic reorganization of the hand representation in cortical area 3b of owl monkeys trained in a frequency-discrimination task. Neurophysiol J 67: 1031–1056.
116. Recanzone GH, Merzenich MM, Jenkins WM. 1992c. Frequency discrimination training engaging a restricted skin surface results in an emergence of a cutaneous response zone in cortical area 3a. J Neurophysiol 67: 1057–1070.
117. Ridding MC, Sheean G, Rothwell JC, Inzelberg R, Kujirai T. 1995. Changes in the balance between motor cortical excitation and inhibition in focal, task specific dystonia. J Neurol Neurosurg Psychiat 59: 493–498.
118. Rijntjes M, Dettmers C, Buchel C, Kiebel S, Frackowiak RS, Weiller C. 1999. A blueprint for movement: functional and anatomical representations in the human motor system. J Neurosci 19: 8043–8048.
119. Roberts TPL, Poeppel D, Rowley HA. 1998. Magnetoencephalography and magnetic source imaging. Neuropsych Neuropsycho Behav Neurol 11:49-64.
120. Roberts TPL, Rowley HA. 1997. Mapping of the sensorimotor cortex: functional MR and magnetic source imaging, ANJR Am J Neuroradio 18:871-80.
121. Sanger TD, Merzenich MM. 2000. Computational model of the role of sensory disorganization in focal task-specific dystonia. J Neurophysiol 84: 2458–2464.
122. Sanger TD, Tarsy DM, Pascual-Leone A. 2001. Abnormalities of spatial and temporal sensory discrimination in writer's cramp. Mov Disord 16: 94–99.
123. Spengler F, Roberts TP, Poeppel D, Byl N, Wang X, Rowley HA, Merzenich MM. 1997. Learning transfer and neuronal plasticity in humans trained in tactile discrimination. Neurosci Lett 232: 151–154.

124. Sterr, a, Miller MM, Elbert T, Rockstroh B, Pantev, C, Taub E. 1998; Perceptual correlates of changes in cortical representation of fingers in blind multifinger Braille readers. J Neurosci 18:4417-4423.
125. Stock SR.1991 Workplace ergonomic factors and the development of musculoskeletal disorders of the neck and upper limbs; A meta-analysis. Am J of Indus Med 19:87-107
126. Stryker, MP, Jenkins, WM, Merzenich, MM 1987 Anesthetic state does not affect the map of the hand representation within are 3b somatosensory cortex in owl monkeys. J Comp Neurol 258:297-303.
127. Tempel LW, Perlmutter JS. 1993. Abnormal cortical responses in patients with writer's cramp. Neurology 43: 2252–2257.
128. Tinazzi M, Frasson E, Bertolasi L, Fiaschi A, Aglioti S. 1999 Temporal discrimination of somesthetic stimuli is impaired in dystonic patients. Neuro-Report 10: 1547–1550
129. Tinazzi, M, Rossa, T, Fiaschi, A. 2003. Role of the somatosensory system in primary dystonia. Mov Disord 18:605-622.
130. Topp KS, Byl NN. 1999. Movement dysfunction following repetitive hand opening and closing: anatomical analysis in Owl monkeys. Mov Disord 14: 295–306
131. Toro C, Deuschl G, Hallett M. 2000. Movement-related electroencephalographic desynchronization in patients with hand cramps: evidence for motor cortical involvement in focal dystonia. Ann Neurol 47: 456–461.
132. Tsui J, Bhatt M, Calne S, Calne DB. 1993. Botulinum toxin in the treatment of writer's cramp: a double-blind study. Neurology 43: 183–185.
133. Tubiana R, Chamagne P. 1983. [Occupational "cramps" of the upper limb.] Ann Chir Main 2: 134–142.
134. Tubiana R. 1998. Focal dystonia. Incidence: classification of severity and results of therapy. In: Winspur I, Wynn Parry CB. (Eds). The Musician's Hand. A Clinical Guide. London, UK: Martin Dunitz LTD; 164–167.
135. Uitti RJ, Vingerhoets FJ, Tsui JK. 1995. Limb dystonia. In: Tsui JK, Calne D. (Eds). Handbook of Dystonia. New York, NY: Marcel Dekker; 143–158.
136. Van Hilten BJ, van de Beek WJT, Hoff JI, Voormolen JHC, Delhaas EM. 2000. Intrathecal baclofen for the treatment of dystonia in patients with reflex sympathetic dystrophy. N Engl J Med. 343: 625–630.
137. Viikari-Juntura E. 1998. Risk factors for upper limb disorders. Clin Orthopaed Relat Res 351:39-43.
138. Vitek, JL. 2002. Pathophysiology of dystonia: a neuronal model. Movement Disorders 17: S49-S62.
139. Wang X, Merzenich MM, Sameshima K, Jenkins WM. 1995. Remodelling of hand representation in adult cortex determined by timing of tactile stimulation. Nature 378(6552): 71–75.
140. Wang X, Merzenich MM, Sameshima K, Jenkins WM. 1994. Afferent input integration and segregation in learning are input timing dependent. Neurosci Abstr 20: 1427.
141. Weiner W. 2001. Can peripheral trauma induce dystonia? No Mov Disord 16:13-22.
142. Wilson FR, Wagner C, Hömberg V. 1993. Bio-mechanical abnormalities in musicians with occupational cramp/focal dystonia. J Hand Ther 6: 298–307.
143. Wilson FR, Wagner C, Hömberg V. Noth J. 1991. Interaction of biomechanical and training factors in musicians with occupational cramp/focal dystonia. Neurology Suppl 41: 291-296.
144. Woolsey CN. 1958. Organization of somatic sensory and motor areas of the cerebral cortex. In Harelow HF, Woolsey CN. (Eds). Biological and Biochemical Bases of Behavior. Madison, University of Wisconsin Press, pp 63-81.

145. Yang TT, Gallen C, Schwartz B, Bloom FE, Ramachandran VS, Cobb S. 1994. Sensory maps in the human brain [letter] Nature 368: 592–593.
146. Xerri C, Coq JO, Merzenich MM, Jenkins WM. 1996. Experience-induced plasticity of cutaneous maps in the primary somatosensory cortex of adult monkeys and rats. J Physiol Paris 90: 277–287.
147. Xerri C, Merzenich MM, Jenkins W, Santucci S. 1999. Representational plasticity in cortical area 3b paralleling tactual-motor skill acquisition in adult monkeys. Cereb Cortex 9: 264–276.
148. Zirh TA, Reich SG, Perry V, Lenz FA. 1998. Thalamic single neuron and electromyographic activities in patients with dystonia. Adv Neurol 78: 27–32.

References for Measurement Documents

149. Ayers AJ, 1989 Sensory Integration and Praxis Tests, Los Angeles, California, Western Psychological Services.
150. Bohannon RW. 1995 Stopwatch for measuring thumb movement time. Perceptual and Motor Skills, 81:211-216.
151. Byl N, Cheney L. 2002. The Byl-Cheney-Boczai sensory discriminator: Reliability, validity, and responsiveness for testing stereognosis. J Hand Therapy Oct-Dec; 315-330
152. Byl N, McKenzie A. 2000e. Treatment effectiveness of patients with a history of repetitive hand use and focal hand dystonia: A planned prospective follow up study. J Hand Ther 13: 289–301.
153. Fung S, Byl N, Melnick M, Callahan P, Selinger A, Ishi K, Devins J, Fischer P, Torburn L, Andrade CK. 1997 Functional Outcomes: The development of a new instrument to monitor the effectiveness of physical therapy. Eur J Phys Med Rehabil 7:31-41.
154. Jamar Dynamometer: TEC, 60 Page Road, Clifton, NJ 07012.
155. Kendall F, McCreary E, Provance P. 1993. Muscles: Testing and Function, Williams & Wilkins, Baltimore, MD.
156. Lafayette Instrument Company.1996. Instructions and Normative Data, Purdue Pegboard, PO Box 5729, Lafayette IN, 47903.
157. Microfet Dynamometer. 1999. Medical Products Division, Hoggan Health Industries Inc.
158. Norkin C, White J. 1995. Measurement of Joint Motion: A Guide to Goniometry, F.A. Davis Company, Philadelphia, PA.

Index

A

Aberrant learning, 235, 236
Acetylcholine (ACh), 74, 75, 174, 227
ACh, *see* Acetylcholine
Action potentials (AP), 11
  activity
    absence of, 15
    fine timing of, 43
    silent cells and sparse, 13
    sparse, 13
  backpropagating, 14
  -dependent signaling, 13
  firing, phasic, 148
  initiation threshold, 3
  M1, 12
  motor effects of, 10
Activities of daily living (ADLs), 248
Activity-based plasticity, 129
Acute limb deafferentation, 211
ADLs, *see* Activities of daily living
Adult motor cortex, *see* Neural plasticity in adult motor cortex
AI, *see* Primary auditory cortex
Airborne media, high frequency information transmitted by, 24
Alert waking states, oscillation amplitude during, 70
Alligator lizards, 27
Alzheimer's disease, 174
α-Amino-3-hydroxy-5-methylisoxazole proprionic acid (AMPA), 116
Amphetamine, 175
AMPS, *see* α-Amino-3-hydroxy-5-methylisoxazole proprionic acid
Amputated limb, activation of stump muscles of, 196
Amputations
  macaque, 196
  motor cortex reorganization after, 190, 195
Analysis of Variance, 244
Anesthetic hands, 103
AP, *see* Action potential
Auditory perception, neural mechanisms of, 47

B

Baclophen, 230
Balance activities, 244
Barrel cells, whisker stimulation and, 118
Barrel cortex
  effects of early SD deprivation on, 119
  experience-induced activity in, 120
  neural modifications in, 66
BCB test, *see* Byl-Cheney- Boczai
BDNF, *see* Brain Derived Neurotrophic Factor
Behavioral compensation, 170
Behavioral gain, 212
Behavioral learning, gap-crossing task, 121
Behavioral process, psychological models of, 140
Behavioral recovery, 200
Behavioral tasks
  models, 145
  primate, 232
Behavioral training, monkeys engaged in, 231
Between-session learning, 98
Bicuculline, 168
Botulinum toxin injections, 230
Braille reading, 101, 102
Brain
  lesions, stroke patients and, 213
  reward circuit, 174
  rewiring, potential for, 253
Brain Derived Neurotrophic Factor (BDNF), 175
  expression, reduction of, 177
  synthesis, blocking of, 176
Brainstem Trigeminal Complex (TC), 117
Byl-Cheney- Boczai (BCB) test, 242

C

CAFÉ 40, 243
Calbindin-immunoreactive cells, 145
Caudal digit-wrist motor cortex, 146
Cell-attached recordings, 6
Central nervous system (CNS), 140, 208
  motor systems, damage to, 167
  rapid changes in, 210
Cercopithicus aethiops, 160
Chronic limb deafferentation, 208
Circuit development, sensory deprivation and, 130

CI therapy, *see* Constraint-induced movement therapy
Clinical tests, summary of, 242–243
CNS, *see* Central nervous system
CO, *see* Cytochrome Oxidase
Cochlear amplification, 27
Cockroaches, airflow discriminated by, 24
Code
  neuronal, key ingredient, 68
  place, 22, 37
Coding
  population, 67
  rate, 148
  temporal, 148
  vibration, 69
Compartmental modeling, layer II/III pyramidal cell, 8
Complete deprivation, restricted deprivation vs., 112
Complex environment, 157
Complex stimuli, 33
Conditioning techniques, 230
Constraint-induced movement therapy (CI therapy), 171, 173, 230
Continuous synaptic activity, 148
Cortical activation, movement related, 172
Cortical activity patterns, 66
Cortical cells, spontaneous firing of, 120
Cortical dysfunction, 229
Cortical function and plasticity in human brain, modulation of, 207–226
  effects of cortical stimulation on motor cortical function and cortical plasticity, 215–217
  somatosensory input modulates motor cortical plasticity, 208–214
    acute limb deafferentation, 210–214
    chronic limb deafferentation, 208–210
    somatosensory stimulation, 214
Cortical modifiability, 71
Cortical neurons, silent, 2
Cortical physiology, central problem of, 12
Cortical plasticity, silent neurons and, 12
Cortical sensory response, synaptic composition of, 8
Corticospinal tract, learning new motor skills and, 173
Cross-correlation index, 72, 73
Cross-digit interval discrimination, 232
Cross-model functioning, 102
Cross-neuron correlations, information in, 68
Current-rodent neuroanatomy, hypothesized, 141
Cytochrome Oxidase (CO)
  patches, 117
  staining, 130
Cytokine changes, measurement of, 239

D

DELAY box, 147
2-Deoxyglucose autoradiography, 127
Desilenced cells, 2
Diaschisis, theory of, 165
Discrimination
  limens (DLs), 82
  task, QA neuron responses during, 84
DLs, *see* Discrimination limens
Dopamine, 174, 227
Dual responses, 162
Dynamic finger motor task, 213
Dystonic cramping, 230

E

Early-mammal neuroanatomy, hypothesized, 141
Electromyography (EMG), 10, 172
EMG, *see* Electromyography
Emotional instability, 230
ESPS's, *see* Excitatory postsynaptic potentials
Excitation-to-inhibition scenario, 8
Excitatory diaschisis, 165
Excitatory neuron classes, 33
Excitatory neurotransmission, effects of sensory deprivation on, 128
Excitatory pathways, lack of, 229
Excitatory postsynaptic potentials (ESPS's), 128
Extracellular unit recording, 3

F

Fast-spiking units (FSUs), 31, 33
Feedback processes, multiple, 146
FHd, *see* Focal hand dystonia
Fine temporal tuning, 47
Finger representation area, monkey, 163
Firing rate(s)
  modulation, QA neuron, 85
  QA neurons, 93
Flutter discrimination
  ambiguous paradigm, 81
  task, psychophysics in, 80
Flutter stimuli, 83
fMRI, *see* Functional magnetic resonance imaging
Focal hand dystonia (FHd), 227–261
  description of, 230
  etiology of, 229
  evidence for aberrant learning, 231

focal hand dystonia, 229–231
human research models, 240–252
case studies, 250–252
intervention, 247–250
relationship of clinical performance and neural structure, 240–246
intervention strategies, 252–253
primate studies, 231–238
rodent animal models, 238–239
summary of animal studies, 239–240
Follicle-sinus complex, receptors in, 116
Food-related odor, 143
Frequency
coding, contact vs., 42
discrimination, QA stimulation for, 94
distribution
periodicity and, 89
QA neurons and, 90
information, redundancy of, 35
specific stimuli, discrimination of, 22
tuning, velocity tuning vs., 38
FSU, *see* Fast-spiking units
Functional magnetic resonance imaging (fMRI), 155, 168, 172, 230

G

GABA
-based inhibitory circuits, 199–200
-immunoreactive axon terminals, 123
-positive spine synapses, 129
receptors, 129, 192, 201
synthetic enzyme for, 129
GABAergic antagonist, 168
GABAergic inhibition, 209, 210
GABAergic transmission, 211
GAD, *see* Glutamic acid decarboxylase
Gap crossing
behavior, single-whisker learning, 63
phases of, 64
physiological analogue of, 67
task, behavioral learning of, 121
Gap detection, 98
Gibson, Eleanor, 97
Global sensory deprivation
anatomical changes produced by, 122
comparison of partial with, 125
in vitro analysis of, 122
Glutamic acid decarboxylase (GAD), 129, 130
Glycoprotein receptor, 175
Golgi-Cox staining, 157
Graphesthesia, 241, 242
Grating discrimination task, 100
Grating set, 98

H

Hand(s)
anesthesia, 103, 211, 212
de-differentiation of, 240
monkey, somatosensory organization of, 237
motor practice, anesthesia of upper arm during, 211
movements, evoked, 170
multiple motor representations of, 173
normal cortical plasticity, 234
normal topography of, 233
presentation, abnormal, 235
representation
M1, 170
normal, 234
use, repetitive, 252
Hebbian-like synaptic mechanisms, 162
Hebbian potentiation, 127
Hebbian synaptic plasticity, 70
Helmholtz principle, 26
Higher harmonics, vibrissa resonance hypothesis and, 39, 41
High frequency stimuli, detection of, 27
High input regimes, 8
High-resolution perception, 55
Human perception, neural mechanisms of, 55
Human research models, FHd, 240
Human tactile perception, high frequency information and, 23
Human tactile system, modular learning in, 65
Human visual cortex, excitability of, 103
Hyperacuity task, 99
Hyper-excitability, 174

I

ICMS mapping, *see* Intracortical microstimulation mapping
INB, *see* Ischemic nerve block
Infiltrating macrophages, 239
Information processing, spikes and, 148
Inhibitory neural tuning, 39
Inhibitory neurotransmission, effects of sensory deprivation on, 129
Inhibitory postsynaptic potentials (ISPS's), 116
Inter-burst conditioning, 72
Intracortical circuits, 118, 119
Intracortical connections, timing-based plasticity of, 71
Intracortical microstimulation (ICMS) mapping, 156
experiment, microelectrode penetration sites in, 161

motor skill training and, 161
technique, 156
Ischemic nerve block (INB), 210, 213
Iso-stimulus intensity, 114
ISPS's, *see* Inhibitory postsynaptic potentials

J

JTT functional motor test, 216

K

Keck Center for Neuroscience, 231
Kinematic training, 230
Kinesthesia, 242
Klüver board, 161, 162, 163, 171

L

LabView® virtual instruments software, 232
Laser scanning photostimulation, 122
Lateral geniculate thalamus (LGN), 142
Layer IV neurons, 75
Layer V corticothalamic neurons, 118
Learned nonuse, 171
Learning
aberrant, 235, 236
behavioral, gap-crossing task, 121
between-session, 98
contribution of silent cells to, 12
effects, 98
-induced plasticity, 156
modular
human tactile system, 65
physiological account for, 66
rat whisker system, 63
motor, 158, 159
new motor activations, 170
perceptual, definition of, 97
reaching tasks, 144
rules, 13, 15
nonspike-based, 15
synaptic, 13
sensorimotor, 150
tactile, 98
mechanoreceptors, 99
transfer of, 99
task, 146
visual, orientation-specificity of, 100
LGN, *see* Lateral geniculate thalamus
Light deprivation, 103
Local anesthesia, 208
Long-term depression (LTD), 111, 157, 174
Long-term memory, 82
Long-term potentiation (LTP), 111, 157, 174, 191, 208
LTD, *see* Long-term depression
LTP, *see* Long-term potentiation
Luria, Alexander, 166

M

M1, *see* Primary motor cortex
Magnetic resonance spectroscopy (MRS), 210
Magneto-encephalographic (MEG) recordings, 101, 230, 246
Mammalian auditory systems, processing of signals by, 37
Mammalian brain anatomy, 1142
Mammalian columnar organization, example of, 63
Mammals
evolution of, 141
movement maps in, 191
Manual Muscle Test, 242
MCA, *see* Middle cerebral artery
Mechanoreceptors, 99
MEG recordings, *see* Magneto-encephalographic recordings
Memory, long-term, 82
Metabolic demands, silent neurons and, 12
N-Methyl-d-aspartate (NMDA), 116
receptor(s), 127, 128
activation, 119
antagonist, 214
blockade, 129
Microvibrissae, 42
Middle cerebral artery (MCA), 175, 176
Minicolumns, 157
Model(s)
behavioral tasks, 145
FHd, 229, 238, 240
psychological, 140
rat vibrissal sensory pathway, 62
synaptic bombardment scenario, 9
Modular learning
human tactile system, 65
physiological account for, 66
Monoamine neurotransmitters, 174
Monocular deprivation, 126
Motor behaviors, rat learning of, 151
Motor control, impaired, 208
Motor Control at Target Test, 243
Motor cortex
decreased excitability in, 217
injury-induced plasticity in, 170
plasticity, 193

reorganization of, 191
transcranial magnetic stimulation of, 199
Motor deficits, 171
Motor deprivation, whisker responsiveness and, 121
Motor learning, 158, 159
Motor loss, long term consequences of, 201
Motor maps
baseline maps and, 163
M1, 167
paired control group, 164
Motor neuron
outputs, loss of, 191
pools, spinal cord, 198
Motor organization, survival-related behaviors and, 149
Motor reaction time, 252
Motor skill(s)
acquisition, 158
corticospinal tract and, 173
training
ICMS maps and, 161
M1 learning induced changes and, 163
Motor system damage, reorganization of motor cortex after, 189–205
amputations, nerve injury, and spinal cord damage, 190–193
loss of limb in primates, 194–197
motor cortex in humans, 199–200
motor cortex in monkeys, 198–199
motor system plasticity after cortical lesions in rats, 193
new connections and motor cortex reorganization, 197–198
plasticity after lesions of motor cortex in humans, 200
Movement
control, 150
maps, 191
MRS, *see* Magnetic resonance spectroscopy
MUA, *see* Multi-unit activity
Multi-digit receptive fields, 103
Multiple-joint movements, 162
Multisensory convergence, visual input and, 104
Multi-unit activity (MUA), 38
Munk, Herman, 165
Muscimol, 169
Musculoskeletal tasks, 244
Mutual Information Theory, 67

N

NASA geological expeditions, 55
Natural stimuli, vibrissae resonance and, 29
Nembutol, 238
Nerve Growth Factor (NGF), 175
Nerve injury
motor cortex reorganization after, 190
peripheral, 189
Nerve root irritation, 229
Neural activity
intrinsic oscillatory patterns of, 51
M1 electrode, 149
suppression of, 192–193
visualization of, 13
Neural code(s)
perception and, 79
stimulus, 80
Neural coding, beginning of study of, 79
Neural firing patterns, 230
Neural frequency tuning, 23
temporal evolution of, 44
vibrissa resonance and, 51
Neural plasticity in adult motor cortex, 155–188
injury-induced plasticity, 164–177
concept of spontaneous recovery, 164–166
directed reorganization of injury-induced plasticity, 170–174
neuromodulation of cortical activity, 174–177
reorganization of M1, 166–170
learning-induced plasticity, 156–164
environmental enrichment and plasticity, 156–158
motor learning, 158–164
Neural processing, task analyses based on, 141
Neural resonance tuning, 39
Neural tuning, trigeminal ganglion neurons, 48
Neural velocity sensitivity, 40
Neuron(s)
conversion of silent to spiking, 13
cortical, silent, 2
corticospinal, 191, 197
firing rates, 7
inhibitory synaptic, 9
inter-burst-conditioned, 73
layer IV, 75
Layer V corticothalamic, 118
M1, 12
model, 9
pooling of, 69
post-synaptic, 174
putative excitatory, 31
putative inhibitory, 31
pyramidal, 157
QA, 83
electric current injected into, 91
firing rates, 85, 88, 93
identification of clusters of, 89

number of spikes fired by, 87
S1, 94
types of responses, 86
SI, fine timing of evoked activity, 52–53
silent, *see* Sensorimotor cortices, silent neurons in
slowly adapting, 31, 92
star-shaped pyramidal layer IV, 5
trigeminal ganglion, 48, 49
Neuronal code, key ingredient, 68
Neuronal discrimination thresholds, 86
Neuronal firing, cyclical oscillations in, 70
Neurophysiological plasticity, first study to document, 171
Neurophysiology, experiments combining psychophysics and, 80
Neuroplasticity, 156
Neurotransmitters, up-regulation of, 228
Neurotrophin-3 (NT-3), 175
NGF, *see* Nerve Growth Factor
NMDA, *see* N-Methyl-d-aspartate
Nonparetic arm, hyper-reliance on, 171
Non skill-learning control, 164
Nonspike-based learning rules, 15
Norepinephrine, 174
NT-3, *see* Neurotrophin-3

O

Orientation tuning, 54
Oxygen, cellular demands for, 172

P

Page Test for Trends, 233
Paired t Test, 249
Paired Wilcoxon Test, 249
Parylene-iridium microelectrodes, 233
Passive fingerpad, threshold discrimination, 99
PCM theory, 126
Perception
audible range in rodents, 30
auditory, neural mechanisms of, 47
high frequency somatosensory, 23
high-resolution, 55
human, neural mechanisms of, 55
neural codes related to, 79
vibrissa-based, 28
Perceptual learning, definition of, 97
Perceptual learning and referral in tactile system, 97–108
inter-manual referral of tactile sensation, 103–104
overview, 97
tactile superiority in the blind, 101–103
transfer of tactile learning, 98–101
early studies, 98
neural basis of tactile learning effects, 101
task specificity, 98
transfer between locations, 98–100
Peripheral nerve
entrapment, 229
injury, 189
stimulation, 214
Peripheral trauma, secondary consequences of, 229
Peri-stimulus time histograms (PSTH's), 74
PET, *see* Positron emission tomography
Phantom limb pain, 209
Phase-locking, 51, 83
Phasic action potential firing, 148
Physical therapy, 251
Place code, 22, 37
Plasticity
activity-based, 129
adaptive synaptic, 157
bursts, spike–timing-dependent, 15
constrained, 158
cortical, silent neurons and, 12
cortical lesions and, 193
definition of, 208
environmental enrichment and, 156
Hebbian synaptic, 70
injury-induced, 170
learning-induced, 156
neurophysiological, 171
normal cortical, 234
protocols, postsynaptic activity and, 15
rapid fluctuations in, 74
rules, spatial, 69
timing-based
intracortical connections, 71
sensory responses, 74
use-dependent, 162, 170, 171, 215
whisker pairing, 125, 126
Poisson control spike trains, 52
PoM, *see* Posterior medial
Population coding, rat barrel cortex, 67
Positron emission tomography (PET), 168, 172, 216
Posterior medial (PoM), 62
Postnatal development, partial sensory deprivation during, 123
Postsynaptic activity, suppressed, 6
Postsynaptic patterns, sensory stimulation, 14
Postsynaptic potential (PSP), 7
amplitudes, sensory responses, 7
responses

opinions, 8
presynaptic spike times and, 9
sensory evoked, 8
Premotor cortex, 142
Presynaptic activity, cortical sensory response and, 8
Primary auditory cortex (AI), 35
Primary motor cortex (M1), 142
connections between somatosensory cortex and, 170
converging sensory input to, 159
damage, compensatory reorganization following, 169
derived motor maps of, 167
effective vs. active cells in, 12
hand representation, 170
intrinsic horizontal connections of, 197
learning induced changes in, 163
location of in placental mammals, 190
microstimulation motor maps of, 194
motor effects of APs in, 10
movement maps in, 191
neurons, effective cells in, 12
recording experiments, 158
re-emergence of thumb representation in, 166
reorganization of, 193
Primary somatosensory cortex (SI), 3, 22, 80, 101, 159
barrel field, 65–66
cortex, whisker-recipient area of, 110
frequency map in, 35
neurons, fine timing of evoked activity, 52–53
QA neurons of, 86, 94
spike rate increases in, 144
temporal coding, vibrissa resonance and, 50
Primate(s)
behavioral tasks, 232
motor cortex reorganization in, 194
motor neuron pools in, 198
Principal whisker (PW), 3
deflection, 4, 14
stimulation, 9
Protein synthesis, 147
Proto-oncogene, 164
PSP, *see* Postsynaptic potential
PSTH's, *see* Peri-stimulus time histograms
Psychological models, 140
Psychophysical discrimination thresholds, 86
Psychophysical performance, stimulus and, 90, 93
Psychophysics, experiments combining neurophysiology and, 80
Purdue Test, 241, 242
Putative inhibitory neurons, 31
PW, *see* Principal whisker
Pyramidal cells, 118

Q

QA cutaneous mechanoreceptors, *see* Quickly-adapting cutaneous mechanoreceptors
QA neuron(s), 83
electric current injected into, 91
firing rates, 85, 88, 93
frequency distribution and, 90
identification of clusters of, 89
number of spikes fired by, 87
responses of during discrimination task, 84
S1, 94
types of responses, 86
QA stimulation, frequency discrimination, 94
Quickly-adapting (QA) cutaneous mechanoreceptors, 81

R

Ranked Sum Wilcoxon Test, 244
Rat barrel cortex plasticity, spatial and temporal rules underlying, 61–78
observations concerning temporal rules temporal plasticity rules, 76
rat vibrissal sensory pathway as general model, 62–63
spatial rules, 63–70
decoding activity of rat barrel cortex populations, 68–69
modular learning in human tactile system, 65–66
modular learning in rat whisker system, 63–65
observations concerning spatial plasticity rules, 69–70
physiological account for modular learning, 66–67
population coding in rat barrel cortex, 67–68
temporal rules, 70–76
fluctuations in cortical excitability, 70–71
locus of modification, 75–76
possible mechanisms for rapid fluctuations in plasticity, 74–75
timing-based plasticity of intracortical connections, 71–74
timing-based plasticity of sensory responses, 74
Rate coding, 148
Reach-to-grasp task, 143, 146, 150
Reaching
rate, 238
repetitive, 239

skilled, 149
task events, perievent histograms, 144
tasks, learning of, 144
Receiver operating characteristic (ROC), 86
Receptive fields, 227
Recording(s)
cell-attached, 6
extracellular unit, 3
magneto-encephalographic, 101
sharp microelectrode, 3
spikelet activity, 7
whole-cell, 4
Recovery
behavioral, 176, 200
function, 164
spontaneous, 167
theory of, 165
Referred percept(s)
anesthetic hand, 103
visual input and, 104
Regular spiking units (RSUs), 31, 33
Repetitive TMS (rTMS), 216
Resonance effects predominance of, 54
Restricted whisker deprivation, cardinal feature of, 113
Retraining, theory of, 166
Retrograde tracer, injections of into squirrel monkeys, 197
ROC, *see* Receiver operating characteristic
Rodents, audible range of perception in, 30
RSUs, *see* Regular spiking units
rTMS, *see* Repetitive TMS

S

SI, *see* Primary somatosensory cortex
SII, *see* Second somatosensory cortex
SA, *see* Spontaneous activity
Saimiri sciureus, 160
SD, *see* Sensory deprivation
Second somatosensory cortex (SII), 101
SEF, *see* Somatosensory evoked field response
Sensations, analogous contralateral referral of, 103
Sensitivity training, 230
Sensorimotor cortices, silent neurons in, 1–19
evidence for predominance of silent cortical neurons in sensorimotor cortices, 2–12
motor effects of APs in primary motor cortex, 10–12
postsynaptic activity in somatosensory cortex, 7–10
presynaptic activity in somatosensory cortex, 3–7
silent neurons and cortical plasticity, 12–15
cortical plasticity and alternative hypotheses for silent neurons, 12–13
data on involvement of silent cells in plasticity, 13
silent neurons and synaptic learning rules, 13–15
Sensorimotor learning, 150
Sensorimotor mirror training, 249
Sensorimotor transformations, role of plasticity in, 139–153
action potentials, 148
behaviors appearing to be organized, 149–151
motor organization in terms of survival-related behaviors, 149–150
sensorimotor learning according to this view, 150–151
conceptualizing models, 140–142
devising task analyses, 141–142
task analyses, 140–141
information from multiple spatial scales processed immediately, 144–145
laboratory animals evaluating information across multiple time scales, 145–147
sensory and motor processing influencing one another, 142–144
Sensory cortex, neuronal populations in, 70
Sensory deprivation (SD), 110
circuit development and, 130
definition of, 111
early, 114, 119
effects of on adult brain, 126
effects of on excitatory neurotransmission, 128
effects of on inhibitory neurotransmission, 129
global, 122
partial, 123, 125
restricted, 124
somatic, 110–111
timing, 113
valid control for, 114
Sensory deprivation, effects of on sensory function of SI barrel cortex, 109–138
effects of early sensory deprivation on barrel cortex, 119–127
anatomical changes produced by global sensory deprivation, 122–123
comparison of partial with global sensory deprivation, 125–126
effects of sensory deprivation on adult brain, 126–127

in vitro analysis of global sensory deprivation during development, 122
in vivo analysis of global sensory deprivation during development, 119–121
partial sensory deprivation during postnatal development, 123–125
methodology of sensory deprivation, 111–115
complete vs. restricted deprivation, 112–113
controls, 114–115
critical or sensitive periods, 111–112
definition of sensory deprivation, 111
plucking vs. trimming of whiskers to produce sensory deprivation, 112
recovery periods, 114
timing, 113–114
molecular mechanisms affected by sensory deprivation, 127–130
excitatory neurotransmission, 128–129
inhibitory neurotransmission, 129–130
overview, 110
whisker to barrel cortex system, 115–119
ascending and recurrent circuits, 117–118
development of normal cortical response properties, 115–117
intracortical circuits, 118–119
Sensory discrimination tasks, training on, 75
Sensory experience, cellular development and, 120
Sensory information
collection of, 76
cortical organization in processing of, 61
Sensory input(s)
coactivation of, 159
degree of plasticity induced by, 72
pattern, timing of, 76
Sensory motor retuning, 230
Sensory processing, modifications of, 111
Sensory representation, neurons sufficient for, 2
Sensory responses, PSP amplitudes, 7
Sensory stimulation
pre- and postsynaptic patterns, 14
vibrissa vibration generated by, 25
Sensory systems, frequency encoding in, 25
Septal cells, whisker stimulation and, 118
Serotonin, 174
Sharp microelectrode recording, 3
Silent cells
desilenced, 2
learning and, 12
Sinus hairs, 112
Sinusoidal stimuli, 30
Skilled reaching, trial of, 149
Slip-stick behavior, high-frequency vibrations and, 28
Slowly adapting neurons, 31
Slowly adapting properties, 91
SMA, *see* Supplementary motor cortex
Somatic clamp experiments, 10
Somatic receptors, deaf individuals hearing sounds through, 24
Somatosensory cortex
connections between M1 and, 170
technique to quantify cellular activity in, 3
Somatosensory discrimination, cortical evidence of, 79–96
artificial induction of activity in S1 underlying flutter discrimination, 89–92
comments, 92–94
neural coding of vibrotactile stimuli in S1, 83–86
neuronal correlates of flutter discrimination in S1, 86–88
overview, 79
psychophysics in flutter discrimination task, 80–82
Somatosensory evoked field response (SEF), 244, 246, 247
Somatosensory isofrequency column, 35
Somatosensory perception, behavioral relevance of high frequency, 23
Somatosensory processing, 143
Somatotopic frequency mapping, 33, 36
Spatial plasticity rules, 69
Spike(s)
-based learning rules, 15
correlations, role of, 67
information processing and, 148
timing, information in, 68
Spikelet activity, 7
Spiking neuron, conversion of silent to, 13
Spinal cord
damage, motor cortex reorganization after, 190
motor neuron pools, 198
Spinal trigeminal nucleus, 130
Spontaneous activity (SA), 115
Spontaneous recovery, 164, 167
SRFs, *see* Surround receptive fields
Stimulus
complex, 33
decoding, 69
flutter, 83
natural, onset of physiological neuronal responses to, 115
neural code of, 80
psychophysical performance, 90, 93
sinusoidal, 30

tactile, spatio-temporal sequence of, 101
Stroke(s)
behavioral recovery following, 200
patients, brain lesions and, 213
Student t test, 233, 244
Supplementary motor cortex (SMA), 168
Surround receptive fields (SRFs), 119, 127
Survival-related behaviors, motor organization and, 149
Synapse(s)
GABA-positive spine, 129
long-term depression of, 111
long-term potentiation of, 111
-models, 9
Synaptic activity
continuous, 148
subthreshold, 148
Synaptic bombardment scenarios, 8, 9
Synaptic conductance changes, 9
Synaptic distributions, simulation of, 9
Synaptic learning rules, 13
Synaptic plasticity, adaptive, 157
Synaptic strengths, cellular mechanisms affecting, 189
Synaptogenesis, 115, 175

T

Tactile learning, 98
mechanoreceptors, 99
transfer of, 99
Tactile spatial capabilities, blind individuals, 101
Tactile system, *see* Perceptual learning and referral in tactile system
Targeted whole-cell recordings, 6
Task(s)
analyses
neural processing, 141
older, 140
gap-crossing, 121
learning, 146
-oriented training, 172
reaching, 144
specificity, 98
Task Specific Motor Control Scores, 251
TC, *see* Brainstem Trigeminal Complex
tDCS, *see* Transcranial direct current stimulation
Temporal coding, 148
SI, 50
trigeminal, 43
vibrissa resonance and, 42
Temporal correlation of inputs, theory of, 162
Tendinonsis, 230
Thalamo-cortico-thalamic processing, 145
Theory of diaschisis, 165
Theory of recovery, 165
Theory of retraining, 166
Theory of temporal correlation of inputs, 162
Theory of vicariation, 166
Thumb representation, re-emergence of, 166
Timing-based plasticity
intracortical connections, 71
sensory responses, 74
TMS, *see* Transcortical magnetic stimulation
Tonotopic mapping, 54
TPTP, *see* Two-photon targeted patching
Training, task-oriented, 172
Transcortical magnetic stimulation (TMS), 172, 210
Transcranial direct current stimulation (tDCS), 215, 216
Trigeminal ganglion neurons, 48, 49
Trigeminal nerve regeneration, 112
Trigeminal temporal coding, 43
Tubiana Dystonia Scale for Musicians, 241
Tumors, behavioral recovery following, 200
Two-photon targeted patching, (TPTP), 6
Two Sample Wilcoxon Test, 244
Type I cutaneous receptors, 144
Tyrosine kinase receptor, 175

U

UCSF Biomagnetic Imaging Laboratory, 241
UCSF Committee on Human Research, 241
UDP, *see* Use-dependent plasticity
Use-dependent plasticity (UDP), 162, 170, 171, 215

V

Velocity tuning, frequency tuning vs., 38
Ventral posterior lateral (VPL), 117
Ventral posterior medial (VPM), 62
barreloids, 117
thalamus, 4
Ventrolateral nucleus of the thalamus (VL), 169
Ventroposterior lateral thalamus (VPI), 142
Vibration-coding mechanism, 69
Vibrissa
-based perception, 28
damping, modulation of, 42
motion, higher harmonics of, 39
oscillations, power spectra of, 29
sensory system, velocity sensitivity of, 50
stimulation frequency, predicted neural response to, 40

Vibrissa resonance hypothesis, 21–60
  active sensing and vibrissa resonance, 41–42
    vibrissa damping, 42
    whisking velocity, 41–42
  behavioral relevance of high frequency somatosensory perception, 23–24
  evidence for, 54
  human implications, 54–55
  overview, 22–23
  rate coding and place coding of frequency-specific tactile information, 28–41
    higher harmonics, 39–41
    neural correlates of vibrissa resonance, 30–33
    somatotopic frequency mapping and isofrequency columns, 33–37
    velocity sensitivity and vibrissa resonance, 38–39
    vibrissa resonance, 28–30
  sensory capabilities of vibrissa sensory system, 24–25
  temporal coding and vibrissa resonance, 42–53
    dynamic evolution of vibrissa resonance tuning, 42–43
    fine timing of action potential activity, 43–51
    intrinsic neural frequency tuning, 51–53
Vicariation, theory of, 166
Visual cortex
  activity of during Braille reading, 102
  low activity in, 116
Visual deprivation, improved acuity and, 102
Visual learning, orientation-specificity of, 100
VL, *see* Ventrolateral nucleus of the thalamus
Volley principle
  instantiation of, 49
  support for, 51
  vibrissa resonance and, 47
Voltage clamp experiments, layer II/III pyramidal cell, 8
Voltage-sensitive dye imaging, 7, 15
Voluntary fine motor control, 252
Voluntary movement generation, 211
von Monakow, Constantine, 165
VPI, *see* Ventroposterior lateral thalamus
VPL, *see* Ventral posterior lateral
VPM, *see* Ventral posterior medial

W

Whisker
  deflections, 62
  movements, intracellular stimulation, 10, 11
  pairing (WP)
    paradigm, 126
    plasticity (WPP), 125, 126
  plucking vs. trimming, 112
  trimming, unilateral, 115
Whisking velocity, modulation of, 22–23, 41, 54
Whole-cell recordings, 4
  derivatives of, 6
  targeted, 6
WPP, *see* Whisker pairing plasticity
WP paradigm, *see* Whisker pairing paradigm

Z

z Test for Correlation Coefficients, 244